STEREOLOGICAL METHODS

Vol. 2

Theoretical Foundations

STEREOLOGICAL METHODS

Vol. 2

Theoretical Foundations

Ewald R. Weibel

Department of Anatomy, University of Berne, Berne, Switzerland

1980

ACADEMIC PRESS

A Subsidiary of Harcourt Brace Jovanovich, Publishers

London · New York · Toronto · Sydney · San Francisco

ACADEMIC PRESS INC. (LONDON) LTD
24/28 Oval Road
London NW1

United States Edition published by
ACADEMIC PRESS INC.
111 Fifth Avenue
New York, New York 10003

Weibel, Ewald R.
Stereological methods
Vol. 2: Theoretical foundations
1. Morphology—Mathematics
2. Stereology
I. Title
574.4'028 QH351 78-75269
ISBN 0-12-742202-1

Filmset in Northern Ireland at The Universities Press (Belfast) Ltd.

Printed in Great Britain by
John Wright & Sons, Ltd., at the Stonebridge Press, Bristol

Preface

A GUIDE TO THIS BOOK

When setting out to write this book several years ago my intention was to produce guide-lines, drawn from practical experience, which should help biologists of varied denominations to "do good stereology", in other words, to use diligently the powerful methods of stereology in the framework of their morphometric endeavours. For this purpose, the text should be general and simple, but I found, in due time, that these two postulates could not be realized together: to keep the exposition simple would have entailed a great loss in generality because of the necessity to restrict the theoretical justification of stereological methods to some well-defined boundary conditions; to make it general requires a rigorous mathematical approach, and this would certainly shy away many biologists.

I therefore decided to split this book into two volumes which are, to a certain extent self-contained, but are also, in part, mutually dependent on each other. The aim of Volume 1 is to present the practical methods of stereology and to provide the user with that minimum of theoretical background which is required in order to prevent serious mistakes. Volume 2 presents the theoretical background for these methods in full. Stereology is actually a branch of applied mathematics. Practical stereological methods are based on usually very simple formulas, but these are so simple only because they were derived by often rather complex mathematical reasoning. A diligent use of these simple methods under sometimes tricky circumstances may make an occasional consideration of their theoretical roots unavoidable.

After a brief general introduction, Volume 1 starts out by an elementary introduction to stereological principles (Chapter 2). Requiring no further mathematical knowledge than that of a high school level it presents plausible justifications of the basic stereological methods, discussing first the fundamental operation of sectioning and its relation to geometrical probability. It then goes on to discuss the basic measurements (volume, surface, numerical density etc.) and the relations between the size of particles and that of their profiles seen on sections. This chapter uses very simple model structures, such as spherical particles; the conclusions drawn are compatible with the more general, and hence more realistic case, as presented in Volume 2, but the insight provided is evidently limited because it is based on special cases. Chapter 3 then deals with the important topic of sampling, important because obviously the study of a spatial structure on a section needs must involve sampling, and a difficult and critical type of sampling. Chapters 4–6 present the practical methods for applying stereological principles, such as point counting methods for estimating aggregate parameters and methods for evaluating the size distribution of particles. This is followed by guide-lines on how one should set about performing a stereological study (Chapter 7) including a brief discussion of instrumentation. In Chapter 8 some special cases of application of stereological methods are presented by various authors.

Volume 2 is devoted to the theoretical foundations of stereology. It is unavoidable that this must make extensive use of mathematical reasoning so that some advanced knowledge of mathematics is required of the reader. However, an attempt was made still to keep this presentation accessible for non-mathematicians, at least on an intuitive basis; indeed full-fledged mathematicians may find the presentation primitive and naive on many accounts, but the book was not written primarily for their benefit, but rather for that of practical stereologists. The introduction (Chapter 1) presents some background information, such as some remarks on the history of stereology and on the specific language and notation used in this volume. The exposition of the foundations of stereological principles (Chapters 2–10) essentially retraces the sequence of topics presented in Volume 1, Chapters 2–6. The focus of Chapter 2 is on sectioning and geometric probability as the foundation of stereology. This forms the basis for the derivation of basic stereological principles in Chapters 3 and 4; the parameters dealt with are volume, surface, and length density as well as curvature and topology for components of arbitrary shape. Chapters 5–8 deal with the methods to estimate number, shape and size distribution of discrete particles and Chapter 9 with the thickness of sheets. In the final Chapter 10 the problem of anisotropic structures and stereology is picked up, a difficult topic of great practical importance, but yet for a large part unsolved.

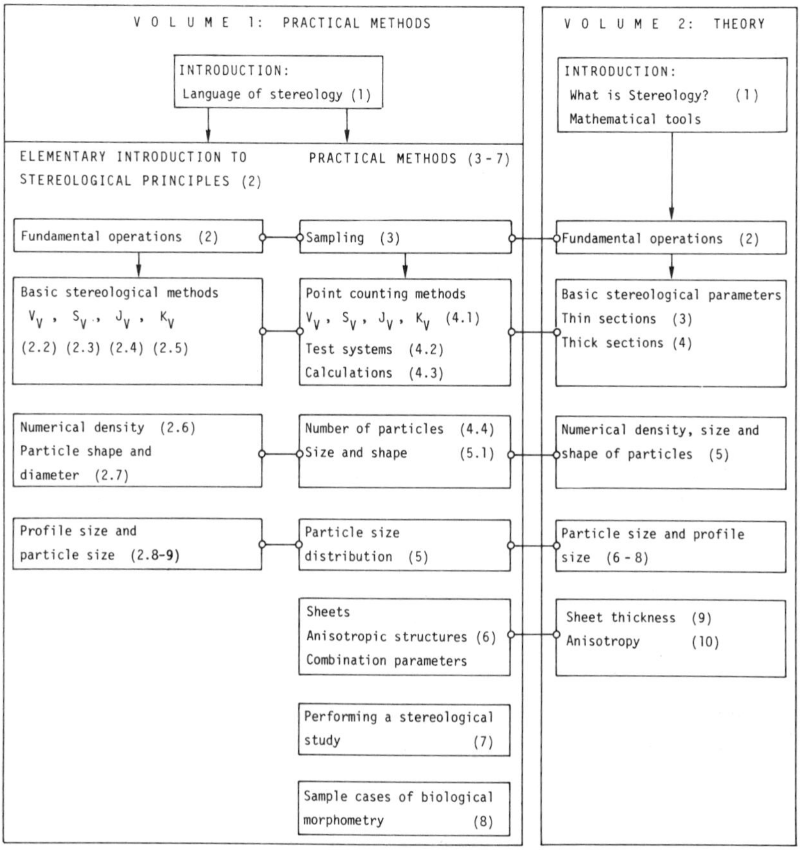

How should this book be used? The above diagram outlines the structure of the book and some of the dependencies. I imagine that most readers from biology will want to start with Volume 1 and consult Volume 2 when the need for deeper understanding arises. A *newcomer to stereology* may first browse or even read through Chapters 1 and 2 of Volume 1 and pick out those bits of information that he finds necessary in order to find out what stereology is about and what it might offer for his own work. He would then probably jump to Chapter 4, reading Section 4.1 to gain an overview and then proceed to those sections of Chapters 4–6 which he may want to put to practical use in his study. When setting out to do the work, he should consult Chapters 3 and 7 which can assist him in proper planning of the study. At one time or another it may be profitable to consult the relevant sections of Chapter 8 where a number of useful practical hints and some specific references to successful use of stereological methods are to be found.

The initiated stereologist may go directly to the presentation of practical methods in Chapters 4–6 and refer to the other chapters when the need comes up. He may familiarize himself with more advanced stereological methods by consulting Volume 2; this gives an adequate background if the development of new methods or substantial modifications of existing methods are required.

If this text is to be used in a course for *students* I would wish that Volume 2 be used to provide them with a sound theoretical basis, presenting the practical side of stereological methods as a result of the theoretical developments, rather than the theory as a justification of some practical methods. As the use of stereological methods becomes more and more sophisticated one cannot do without a sound theoretical basis.

Many a biologist will in time find it necessary to *consult a mathematician* or statistician in trying to solve some tricky problems. If this mathematician is not familiar with stereology—which will often be the case—Volume 2 could serve as a basis for discussion, rather than Volume 1. This would prevent the frequent event that one wastes a lot of time and energy at "reinventing the wheel". The mathematician will also find more appropriate references to the literature in Volume 2 than in Volume 1.

I hope that this book will help a variety of potential readers in "doing good stereology".

January, 1980 Ewald R. Weibel

Errata, Stereological Methods, Volume 1

The following equations should read as follows:

4.22: $$a_T = \frac{\sqrt{3}}{2} \cdot d^2$$

4.61: $$\text{S.D.}^2(R) = \frac{n}{(n-1)(\Sigma P_c)^2}\{\Sigma X_a^2 + \hat{R}^2 \Sigma P_c^2 - 2\hat{R}\Sigma X_a \cdot P_c\}$$

4.93: $$S_V = \frac{S}{V_0} = \frac{S}{V_s \cdot f^{-3}} = S'_V \cdot f_s^3$$

In Fig. 6.5 the midpoint of class 3 is 1.74 and not 1.95.

Acknowledgements

The writing of a book of this kind is a lonely task, but it cannot be done alone. I would therefore like to express my deep gratitude first to all those who have helped me acquire the notions presented in this book, and then those who assisted me in the actual preparation of the text.

The seed was sown two decades ago when I learned to think in terms of mathematical models while working with Domingo Gomez at Columbia University and later on with Bruce W. Knight at Rockefeller University in New York. I was then fortunate to share the decisive moments in the early years of the International Society for Stereology with experienced "stereologists" from various disciplines, and it is particularly through the interaction with Hans Elias, Robert T. De Hoff, Ervin E. Underwood, Herbert Haug, Hans Giger, John E. Hilliard, August Hennig and many other colleagues that I could gradually get a grasp of the basic notions of stereology, that I could learn to correct some of the mistakes I had made and to avoid many of the pit-falls. A decisive event for me was the acquaintance and then friendship with Roger E. Miles to whom Volume 2 is dedicated; it will become obvious from the text, particularly in Volume 2, how much I owe to him for having taught me what mathematical stereology truly is. I hope to have done him justice in my attempts to "translate" some of his thinking into a language accessible for biologists.

Among those who helped me in the process of assembling this text over a period of several years, my secretary Ms Gertrud Reber deserves to be mentioned and thanked in the first place; without her devoted and meticulous assistance this task could not have been accomplished. No lesser gratitude goes to Mr Karl Babl who did all the graphical work for this book, diligently assisted by Ms Marianne Rüfenacht.

The writing of Volume 2 presented me with a number of problems difficult to solve for a non-mathematician. I am most grateful to Dr Luis-Manuel Cruz-Orive for his invaluable assistance in this, particularly in the chapters on size distribution and in preparing the chapter on anisotropy. He also critically read the manuscript and thus helped to improve it.

Finally I should thank the institutions which have generously supported our stereological work, particularly the Swiss National Science Foundation, the University of Berne, and the Roche Research Foundation.

Contents

To ROGER EDMUND MILES

who, from the beautiful building blocks of integral geometry, built a solid mathematical framework for stereology.

CHAPTER 1

Introduction

1.1. What is stereology?

Literally, "stereology" means "the knowledge of space", a meaning which is much more general and broad than what Hans Elias had in mind when he introduced the term in 1961: he was looking for a label for a newly emerging branch of morphological science where positive attempts were made to quantitatively interpret two-dimensional microscopic observations in terms of the three-dimensional structures to which they relate. Microscopic images are necessarily flat projections, and this is particularly troublesome if one observes a section of a three-dimensional specimen. If the internal make-up of this specimen is unknown, as it very often is, one may have very few leads for arriving at a clear vision of its true three-dimensional structure. This has resulted in many misconceptions of natural structures because we are psychologically not prepared to think in terms of sections.

This is where stereology is expected to be of value: by giving the process of sectioning due consideration, methods can be developed which allow observations made on sections to be interpreted in terms of the spatial structure of the specimen. Particularly, a quantitative description of the structure should become possible if the probability of obtaining certain observable section images is considered. Thus it is not astonishing that stereology will be based on mathematical methods to a very large extent, as it is reflected in the following *definition* of the term in its original meaning:

Stereology
is
a body of mathematical methods
relating
three-dimensional parameters defining the structure
to
two-dimensional measurements obtainable on sections of the structure

This definition is, however, too narrow in that it only considers the step from a "two-dimensional section" to a "three-dimensional structure". This resulted from the most common practical situation where the microscopist uses flat sections to study spatial structures. However, this restriction is not only unfortunate, it is unnecessary: the same general principles hold, in fact, if we cut a planar structure or a spatial structure with a line. Miles (1972) has established what he called a "multidimensional perspective for stereology", meaning that we may consider any n-dimensional structure and study the result of sectioning it with any s-dimensional section: the principles are always the same. One can, hence, easily modify the definition of stereology by replacing "three- and two-dimensional" by "n- and s-dimensional".

In terms of a scientific discipline there are two sides to stereology: on the one hand it offers the microscopist a *practical methodology* by which he can infer, by calculation, parameters of structure from some measurements performed on sections; this was dealt with in detail in Volume 1. On the other hand there is an intrinsic *theoretical* interest in stereology which has begun to attract mathematicians and which, in fact, is of fundamental importance for establishing sound foundations for *stereology as a science.* This is the topic of this second volume.

Although much of the original work in deriving stereological methods has been done in a pragmatic way, trying to look for solutions to particular problems with which the microscopist was faced, this is not the optimal approach if methods of some generality are to be derived. The proper approach is that of the mathematician: in trying to solve a problem he will assume and define a model which is convenient for his analytical approach and which is as general as possible; in the present case such a model will have properties of a structure. He will then also assume and define a sectioning process by which the model structure is cut. If his choice is appropriate he should now be able to predict the result of the sectioning process, the "section image", in quantitative terms. This is the approach we shall pursue in this volume.

To the practical stereologist this may sound esoteric, or even arbitrary, but it is not. Indeed, we cannot expect to derive and prove rigorously any method of a mathematical nature without choosing this heuristic approach. The practical advantage is that the model structures and the sectioning process can and must be precisely defined: this alone makes it possible for the practical stereologist to test whether a particular method is applicable to a given practical case.

One of the fundamental peculiarities of stereological methods is that they should allow us to infer certain parameters of the structure from incomplete information, e.g. from information gathered on sections of a dimension lower than that of the structure. This has two consequences: (1) certain parameters of structure—e.g. connectivity between parts—are not accessible to a straightforward stereological evaluation; (2) the section being a sample of a particular type, stereological methods are statistical in nature and measurements obtained on these samples yield estimates which are averages and are affected by some sampling error. This limitation actually requires that the unit (or object part of the structure) on which the parameter is defined occurs in large number within the structure, and model structures must be made to comply with this rule. Thus it is fairly easy to estimate the total surface, or the average diameter, or even the size distribution of a population of particles, but it is most difficult, and often practically impossible to estimate the surface or the diameter of a single object by stereological methods, unless we are able to cut a large number of random sections from this one and only object. We will therefore suppose that the structures subjected to stereological analysis are made of a very large number of similar units, similar at least in the sense of the parameter whose value is to be estimated. But this requirement is, actually, closely related to the definition of the term "structure" as "something made of interdependent parts in a definite pattern of organization" (Merriam–Webster).

Likewise there are some limitations on what we mean by a "section", for which it would be more appropriate to use the term "probe". A *section* or *probe* in the sense of stereology is a statistical sample of the structure, applied to the structure from outside. One important notion—implicit in the term "statistical sample"—is that this probe be independent of the structure, i.e. not influenced by the make-up of the structure. The section image we observe must be a chance intersection between structure and probe. Great emphasis must therefore be placed on properly defining the meaning of random sectioning in each case.

Finally, stereological methods are mathematical methods, and sound methods can only be derived by rigorous mathematical reasoning. It is unavoidable that this will, on occasion, require the use of advanced mathematics difficult to follow for the unexperienced. In fact, it is the job of the mathematician to do this for us, and we non-mathematicians must trust him. In this second volume I have tried to present some of the most important derivations in such a way that a non-mathematician with a minimal background in mathematics should be able to follow the reasoning, even if he cannot reproduce the derivations on his own. Perhaps this presentation will allow him to build a bridge for communication with mathematicians unfamiliar with stereology, which often proves very useful.

1.2. Landmarks in the history of stereology

In the spring of 1963 a small group of scientists from fields as wide apart as metallurgy and biology, geology and medicine met in Vienna for what was ambitiously called the "First International Congress for Stereology". Apart from them nobody knew what "stereology", a newly coined term, was; those practising it were few and somehow outsiders.

The Vienna meeting was the result of an informal meeting of a dozen people whom Hans Elias had gathered on the Feldberg in Germany in 1961 because of their common interest in quantitative problems in histology. A note on this meeting—at which the term "stereology" was invented—published in Science led to a response by metallurgists and geologists who shared with the Feldberg biologists the problems of section analysis. Out of this grew the first interdisciplinary gathering in Vienna where the International Society for Stereology was formally founded, with Hans Elias, its main promoter, as first President.

Stereological methods were not invented in 1963. Their history rather begins in 1777 when Jean Louis Leclerc, Comte de Buffon, formulated the first problem of geometric probability theory—and solved it: he asked what the probability was that a randomly tossed needle of length l intersects a grid of parallel lines of distance d. As will be developed in a subsequent chapter (Section 3.2.3) the answer to this question was that

$$\mathrm{Pr} = (2/\pi) \cdot (l/d)$$

which means that the probability of having an intersection depends on the ratio l/d reduced by a factor $2/\pi$ which results from integration of the angular orientation of the needle.

Evidently, this so-called Buffon principle was not yet a stereological principle, but it contained the conceptual elements of the latter: there was a *structural object*—the needle—characterized by its dimension; there was a geometric *reference system*—the grid—also quantitatively defined; the random interaction between the two led to an *observable result*—the intersection—which occurred with a predictable *probability*. This characterizes all stereological principles.

The definition of stereology states that "stereology is a body of mathematical methods which relate parameters defining three-dimensional structures to measurements obtainable on two-dimensional sections". From this it is evident that the structural objects are the elements of a three-dimensional material or tissue and that the fundamental reference system is a plane section cut at random across the material. The observable results are the transections or profiles of the structure elements seen on the section. In view of this definition the origin of stereology must be located in 1847 when the French geologist Auguste Delesse presented his classical paper entitled "Procédé mécanique pour déterminer la composition des roches" and demonstrated that the areal fraction of a section covered by transections of a certain component was equal to the fraction of the rock volume occupied by this component. In present-day terminology this would mean that the volume density of a component is equal to the areal density of its profiles on a random section. We are going to develop this fundamental principle in detail in a subsequent chapter (Section 3.1.1). Delesse proposed to trace the section image on a sheet of heavy paper, to cut out its component profiles and to weigh them. This was evidently an awkward procedure, but it is still used today!

In 1898 this first stereological principle was modified and extended by A. Rosiwal, again a geologist. He demonstrated that one could randomly draw a line on to the section and measure the length of the line segments included in profiles of a certain component. In doing this, one can easily determine the fraction of the test line included in this component, and Rosiwal showed that this is again equal to the volume fraction occupied by the component. This evidently greatly simplified the analytical procedure as compared to Delesse's "cut out and weigh" method.

It took another three decades until the last step in this development was made by the geologists Thomson and Glagolev when they showed independently, in 1930 and 1933 respectively, that one could use a test grid made of a set of points to estimate volume density. If this grid of points is randomly deposited on the section the fraction of the points falling on profiles of the component of interest is equal to the fraction of the volume it occupies. This method was evidently very simple to apply, which was a progress. But it had another virtue which was more theoretical in nature: it showed clearly the statistical or probabilistic nature of stereological principles. It is conceptually quite close to the principle of Buffon: the structural elements are those parts of the structure which belong to a certain component; the reference system is a set of test points, which is randomly tossed onto a section, i.e. actually into the structure; the observable result is the location of the points within the various components; the probability that a point lies in a particular component is predictably given by the fraction of the total volume of the structure occupied by this component, as will be discussed in Section 2.2.

For biologists another historical landmark is important. In 1943 Harold Chalkley at the National Cancer Institute in Bethesda independently proposed an identical point counting method and thus introduced stereological methods into biological research.

Only relatively recently was it recognized that the surface area of spatial objects could also be estimated on sections by similar procedures. In 1945 the Russian metallurgist Sarkis Saltykov showed that (1) the boundary length of profiles on sections was proportional to the surface area of embedded objects and (2) that this surface can be estimated from the number of intersections the boundaries form with a random test line system. In the same year, Tomkeieff independently developed an identical method, but formulated it somewhat differently. Both papers remained largely unknown, so that the fundamental principle of stereology, which states that the surface density, S_V, is equal to twice the intersection density on test lines, I_L, could be rediscovered several times, most notably by Smith and Guttman in 1953 and by Hennig in 1956. It should be noted that this principle is again very closely related to the Buffon principle. As we shall show later on, the length of the boundaries of profiles appearing on sections can be decomposed into short straight lines of equal length which correspond to Buffon's needles. The number of "needles" which intersect the test line grid, related to the number of needles making up the boundary length must evidently correspond to the probability of obtaining an intersection between boundary and test lines, and this follows from the considerations of the Buffon principle.

A very similar method was originally proposed in 1949 by Chalkley, Cornfield and Park for the estimation of the surface-to-volume ratio of objects studied on sections. This method had much resemblance to the Buffon principle in that these authors proposed to toss short test lines of known length on to a section and to obtain two counts: the number of intersections between profile boundary and test line and the number of end points of the test lines lying within the profiles. This is evidently a combination of point counting for volume and intersection counting for surface estimation. It will be shown later that such combination methods allow the determination of useful parameters, such as the thickness of sheets (Weibel and Knight, 1964). In the context of pointing out historical landmarks one additional method of this kind should be mentioned, namely that of estimating mean curvature of surfaces from a count of intersections and tangents. This principle was independently and simultaneously developed by John Cahn and Robert De Hoff in 1967.

Both Saltykov (1945) and Smith and Guttman (1953) noticed that the principle establishing the probabilistic relationship between the area of a surface and the length of a line, expressed by the number of intersections formed when they interact in a random fashion, could also be used to estimate the length of linear features of the structure—edges or filaments. The reference system is now a set of planes, the sections, and one has to count the number of times the lineal features transect the plane of defined area.

The principles so far described in their historical perspective all aimed at an estimation of composition of the structure in terms of different materials but they were not concerned with the individual objects building the structure. A second group of stereological principles explicitly considers the individual object or element of structure when they aim at determining the number and size of particles. It is intuitively plausible that section profiles will mostly have a smaller size than the object from which they were derived. Stereological considerations will therefore have to define methods by which we can transform measurements of profile size into dimensions describing the size of the objects. This problem was first tackled in 1925 and 1926 by the Swedish scientist S. D. Wicksell when he derived the relationship between the size distribution of spherical and ellipsoidal particles and that of their profiles on section. Wicksell presented a very profound theoretical treatment of the problem which is still worth studying; but he also proposed practical solutions for determining particle size from measurements of profile diameter.

Wicksell had derived his interest in this principle from a biological problem presented to him by an anatomical colleague. He called it, "The corpuscle problem. A mathematical study of a biometric problem", and published two papers in a biometric journal. This is probably why these classical papers remained unknown to the materials scientists who had similar problems to solve. Beginning with studies by Scheil (1931), Schwartz (1934) and Krumbein (1935), a whole sequence of papers treated this problem anew, but from the point of view of the materials scientist, and derived methods in part quite similar to those of Wicksell. Since these early years a large number of different procedures have been proposed; this is due to the fact that the transformation of profile to particle size distributions depends on a number of assumptions so that a perfect method applicable to the general case has not yet been found.

The description of a material or tissue in terms of the particles it contains also requires the estimation of particle number per unit volume, and this should optimally be obtained from profile counts on sections. The fundamental stereological principle states that the number of profiles found on the unit area of section is equal to the number of particles in the unit volume multiplied with the mean particle diameter. This formula is already contained in Wicksell's original papers on spherical (1925) and ellipsoidal particles (1926); it is however mostly attributed to De Hoff and Rhines (1961) who presented a rigorous derivation.

The historical development of stereological principles described so far was characterized by a pragmatic approach which explains why its path was in some way erratic, certainly not systematic. Alongside this pragmatic development a branch of mathematics evolved which worked out the theoretical basis of geometric probability closely related to stereology. For too many years there was too little contact between those biologists or materials scientists who developed stereological methods for immediate use in their research and the mathematicians rigorously studying geometric probability. This is unfortunate as can be shown in the case of volume density estimation. We should remember that, following Delesse's fundamental development, Rosiwal introduced in 1898 the method of "linear integration" and Glagolev proposed in 1933 the method of point sampling. This last method was, however, already rigorously introduced on mathematical grounds by M. W. Crofton in his article on probability in the Encyclopedia Britannica in 1885! In recent years some most valuable contributions have come from the fields of geometric probability theory and integral geometry through the work of Santaló (1953, 1976), Hadwiger (1955), Kendall and Moran (1963), Giger (1967, 1970), Moran (1972) and Miles (1972), as is perhaps best reflected in the proceedings commemorating the 200th anniversary of Buffon's needle problem (Miles and Serra, 1978).

At the time of writing, the process of giving stereology a sound mathematical basis is in full swing. Indeed the many new insights have made it necessary to revise and modify the presentation of the theoretical foundations of stereological methods all along, and it is likely that the coming years will broaden and deepen our understanding of these methods even more. The account given in these pages attempts to present a current view of the theoretical foundations of stereology, but this has, definitely, not yet become "historical"!

Finally, the practical application of stereological methods greatly depends on the development of adequate technology. The first principle that was used for instrumental application was Rosiwal's "lineal integration" method for volume fraction analysis. In 1916 Shand developed the first scanning stage for a light microscope which greatly facilitated this work; the specimen was driven across the microscope field containing a hair-line cross by means of two co-linear micrometers, one being assigned to the particular phase under investigation, the other one to all the other phases. The investigator had to move the specimen with one micrometer until the hair-line cross changed into the other phase; then he used the other micrometer, and so on. The total lengths of traverse of the hair-line cross through the two phases could be directly read off the micrometers. Improved variants of this original instrument, mainly those allowing for the simultaneous analysis of various phases, have been extensively used in materials sciences. In Volume 1, Section 7.4, we discussed in detail the tools for this type of analysis, including point counting and profile sizing, in which interpretation of the image is left to a human observer.

It is evident that automatic methods of image analysis were introduced into stereology as soon as they became available. The basis of these instruments is the use of a linear scan combined with a sensitive photodetector which records the phases appearing in a small spot of the microscope image on the basis of contrast differences. The first instrument of this kind was developed from the scanning stage (Lagerkrantz, 1948). A more powerful variant was the flying spot microscope, first developed by J. Z. Young and F. Roberts in 1951; here a fine spot of light produced by a cathode ray oscilloscope was projected in demagnified form onto the stationary specimen through eyepiece and objective and rapidly scanned. A photomultiplier picked up the signal emitted from the specimen and this signal was transferred to another cathode ray tube which was synchronized with the light-emitting tube. Again, the length of the scan traverse in a given component can be recorded to yield, for example, volume density from the Rosiwal principle. Many variants of these scanning instruments have been produced and used (cf. Fischmeister, 1968).

The most recent development is the use of television cameras to cut up the microscope image into a dense lattice of fine lines or points; this was first realized in the Quantimet B by Metals Research in England, which was put on the market in 1967. These instruments are still in full development and hold much promise, as discussed in Volume 1, Chapter 7.

In connection with the development of such instruments important progress has been made in the field of mathematical image analysis and pattern recognition, as exemplified by the work of G. A. Moore (1972) and K. D. Preston (1973), or by the concepts of mathematical morphology pioneered by G. Matheron (1967) and J. Serra (1969, 1978).

Point counting methods are basically so simple to apply that they have not called for much sophisticated instrumentation. However, their application is quite time-consuming; the advent of efficient mini-computers in recent years has greatly contributed to rendering these methods more efficient.

In summarizing the historical development of stereology during the past two centuries since Buffon's first formulation of a problem of geometrical probability, we can note a long early phase, spanning from 1847 to about 1960, in which basic stereological principles have been developed in a pragmatic way in materials sciences and in biology. Alongside, and for the most part without any interaction with stereology, mathematics has developed the sophisticated fields of geometric probability and integral geometry. Since 1960 the main efforts have been on trying to develop a unified theory of stereology, to strengthen its mathematical foundations, and to improve the methods of their practical applicability. In this last decade one also notices a surge of application of stereology both in the materials sciences and in biology.

1.3. The language and symbolism of stereology

As is the case with most scientific disciplines stereology has developed its own language, particularly by giving some terms a restrictive connotative meaning. Some of these terms need to be explained.

A *structure* is generally defined as "something made up of interdependent parts in a definite pattern of organization" (Merriam–Webster). This definition holds for stereology as well: a structure is made of at least two parts or components, and to clarify the pattern of organization—degree of interconnections between the parts, their relative size etc.—is the declared purpose of stereology.

The *components* of the structure can be described in various ways. First, the different components must be clearly separated and identifiable; regions of transition from one to the other component (e.g. zones of diffusion) are difficult to handle. We call a *phase* the aggregate of all parts which are identical in nature; e.g. all mitochondria of a cell constitute its mitochondrial phase. If the component is made of discrete elements that could be isolated as units we will call these *objects* or *particles*. The implications of this concept of structure on mathematical reasoning in terms of set theory will be discussed in the next section.

The particles may have an arbitrary shape, but one specially important class of particles are *convex solids*; they are defined as follows: the line segment connecting any two points of the solid must lie totally within the solid, in other words, a line traversing the solid forms only one chord or intercept and likewise a section hitting the solid forms only one profile.

We shall, in the following, often describe a structure quantitatively as

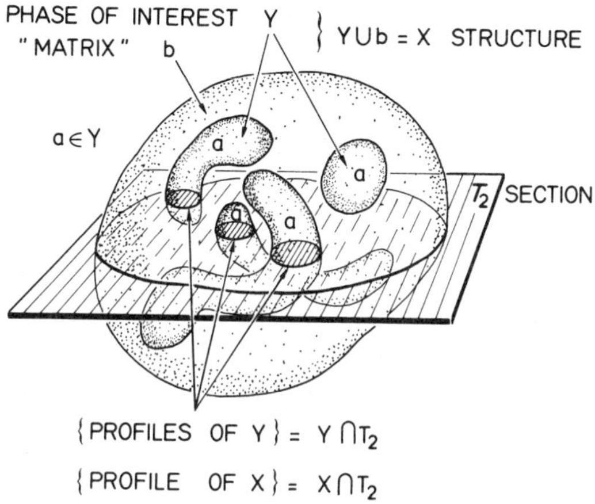

FIG. 1.1: Model structure containing component objects *a* embedded in matrix *b* cut by plane section T_2.

the "containing space", i.e. the space containing the components of interest, often together with some "uninteresting" phases. A structure of this kind is shown in Fig. 1.1. The fundamental quantitative descriptor of the make-up of the structure is the *density* of the various components within the structure, i.e. within the containing space. A "density" is defined as "the quantity per unit volume, unit area or unit length" (Merriam–Webster). The basic quantities we shall be interested in are the volume of the component phases, their surface, their length or the number of their particulate elements, so that the terms *volume density*, *surface density*, *length density*, *numerical density* describe these quantitative properties of the structure in relation to the structure space. The volume density is the volume of the phase in the unit volume of the structure, the surface density the surface area of the phase contained in the unit volume of the structure, etc.

The concept of density is particularly useful in stereology because of the very nature of the sampling process used: sectioning. A *section* is defined as a plane intersecting the structure; it may hence also intersect some of the components of the structure. In terms of set theory (see Section 1.4) the "image" of the structure on the section plane, called "*profile*", is the intersection between the two sets "section" and "structure". It is intuitively understandable that a relationship must exist between the density of component profiles within the profile of the structure (the "quantity of profile per unit area") and the density of components within the structure.

Since we considered a section to be a set intersecting the set "structure" it is evident that the "section" need not necessarily be a plane; it can just as well be any other *geometric probe*, such as a *test slice*, a *test line* and a *test point set*. In either case, the image of the structural composition on these probes, the intersection between section and structure, again quantitatively reflects the make-up of the structure: the density of *intercept length* of some component on the (unit length) of test line is related to its volume density, and so is the "density of points" (per unit point set) falling into the component related to its volume density in the structure.

This narrative description of the nature of stereological procedures reveals one important point: all stereological measurements are in principle obtained as relative measurements, more precisely as a ratio of at least two joint measurements: one relating to the components, the other to the structure as a whole; the latter is called the "reference system". Stereological principles establish the precise relationship between such ratios measured on sections, and the corresponding ratios in the spatial structure.

This fundamental nature of stereological principles is reflected in the set of *stereological symbols* that have come into general usage over the past decade: the ratios are noted as a double symbol of two capital letters: the first capital letter defines the component parameter, the second, usually written as a subscript, the parameter of the containing space or, more generally, of the reference system. For example, the numerical density of some particulate component (b) in the containing space (c) is written as

$$N_V = N(b)/V(c) \tag{1.1}$$

or the areal density of the profiles of (b) in the profile of the structure (c) as

$$A_A = A(b)/A(c) \tag{1.2}$$

The list of the symbols used in this text is given in Table 1.1, which is

Table 1.1
List of stereological symbols

Definition	Symbol	Alternative symbols	Dimension
1. *Parameters of structure*			
Volume (of component or structure)	V		cm^3
Surface area (of components)	S		cm^2
Length of linear feature in space	J	L	cm
Mean curvature, integral	K	M, H	cm
Total (Gaussian) curvature, integral	G		cm^0

Table 1.1 (*Continued*)

Definition	Symbol	Alternative symbols	Dimension
Number of objects	N		cm^0
Diameter (of spheres)	D		cm
Caliper diameter	H	D	cm
Sheet thickness	τ		cm
2. *Parameters of profiles on section area*			
Area of profiles	A		cm^2
Boundary length of profiles (perimeter)	B	L, P, C	cm
Transections of linear features	Q		cm^0
Tangents to profile boundary	T		cm^0
Number of profiles	N	n	cm^0
Diameter (caliper) of profiles	d, h		cm
Curvature of plane figure	C		
3. *Parameter on test lines*			
Length (total) of profiles on test line	L		cm
Intersections with surface or profile boundary	I	P, N, C	cm^0
Length of intercept or chord	l	L_2, L_3	cm
Number of intercepts	n	N	cm^0
4. *Parameters on test point sets*			
Point number on profiles or structure	P		cm^0
5. *Parameters of test sets*			
Test volume	V_T		cm^3
Test area	A_T		cm^2
Test line length	L_T		cm
Test point number	P_T		cm^0
Section (slice) thickness	t	T	cm
6. *Densities in space*			
Volume density	V_V		cm^0
Surface density	S_V		cm^{-1}
Length density in space	J_V	L_V	cm^{-2}
Mean curvature density	K_V	M_V	cm^{-2}
Total curvature density	G_V		cm^{-3}
Numerical density	N_V		cm^{-3}
7. *Densities on test sets*			
Areal density	A_A		cm^0
Boundary length density	B_A		cm^{-1}
Transection density on area	Q_A		cm^{-2}
Numerical density of profiles on area	N_A		cm^{-2}
Tangent density on area	T_A		cm^{-2}
Intercept length density on line	L_L		cm^0
Intersection density on line	I_L		cm^{-1}
Intercept number density	N_L		cm^{-1}
Point density	P_P		cm^0

repeated in slightly expanded form in Appendix 1. There has been a tendency for unification of symbols in the International Society for Stereology, so as to facilitate reading of articles by various authors. Whilst this endeavour is certainly laudable we have found it impossible to restrict the symbols to a few selected letters because this can lead to ambiguities. For example, it is inadequate to use the letter P ("points") for both point hits by a point lattice *and* intersection points between a test line and a surface; we have therefore introduced the symbol I for "intersections" and have retained P for "point hits". Likewise the use of L for both test line length and length of a profile boundary leads to ambiguities, so that we have substituted B for "boundary length". The second column in Table 1.1 lists alternative symbols used by other authors.

A particular problem concerns the identification of the phases or components to which the parameter symbol relates. We have chosen to use lower-case arabic letters for this purpose and to attach them to the parameter symbol either as a subscript or in parenthesis, such as

$$N_b = N(b) \tag{1.3}$$

for the number of objects of phase b. The numerical density is then written as

$$N_{Vb} = N_V(b) = N(b)/V(c) \tag{1.4}$$

If the containing space or reference system needs to be identified, its symbol is added following a comma:

$$N_{Vb,c} = N_V(b, c) = N(b)/V(c) \tag{1.5}$$

In such cases, one could evidently just as well write out the ratio in full; the main advantage of using the stereological double symbol, followed by phase identifiers, is that it reminds us that stereological estimates are never absolute but always relative estimates: we cannot obtain $N(b)$ by stereological methods, but only $N_V(b, c)$! In the next section we shall see that mathematical rigour and clarity often require even more complex symbols to be introduced in order to make the mathematical statements unambiguous.

1.4. Introducing the concepts of set theory and the corresponding symbolism

Some of the subsequent developments of stereological principles are based on "modern mathematics" which makes heavy use of the concept of sets and their interactions. This is not the place to write at length on this basic topic; the reader is referred to any of the good texts dealing specifically with set theory, e.g. Green (1965). We shall merely explain the use of set theory made in the present text and the terminology and symbolism that needs to be introduced for that purpose.

The concept of a *set*—defined simply as a collection of "things" called *elements*—is evidently natural in stereology: the definition of a structure given in the preceding section is clearly that of a set made of a number of elements or objects. In Fig. 1.1 the structure contains a number of elements or objects labelled a; all these objects taken together form the phase of interest Y, and one says that a belongs to the set Y and uses the symbol $\in$ for "belongs to", so that

$$a \in Y$$

is a notation for the statement "a is an *element* of the set Y". The negation of this, namely that a does not belong to Y, is denoted by $a \notin Y$. On the other hand, we observe that the objects of set Y are all contained in the structure X, so that Y is a *subset* of the set X, denoted by

$$Y \subset X$$

which means "Y is contained in X", or, alternatively,

$$X \supset Y$$

meaning that "X contains Y as a subset".

A structure X is, by definition, made of various subsets, at least of two. In general, one of these represents the phase of interest which we call Y, whereas the other components are considered the "matrix" b, in which Y is suspended. The total structure X is then defined as a set by the *union* of Y with b (Fig. 1.1):

$$X = Y \cup b \tag{1.6}$$

where the symbol $\cup$ means "union", i.e. that Y and b are joined together. It is clear that b is the difference between the sets X and Y, and is called the *complement* of Y. We should note that, in terms of set theory, this is a special case because we have assumed Y and b to be clearly separated phases. In set theory this is not a condition (Fig. 1.2): if two sets, a and b, overlap, their union $a \cup b$ describes the set for which each element (each point, for example) is contained either in set a or set b or in both (Fig. 1.2). This kind of set does not appear compatible with the condition that the phases must be clearly separated for stereological principles to be applicable; however, we shall make use of it on occasion when attempting to build structures of some generality.

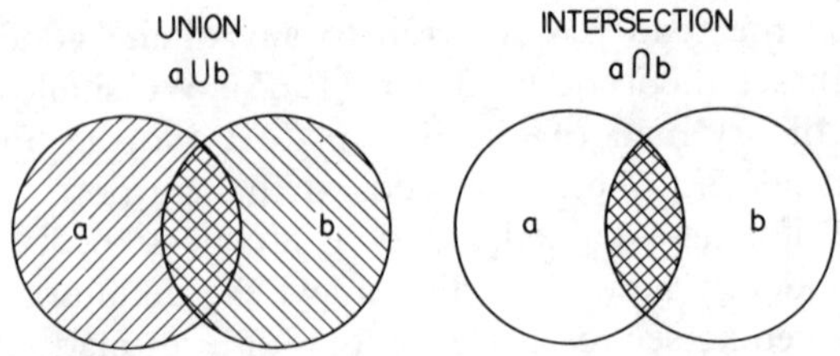

FIG. 1.2: Venn diagrams explaining union and intersection between sets a and b.

Looking at Fig. 1.2 we see that part of the set $a \cup b$ is common to both subsets a and b; this part is called the *intersection* of a and b (Fig. 1.2) written as

$$a \cap b$$

where $\cap$ means "intersection". This is an important operation in stereology, because it allows us to express the "image on a section" in terms of set theory. Consider the specimen X cut by a section T (Fig. 1.1) where both X and T are sets; the part of the specimen contained in the section is now clearly the *intersection* of X and T:

$$X \cap T$$

And, furthermore, if X contains a phase of interest Y the part of Y contained in the section is

$$Y \cap T$$

This now opens the possibility of introducing some quite unambiguous notation for stereological parameters. Thus, if X_3 is a three-dimensional structure containing a three-dimensional object phase Y_3, and T_2 is a two-dimensional plane section, then the intersection between X_3 and T_2 is a plane whose area is

$$A(X_3 \cap T_2)$$

whereas the "profile" area of Y_3 is

$$A(Y_3 \cap T_2)$$

and accordingly

$$A_A = A(Y_3 \cap T_2)/A(X_3 \cap T_2) \tag{1.7}$$

symbolizes the areal density of profiles of Y_3 on the profile of X_3, both with respect to the intersecting set T_2, the section. Similarly, if Y_2 is the set of points describing the interface between the phase of interest and the matrix, its intersection with T_2 is a trace length

$$B(Y_2 \cap T_2)$$

and

$$B_A = B(Y_2 \cap T_2)/A(X_3 \cap T_2) \tag{1.8}$$

is the length density of the surface trace on the section area. In both instances, this notation makes it absolutely clear that areal or length density are expressed in relation to the profile of the specimen, $X_3 \cap T_2$, and not, simply, with respect to the section plane. This notation introduces rigour and will therefore be extensively used throughout this text.

When dealing with geometric probability we will have to introduce an additional symbol which, at first sight, bears some similarity with $\cap$. We shall talk about the probability, $\Pr\{\cdot\}$, that a probe (or section) T hits the specimen X or the object phase Y, and use the symbol $\uparrow$ for "*hit*", as introduced by Miles (1972). Thus

$$\Pr\{T \uparrow Y\}$$

means "the probability that the probe T *hits* the object Y". The distinction between $\uparrow$ and $\cap$ is important because $\{T \cap Y\}$ can be an "empty set" $\varnothing$, i.e. the statement includes the possibility $\{T \cap Y\} = \varnothing$ where there is no intersection. In contrast $\{T \uparrow Y\}$ positively requires that T hits Y and that an intersection is formed.

An important logical symbol introduced in this connection is the *conditional* bar, meaning that a certain event or statement, a, is only valid under the restrictive condition that another event, b, also occurs:

$$(a \mid b)$$

The first example to be mentioned is that the probability $\{T \uparrow Y\}$ can only be worked out under the condition that T also hits X, so that we must seek the conditional probability

$$\Pr\{T \uparrow Y \mid T \uparrow X\}$$

because any section that does not hit the specimen may not be considered in deriving this probability. Another example relates to the orientation of the specimen with respect to the section plane; if the orientation is fixed at angles (θ, ϕ) we may derive the conditional probability

$$\Pr\{T \uparrow X \mid \theta, \phi\}$$

which we will find to be related to the caliper diameter of X

$$H(X \mid \theta, \phi)$$

measured under the condition that the specimen is held at orientations θ and ϕ.

Closely related to the conditional probability are such terms as the conditional probability density function $f(a \mid b)$ or the conditional average $E\{a \mid b\}$. The reader will find some of the more frequently recurring symbols explained in Appendix 1.

For the uninitiated this kind of symbolism appears clumsy and difficult to grasp, at first, but it will soon be found that it is convenient because it clearly exposes some of the conditions necessary for the derivation of stereological principles.

CHAPTER 2

Fundamental operations in stereology

2.1. Sectioning

The fundamental operation of stereology is to section a block of material and to perform some observations on the internal region thus exposed. Stereological formulae permit to transform these observations back to the three-dimensional material. Object properties of the material leave traces on the section and it is from these that we extract the required information. We must therefore first define the term "section" in a general way, and then examine the effects of sectioning on the dimensionality of traces.

Our common understanding of the term "section" implies a solid to be cut open so as to expose an internal plane. This internal plane is our region of observation; it is a sample of the solid. This sample is a two-dimensional space, whereas the solid is a three-dimensional space. It is hence immediately evident that the trace of objects on a two-dimensional section can at most be two-dimensional.

Let us look at a polyhedron as an object (Fig. 2.1); it can be characterized by its volume, V, by the surface area of its faces, A, by the length of its edges, L, and by the number of its corner points, P. Let a plane intersect this polyhedron; the resulting trace will be a polygon, which we can characterize by its area, A', the length of its edge, L', and the number of corner points, P'. But it is quite evident that A' is a trace of V, L' a trace of A, and P' a trace of L, whereas P is not represented on the trace, since the probability that a true plane hits a true corner point is zero, as will be shown later. In general terms, we can conclude that, when sectioning a solid with a plane, the dimensions of the parameters describing the trace are by one smaller than the equivalent parameters describing

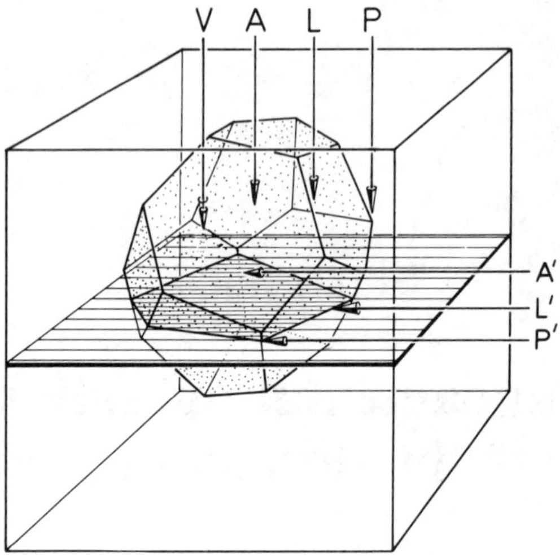

FIG. 2.1: Model structure in form of a polyhedral object contained in cube. A plane section exposes a polygonal profile of the object.

the solid. In mathematical terms calling the dimension of an object parameter $d(Y)$ and that of the corresponding trace parameter $d(Z)$ we have

$$d(Z) = d(Y) - 1 \tag{2.1}$$

This can be generalized if we accept a broader definition of the term "section". Using the concepts of set theory we had defined a structure as the union of its various non-overlapping subsets represented by the objects or phases in a two-phase system (Fig. 1.1):

$$X = Y \cup W \tag{2.2}$$

Similarly we have defined the section as a set, T, so that the trace, Z, would then be the intersection (in terms of set theory) between the set T and the set X. The total trace of the structure would be

$$Z(X) = T \cap X \tag{2.3}$$

and the trace of phase Y

$$Z(Y) = T \cap Y \tag{2.4}$$

The meaning of these expressions is illustrated in Fig. 2.2 using a simple three-dimensional "real" model. In very general terms we will consider the sets Y, W, X and T as point sets, i.e. collections of densely spaced points. In Y, W and X they are grouped in three dimensions to form

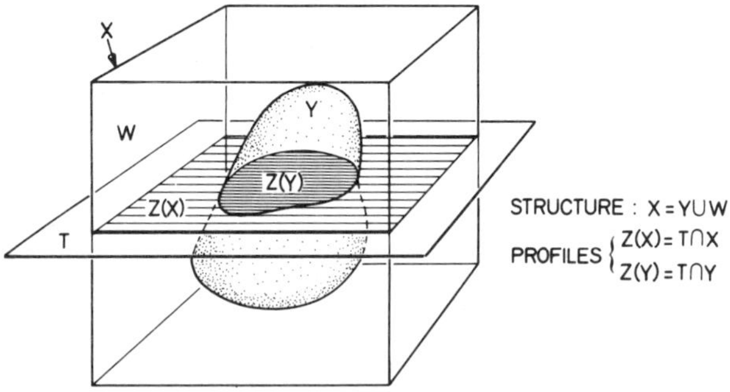

FIG. 2.2: Relationship between structure, section and profiles (or traces) interpreted on the basis of set theory. The intersection between structure and section plane produces a two-dimensional profile or sample.

solids, in the section T in two dimensions to form a plane. The intersection means that we are picking out those points of the solid set which are at the same time included in the plane set.

Now, set theory does not specify in which way the points must be grouped; we have complete freedom. Let us therefore play a bit with this general notion and first vary the arrangement of the point sets building the structure, and then those building the section, in order to see what types of traces we may expect. Since we are, for the time being, only interested in working out the way in which the section interacts with the object we shall only look at one object, set Y, and ignore the embedding matrix. And we assume again that this object is a polyhedron (Fig. 2.3). In terms of integral geometry this can be considered a convex solid or a closed point set, closed because it has an outer surface. We can now define four point sets which make up the polyhedron (Fig. 2.3):

Y_3 = {the points internal to the surface}. It is evident that the number of the points of set Y_3 is proportional to the volume of the polyhedron.

Y_2 = {the points forming the polygonal surface elements}. The number of points in Y_2 is proportional to the surface area of the polyhedron.

Y_1 = {the points forming the edges}. The number of points in Y_1 is proportional to the edge length of the polyhedron.

Y_0 = {the points on the corners}. Their number defines the number of vertices of the polyhedron.

The following expression holds:

$$Y = Y_3 \cup Y_2 \cup Y_1 \cup Y_0 \tag{2.5}$$

In this sense Y is a closed set, whereas Y_3 is the corresponding open set.

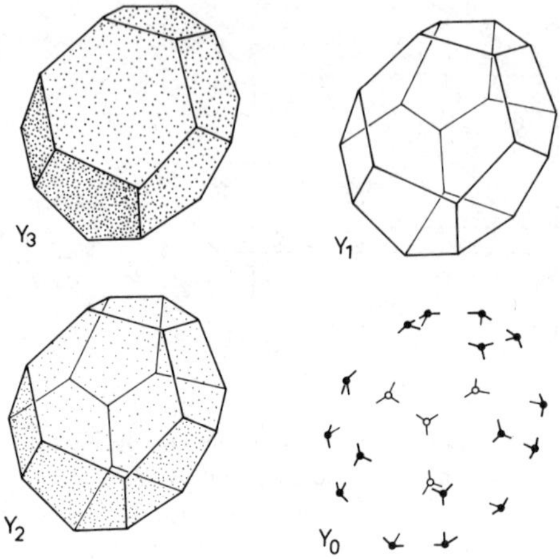

FIG. 2.3: Decomposition of polyhedron into four subsets.

It should be noted that, if Y is embedded in W and is therefore a subset of X, Y_2 will be the point set at the interface between Y and W. In considering Y_2 to be part of Y we must leave W an open set; in other words, W has no surface set. This actually has the following consequence: we may define Y_2 just as well as a subset of the structure X:

$$Y_2 \subset X$$

and if the number of points making up X is proportional to the volume of the structure, then the ratio of the points in Y_2 to those in X is the surface density. Similarly,

$$Y_1 \subset X$$

$$Y_0 \subset X$$

can define the length density of edges or the numerical density of corners, respectively, just as

$$Y_3 \subset X$$

defines the volume density of the object in the structure.

After these parenthetical remarks on the relationships between the subsets of Y and the structure as a whole let us now define the "section" in similar terms. As with the polyhedron, the points making up the set "section", T, can—at our discretion, because we as investigators impose the section—be arranged as a 3-, 2-, 1- or 0-dimensional set. We will however demand that this set be always "flat" (a condition which is not really necessary, but convenient!). Therefore

$$T_3 = \{\text{plane-parallel slice of tissue}\}$$
$$T_2 = \{\text{plane}\}$$
$$T_1 = \{\text{straight lines on plane}\}$$
$$T_0 = \{\text{plane lattice of points}\}.$$

Let us now see how these sets T_3, T_2, T_1 and T_0 intersect the sets Y_3, Y_2, Y_1 and Y_0; to do this we must simplify the latter sets. The definition of Y_2 says, in a general sense, that its component points are arranged on a surface; if, for reasons of simplicity, we consider Y_2 to be represented by a plane it is evident that $Y_2 \cap T_2$ must be a line, the result of two intersecting planes. The trace will therefore be

$$Y_2 \cap T_2 = Z_1 \tag{2.6}$$

where Z_1 denotes a one-dimensional trace. The reader can easily verify for himself that Y_1 can be represented by a segment of a straight line, so that

$$Y_1 \cap T_2 = Z_0 \tag{2.7}$$

or, the intersection of a line with a plane is a point. Furthermore, Y_3 is a solid, so that

$$Y_3 \cap T_2 = Z_2 \tag{2.8}$$

a plane, and so on. We can, however, also observe that Y_0 is a cloud of points; its intersection with a plane section $Y_0 \cap T_2$ leaves no trace since the probability that the random plane T_2 hits a true point is 0, as will be shown below: the set $Y \cap T_2 = \varnothing$ is void.

There is evidently a dependence of the dimension of the trace Z on those of the intersecting sets, object Y and section T. This is shown in Table 2.1. Denoting the dimension of the object set Y with $d(Y)$, that of the section with $d(T)$ and that of the trace with $d(Z) = d(Y \cap T)$, we will find, for example,

$$d(Y_3 \cap T_2) = 2 = d(Z_2) \tag{2.9}$$

which is actually

$$d(Y_3) + d(T_2) - 3 = d(Z_2) \tag{2.10}$$

The reader can easily verify that this holds for all the intersections considered in Table 2.1, so that we can generalize:

$$d(Z) = d(Y) + d(T) - 3 \tag{2.11}$$

Table 2.1

Parameters (a) and dimensions (b) of the intersections Z between structure Y and section T

(a)

Trace $Z = Y \cap T$ ↘	Section T: $V(T_3)$	$A(T_2)$	$L(T_1)$	$P(T_0)$
$V(Y_3)$	$V'(Y_3 \cap T_3)$	$A'(Y_3 \cap T_2)$	$L'(Y_3 \cap T_1)$	$P'(Y_3 \cap T_0)$
$A(Y_2)$	$A'(Y_2 \cap T_3)$	$L'(Y_2 \cap T_2)$	$P'(Y_2 \cap T_1)$	$\varnothing$
$L(Y_1)$	$L'(Y_1 \cap T_3)$	$P'(Y_1 \cap T_2)$	$\varnothing$	$\varnothing$
$P(Y_0)$	$P'(Y_0 \cap T_3)$	$\varnothing$	$\varnothing$	$\varnothing$

(b)

$d(Z) = d(Y \cap T)$ ↘		$d(T)$: 3	2	1	0
$d(Y)$	3	3	2	1	0
	2	2	1	0	—
	1	1	0	—	—
	0	0	—	—	—

whereby we must specify the condition that

$$d(Y) + d(T) \geqslant 3 \tag{2.12}$$

for a trace to be formed. All intersections which do not conform to this condition will not form any traces; the set is void. (Table 2.1).

We can generalize further if we observe that the number 3 entering eq. (2.11) is the dimension of the space in which we have operated; we have purposely used the "real world" of three dimensions for the development of this argument. In accepting the existence of (Euclidean) spaces of n dimensions the general formula becomes

$$d(Z) = d(Y) + d(T) - n \tag{2.13}$$

with the limiting condition that

$$d(Y) + d(T) \geqslant n \tag{2.14}$$

for traces to be formed by intersection. We had originally developed this fundamental relationship of stereology by purely intuitive reasoning (Weibel, 1967); it has been proved rigorously by Sommerville (1958) and by Miles (1969, 1972). The reader can easily verify it for the "real space" of two dimensions.

The practical value of such abstract exercises is that we can now immediately predict what type of trace we can expect when we use a probe of a certain dimensionality to test an object parameter of defined dimensionality. Testing the surface area of our objects (Y_2) with a plane probe (T_2) will give a linear trace (Z_1) whose length we can measure. Testing the surface with a linear probe (T_1) will yield a trace Z_0, i.e. intersection points which we can count. But using a point set (T_0) we will not be able to observe a trace of the surface set; point sets can therefore not be used to test surface areas.

In perusing Table 2.1 we will observe that for every object parameter (Y) there is a "section" or probe (T) which produces a trace Z_0, i.e. a set of points which we can count. This is the case when

$$d(Y)+d(T)=n \tag{2.15}$$

When this condition is met we can use point counting methods of stereology which are very efficient because the process of measuring is reduced to counting trace points with respect to a well-defined probe or test system (see Volume 1).

We have shown that metric properties of the objects are represented in the section by a trace and can therefore be studied on sections. It must be mentioned, however, before concluding this chapter, that a section will almost always destroy the topological properties of the object, at least to the point that no statement on the topology of the object is possible from a single section. If the object is a convex solid—i.e. if a line connecting any two points of the solid remains entirely within the solid—then any section will produce one trace only. As soon as the solid is not convex, or if its topological genus is greater than zero, then multiple traces may arise from one and the same object (Fig. 2.4). There is no way of extracting information on the three-dimensional connectivity of traces on a single two-dimensional section, for example. Dense serial sectioning through the entire structure is needed for that (De Hoff, 1967).

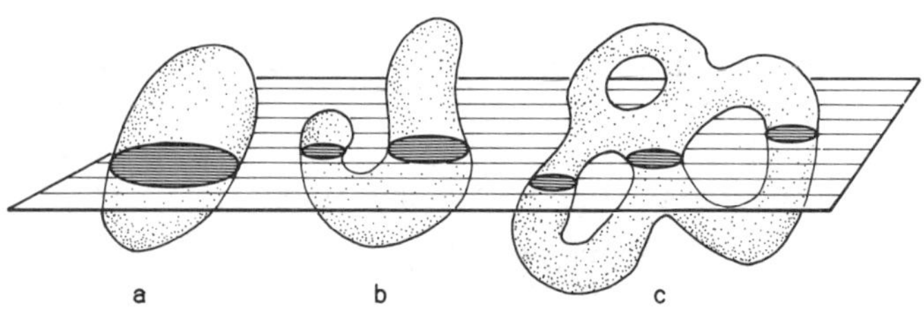

FIG. 2.4: Sectioning a convex solid (a) produces a single profile, whereas multiple profiles may be produced by sectioning a non-convex solid (b) or one of topological genus 3 (c).

2.2 Geometric probability

Stereological procedures are statistical in nature. We demand a representative sample of profiles to be studied, in order for stereological methods to be applicable. In deriving these methods we must therefore give due consideration to the probability with which a certain profile is formed.

Probability is usually defined as the ratio of the number of expected or "favourable" outcomes of some kind of operation to the total number of possible outcomes. It is generally associated with a process of repetition and enumeration, for example if we say that in tossing a coin the probability of having heads up is 1/2 or 50% we mean that out of 100 tosses 50 should land heads up. Or, in throwing a dice, the probability of having 4 up is 1/6 means that every sixth throw should, on the average, be a 4.

In probability experiments of this kind the outcomes can be grouped into at least two sets with different outcomes: for coins there are the sets

$$H = \{\text{heads up}\}$$

$$T = \{\text{tails up}\}$$

or for the dice experiment the sets with 1, 2, 3, 4, 5 or 6 up:

$$\{1\}, \{2\}, \{3\}, \{4\}, \{5\}, \{6\}$$

We can now assign each of these sets a certain probability $\Pr\{H\}$, $\Pr\{T\}$, $\Pr\{1\}$, $\Pr\{2\}$ etc. These probabilities should satisfy the following fundamental conditions:

(1) the probability should be so chosen that it is

$$0 \leq \Pr\{i\} \leq 1 \tag{2.16}$$

(2) the sum of the probabilities in one experiment must be 1:

$$\sum_i \Pr\{i\} = 1 \tag{2.17}$$

In the two elementary experiments mentioned above the situation is very simple: since we postulate that heads and tails are equally likely to be shown up, and that all the numbers on the dice faces have equal likelihood to be up, the probabilities in each set must be equal; therefore

$$\Pr\{H\} = \Pr\{T\} = \tfrac{1}{2}$$

so that

$$\Pr\{H\} + \Pr\{T\} = 1$$

or, in the dice experiment:

$$\Pr\{1\} = \Pr\{2\} = \Pr\{3\} \ldots.$$

$$\sum_{i=1}^{6} \Pr\{i\} = 1$$

so that it is evident that

$$\Pr\{i\} = \tfrac{1}{6}$$

In these two cases we have, in fact, treated the entire experiment as a set of events E composed of several subsets with different outcomes:

$$E\{\text{coin}\} = H \cup T$$

$$E\{\text{dice}\} = \{1\} \cup \{2\} \cup \{3\} \cup \{4\} \cup \{5\} \cup \{6\}$$

We have counted the number $N(E)$ of subsets in E and have calculated the probability of the subset of events as

$$\Pr\{i\} = 1/N(E) \tag{2.18}$$

Let us add one more rule of probability theory. In the dice experiment we may ask what is the probability that either 1 or 3 are up, in other words, we would like to know the probability of the united sets $\{1\}$ and $\{3\}$; it is

$$\Pr(\{1\} \cup \{3\}) = \Pr\{1\} + \Pr\{3\} = \tfrac{2}{6}$$

In very general terms

$$\Pr\{a \cup b\} = \Pr\{a\} + \Pr\{b\} - \Pr\{a \cap b\} \tag{2.19}$$

whereby the last term compensates for the fact that if there is an overlap between $\{a\}$ and $\{b\}$ these events may be counted twice. If there is no overlap $\Pr\{a \cap b\} = 0$.

This excursion into elementary probability theory was motivated by the aim to introduce the notion that in probability theory we must choose a *reference set* with respect to which we can express the probabilities of the subsets showing specific outcomes. The reference set must evidently comprise all the possible outcomes. It is furthermore necessary to find what is called the *element of probability measure* for which the probability is given by $1/N$, N being the number of such elements making up the entire reference set. One can then obtain other probability statements by finding out how many of these elements of probability measure show the same outcome of the probability experiment.

This is no different in geometric probability, as a simple but fundamental example will show. We choose the basic model structure of one convex object Y embedded in fixed orientation in a cube X of side l (Fig. 2.5). In

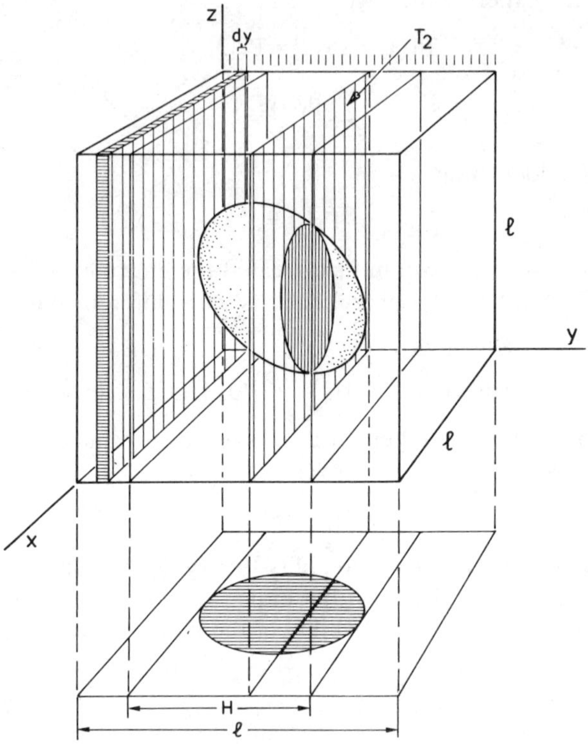

FIG. 2.5: Model for deriving probability of cutting an object by a random section plane T_2. The object is enclosed in a cube of side l.

the notation of set theory introduced above this structure is $\{X \supset Y\}$. Let us place this cube into an x, y, z coordinate system. The experiment is to cut this cube with a plane section T_2 which is parallel to the x–z plane. It is required to find the probability that the plane intersects the object: $\Pr\{Y \uparrow T_2 | Y \subset X\}$. In this experiment two outcomes are possible:

$$\{a\} = \{\text{section cuts object}\} = \{(T_2 \cap Y \mid T_2 \uparrow X) = Z_2\} \tag{2.20}$$

$$\{b\} = \{\text{section misses object}\} = \{T_2 \cap Y = \emptyset\} \tag{2.21}$$

In set a there will be a trace Z_2 of the object on the section, in set b there will be none. Evidently, the set of all possible outcomes is

$$E = \{a\} \cup \{b\} \tag{2.22}$$

so that

$$\Pr\{a\} + \Pr\{b\} = 1 \tag{2.23}$$

But here we cannot state that both outcomes a and b are equally likely; this evidently depends somehow on the size of the embedded object relative to the size of the containing space: a large one will be more frequently cut than a small one. It is therefore necessary that we find a probability measure, one of the most tricky tasks of geometric probability.

Let us proceed empirically and study the model of Fig. 2.5. As the section plane is moved from left to right it will first miss the object, then intersect it, then miss it again. Let us now cut up this structure into a large number of very thin slices of thickness dy; each of these slices is a subset of E, and we can determine for each of them whether they contain part of the object (a^*) or not (b^*). Since each slice has the same thickness dy we may postulate that a random section (parallel to the x–z plane) will hit each slice with equal likelihood so that the probability assigned to each slice is

$$p(a^*) = p(b^*) = 1/N(E) \tag{2.24}$$

on the basis of the general considerations on probability outlined above. The slice therefore is a useful element of probability measure. In this model structure $N(E)$ is evidently

$$N(E) = l/dy \tag{2.25}$$

so that we obtain for the element of probability measure

$$p(a^*) = p(b^*) = dy/l \tag{2.26}$$

In a similar way we can ask how many elements (a^*) are contained in subset $\{a\}$; it is evidently the number of slices of thickness dy that can fit between the left and right tangent planes to the object. If the distance between these tangent planes is H, then

$$N\{a\} = H/dy \tag{2.27}$$

But now, a is the union of all the (a^*) so that

$$\Pr\{a\} = \sum p(a^*) = N\{a\} \cdot p(a^*) = \frac{H}{dy} \cdot \frac{dy}{l} \tag{2.28}$$

from which we conclude that the probability that a random section parallel to the x–z plane will cut the object is

$$\Pr\{a\} = H/l \tag{2.29}$$

the ratio of the tangent diameter to the extent of the structure. Note that H and l are both measured perpendicularly to the section plane (Fig. 2.5), an important concept to which we shall return. It suffices at present to note that we have defined a probability measure as the ratio of two compatible geometric properties: the distances between planes parallel to the section.

We can outline this procedure on another basic example from stereology. Take the same model structure (Fig. 2.6); but instead of cutting it with a plane we probe it with a vertical test line parallel to the z-axis and ask what is the probability that the line intersects the object; in set theory notation:

$$\Pr\{a\} = \Pr\{T_1 \cap Y \mid T_1 \uparrow X = Z_1\} \tag{2.30}$$

In the complementary probability $\Pr\{b\}$ there will be no intersections with the object.

The following is immediately evident: if we project the structure and its content vertically on a horizontal plane, then those (vertical) lines which pass through the projection image of the object will also traverse the

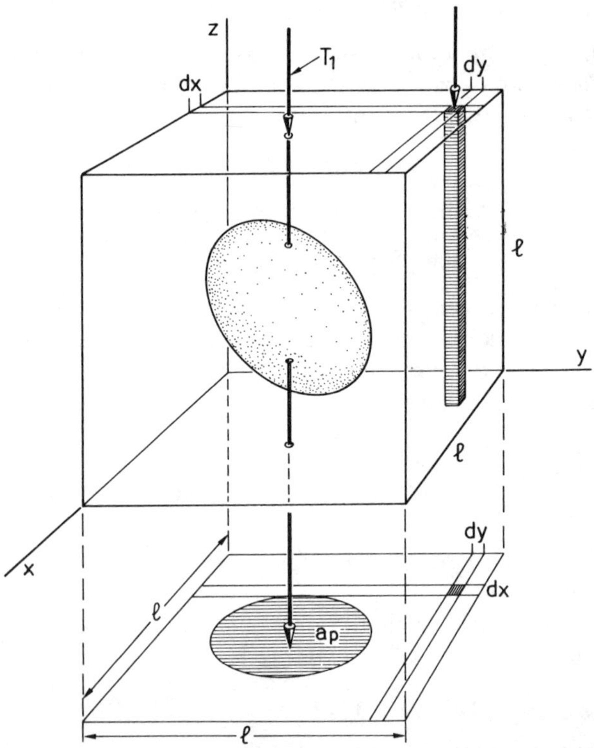

FIG. 2.6: Model for deriving probability that a random line T_1 cuts across an object.

object. We can proceed as before: we cut up the entire structure E into little subsets, this time not slices but prisms of side dx and dy. They are all equal in size (volume $= dx \cdot dy \cdot l$) and the face they offer to the test lines is of equal area $dx\,dy$ for all of them. Hence, we may again postulate equal likelihood for each of them to be penetrated by test lines; they are elements of probability measure. But some prisms contain the object (a^*) some not (b^*), as before. The elementary probabilities are

$$p(a^*) = p(b^*) = 1/N(E) = dx\,dy/l^2 \qquad (2.31)$$

where l^2 is the total area of the cube face penetrated by test lines. In analogy to the previous example we find the number of elementary prisms which contain the object from the ratio of the projection area of the object a_p to the area of the element of probability measure:

$$N\{a\} = a_p/(dx\,dy) \qquad (2.32)$$

so that, finally, the probability for a test line to traverse the object is

$$\Pr\{a\} = N\{a\} \cdot p(a^*) = \frac{a_p}{dx\,dy} \cdot \frac{dx\,dy}{l^2} \qquad (2.33)$$

or

$$\Pr\{a\} = a_p/l^2 \qquad (2.34)$$

In this case we again find the probability from the ratio of two compatible geometric properties, namely the ratio of two plane projection areas of structure and object which are perpendicular to the direction of the test lines (Fig. 2.6).

Let us now finally ask what the probability is that a point, T_0, intersects the object Y, i.e. lies within the object, if it is randomly placed into the structure (Fig. 2.7); in set theory notation this is

$$\Pr\{a\} = \Pr\{T_0 \cap Y_3 \mid T_0 \uparrow X = Z_0\} \qquad (2.35)$$

where again the complementary probability $\Pr\{b\}$ marks no intersection. The element of probability measure is now a little cube of volume $dx \cdot dy \cdot dz$; for each of them, the probability to be hit by the random point is equal, whether it is part of the object (a^*) or part of the matrix (b^*), so that

$$p(a^*) = p(b^*) = 1/N(E) = dx\,dy\,dz/l^3 \qquad (2.36)$$

The number of elementary cubes which fits into the object is evidently proportional to the volume of the object $V(Y)$:

$$N\{a\} = V(Y)/(dx\,dy\,dz) \qquad (2.37)$$

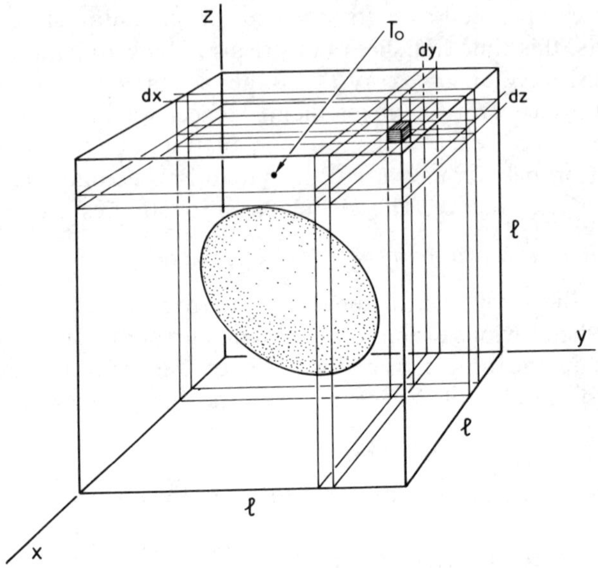

FIG. 2.7: Model for deriving probability that a random point T_0 lies within object.

from which we derive

$$\Pr\{a\} = N\{a\} \cdot p(a^*) = V(Y)/l^3 \tag{2.38}$$

But since $l^3 = V(X)$ is the volume of the structure we have found that

$$\Pr\{a\} = V(Y)/V(X) = V_V \tag{2.39}$$

is directly given by the volume density of the object in the structure. This is evidently a straightforward derivation of the point-counting principle for estimating volume density: $\Pr\{a\}$ can be experimentally obtained by throwing into the cube a large number of random test points, P_T, and counting the number of points, P_a, within the object:

$$\Pr\{a\} = P_a/P_T = P_{Pa} = V_V\ (Y \subset X) \tag{2.40}$$

This is perhaps the place to briefly answer the frequent question why the probability that a plane hits a point in space is 0. Take the same model situation as for cutting an object by a plane, only let the object be a point Y_0 (Fig. 2.8). Sectoning parallel to the x–z plane, the element of probability measure is again a slice of thickness dy. It is evident that only one slice can contain the point Y_0; therefore the probability of cutting that slice, a^*, is

$$\Pr\{T_2 \uparrow a^* \mid a^* \supset Y_0\} = dy/l \tag{2.41}$$

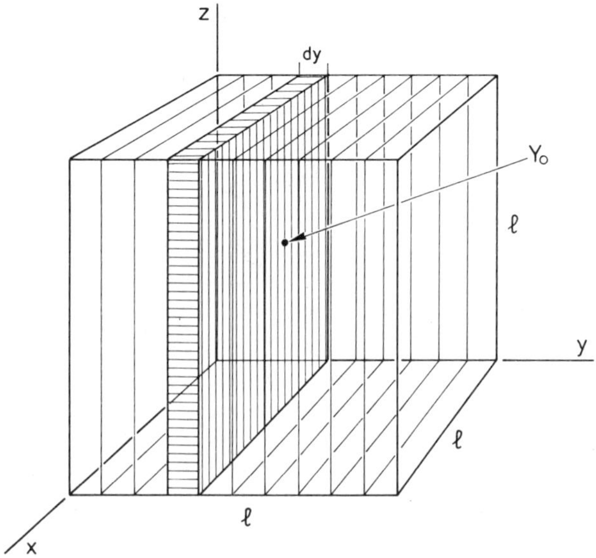

FIG. 2.8: Model for demonstration that the probability to hit a point in space, Y_0, by a random section plane is zero.

If we now let

$$dy \to 0$$

it is evident that also

$$\Pr\{T_2 \uparrow Y_0\} \to 0 \tag{2.42}$$

One can proceed similarly to show that the probability that two lines intersect, or that a line intersects a point are 0.

With these basic examples we have learned a few tricks for handling problems of geometric probability, as they come up in stereology. Most important is the definition of a correct reference set for which elementary probability statements can be made. The definition of this reference set must however take into consideration the experimental procedure for which one wants to define a probability of certain outcomes. By changing the reference set we could obtain different probabilities for what appears to be the same experiment. For example, in the second experiment where we asked for the probability that a test line intersects the object we could have used lines which pass through the origin of the coordinate system, rather than parallel lines (Fig. 2.9); the probability $\Pr\{a\}$ would have been different.

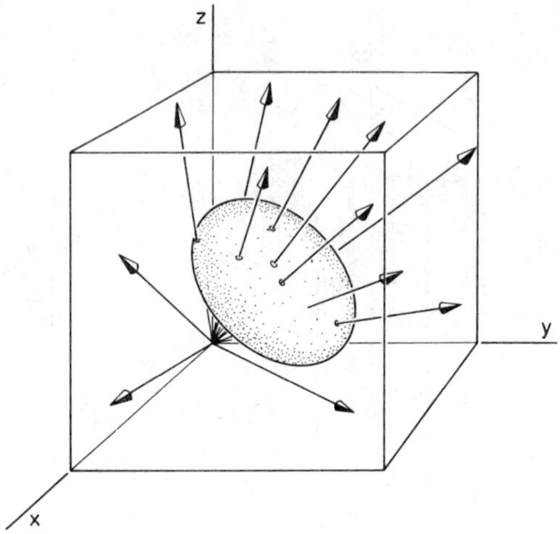

FIG. 2.9: Alternative model for defining probability of intersection between object and random lines.

Bertrand's paradox

Probably the nicest example of how such paradoxical probability statements can be obtained by changing the reference set is "Bertrand's paradox" formulated in 1899 by J. Bertrand and described by Kendall and Moran (1963) in their remarkable and highly recommended monograph on geometric probability. Take a circle of radius 1 with an equilateral triangle inscribed (Fig. 2.10). The problem is to find the probability $\Pr\{l > \sqrt{3}\}$ that a random chord has a length greater than the side of the triangle whose length is $\sqrt{3}$, as can be easily found by elementary geometry. Three different results may be obtained as follows:

Experiment 1 (Fig. 2.10b): Each chord intersects the circle at two points. Assume one to be fixed at *A* and determine the location of the other. For a random chord we would postulate that each point on the circle circumference is equally likely to be chosen as the second intersection point. Only those chords with their end point between *B* and *C* will have length $>\sqrt{3}$. The circle segment *BC* is exactly $\frac{1}{3}$ of the circle circumference, hence

$$\Pr_1\{l > \sqrt{3}\} = \tfrac{1}{3}$$

Experiment 2 (Fig. 2.10c): We cut the circle with a grid of parallel lines; those chords whose distance from the centre of the circle is $<\frac{1}{2}$ will have chord length $>\sqrt{3}$, hence

$$\Pr_2\{l > \sqrt{3}\} = \tfrac{1}{2}$$

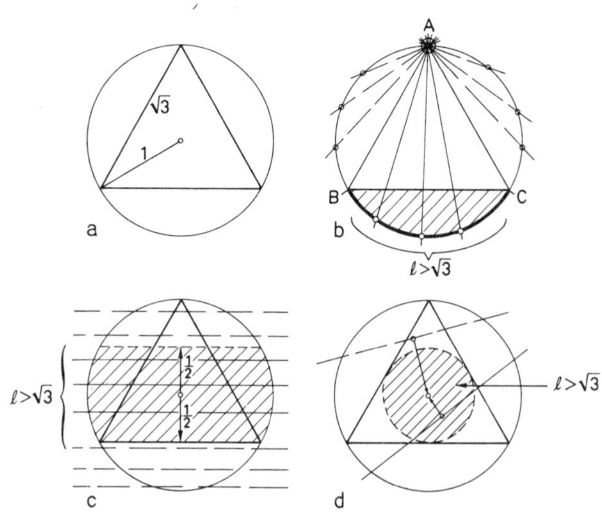

FIG. 2.10: Bertrand's paradox: three different ways of defining the probability that a random chord on the circle will be longer than the side of an inscribed equilateral triangle.

Experiment 3 (Fig. 2.10d): Each chord is defined by its centre point which is actually the termination of the normal to the chord which passes through the centre of the circle. All those chords whose centre point lies within a circle of radius $\frac{1}{2}$ will have length $>\sqrt{3}$, hence

$$\mathrm{Pr}_3\{l>\sqrt{3}\}=\tfrac{1}{4}$$

It is in general said that all three solutions of Bertrand's paradox are correct, which establishes the paradox. But is there really no possibility to decide which of the solutions is "truly correct", that is which solution is the most or only relevant one for stereology? Let us go through the reasoning again.

Imagine a circle drawn on a large plane and drop an infinitely long line on that plane at random (Fig. 2.11a). What is the parameter determining the probability that the line will intersect the circle? From what we have said before it is the caliper diameter of the circle, i.e. the diameter measured at right angles to the test line. Why is this? Because the only displacement of the circle with respect to the line which might change the event "intersection" to "non-intersection" is a shift "away from the line", that is a displacement at right angles to the line (Fig. 2.11b). If we shift the circle along the line, or if it is rotated about its centre, its position relative to the line is unchanged; and all other displacements can be broken up into rotations, longitudinal shifts and vertical or orthogonal

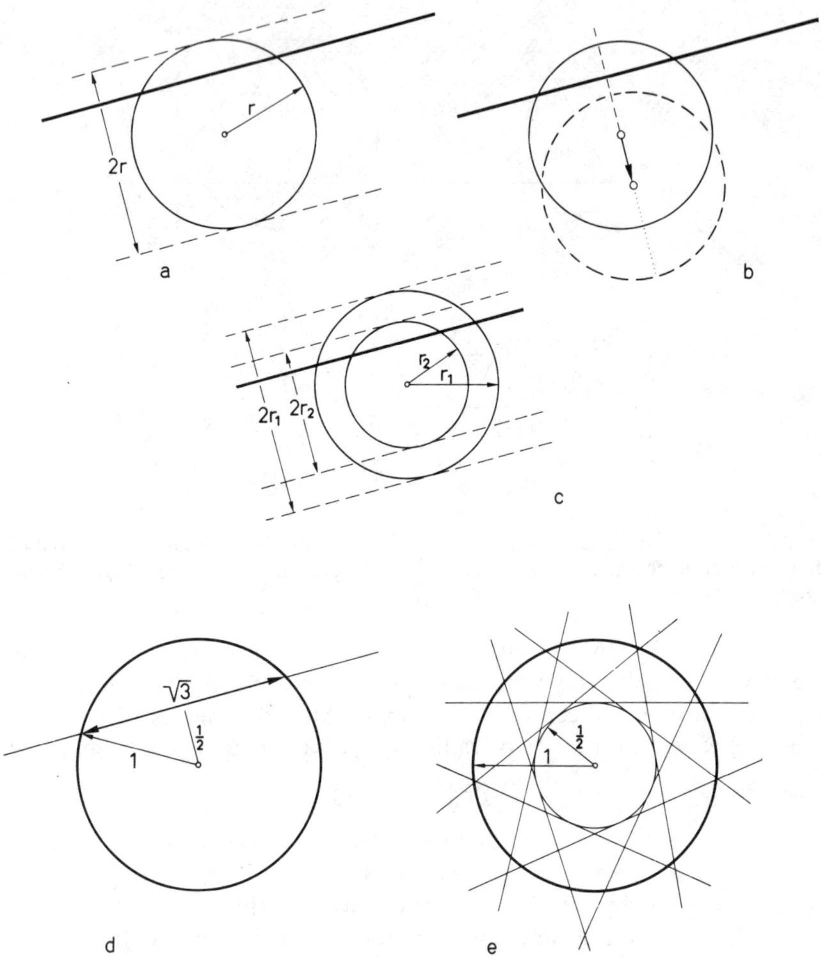

FIG. 2.11: Solution of Bertrand's paradox.

displacements, whereby only the latter component matters. This is nothing but a further illustration for the general statement derived below that a geometric probability is measured on the orthogonal complement to the test line, which, in a two-dimensional system, is a distance at right angles.

Now consider two concentric circles of radius r_1 and r_2 (Fig. 2.11c). What is the probability that a random line T_1 which intersects the outer circle, a_1, also intersects the inner circle, a_2? From the above reasoning it is evident that the probability for a random line to intersect the outer circle is proportional to its orthogonal caliper diameter which is $2r_1$, and, evidently, the probability for cutting the inner circle is proportional to *its* orthogonal caliper diameter, $2r_2$. If we consider all lines which intersect the outer circle it follows immediately that the probability of also intersecting the inner circle is the ratio of the two caliper diameters:

$$\Pr\{T_1 \uparrow a_2 \mid T_1 \uparrow a_1\} = r_2/r_1 \tag{2.43}$$

Here again only orthogonal displacements determine the relative position of circles and line, whereas longitudinal shifts or rotations are irrelevant.

This now allows us to approach Bertrand's problem with a general statement about geometric probability at hand. This general statement is based on the condition that the relative position of line and circle should be unrestrained, i.e. that rotations, longitudinal shifts and orthogonal displacements are allowed to occur freely; its result was that only orthogonal displacements influence the relative position of line and circle.

Bertrand's problem seeks the probability that a line intersecting a circle of radius 1 at random forms a chord of length $l > \sqrt{3}$. A chord of length $l = \sqrt{3}$ is formed if the line intersects the circle at a distance $\frac{1}{2}$ from the centre, as can readily be seen from Fig. 2.11d. Any line lying closer to the centre forms a chord $l > \sqrt{3}$, any one further away forms shorter chords. If we now allow for free rotation it becomes evident that all lines having $l = \sqrt{3}$ form the envelope to a concentric circle of radius $\frac{1}{2}$ (Fig. 2.11e); all lines intersecting this inner circle will be closer to the centre and form chords of length $l > \sqrt{3}$. From the above general statement of probability of cutting two concentric circles with a random line we immediately derive that it is the ratio of the radii of inner and outer circle which determines this probability. In the specific case of Bertrand's problem $r_1 = 1$ and $r_2 = \frac{1}{2}$. Therefore we derive

$$\Pr\{l > \sqrt{3}\} = r_2/r_1 = \tfrac{1}{2} \tag{2.44}$$

This result can be easily verified experimentally by drawing two concentric circles of radius 1 and $\frac{1}{2}$ on a transparent disc and tossing it many times onto a long line drawn on a board. In such a trial it was found that out of 100 tosses the inner circle was intersected 49 times.

It is seen that this solution agrees with solution 2 of Bertrand's paradox. Looking at that solution again we note that it was also based on the ratio of orthogonal distances and was hence compatible with the present reasoning. It did not, however, explicitly consider rotation, but we noted that this is irrelevant anyway.

What is then wrong with the two other solutions? The "random lines" were not unrestrained. In solution 1 the lines were assumed to intersect one specific point on the circle which amounts to a severe restraint, mainly on the othogonal displacement of the line. In solution 3 the longitudinal shift of the lines was restrained by fixing a "centre point" and taking its location as a measure of probability. Because of these restraints true randomness of the line was not assured, so that, with respect to stereology at least, these two solutions must be considered false.

Generalization

Returning now to the more realistic stereological examples of cutting an object contained in a structure with a plane or a line we can now relax one of the simplifications introduced in modelling the structure: instead of embedding the object in a cube (Figs 2.5–7) we now consider a structure of arbitrary (in general convex) shape, X, which contains an object Y (again convex), as shown in Figs 2.12 and 2.13. What is the probability of cutting Y with a random plane $T_s = T_2$, under the condition, of course, that the plane cuts the containing solid X (Fig. 2.12a)? We have learned that, with respect to probabilities of cutting with a plane, the reference set is made of all planes which can cut X, and these are proportional to the distance between the two tangent planes to X parallel to the section plane T_2 (Fig. 2.12b); the "measure" of X in this respect shall be denoted $|X|_1$ because it is one-dimensional. The probability of cutting Y with the plane is similarly proportional to the tangent diameter of Y, measured perpendicularly to the section plane; we shall call it $|Y|_1$. It now follows immediately from the arguments given in relation to deriving eq. (2.29) that the probability of cutting Y, if X is cut by a plane T_2, is

$$\Pr\{T_2 \uparrow Y \mid T_2 \uparrow X\} = |Y|_1/|X|_1 \tag{2.45}$$

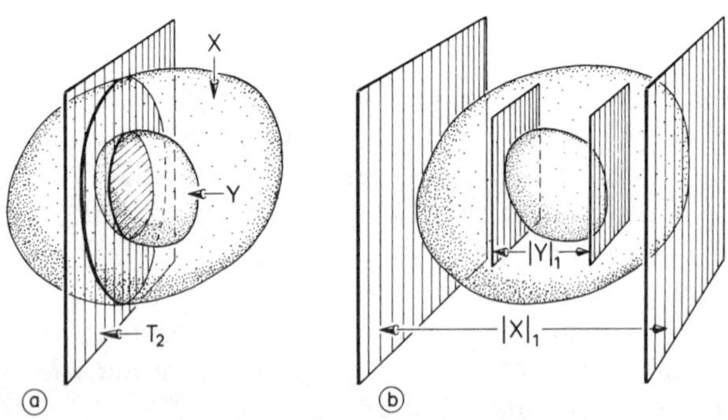

FIG. 2.12: Probability of cutting object Y within structure X with a plane depends on the ratio of the respective caliper diameters, the measures of object and structure on the orthogonal complement to the section.

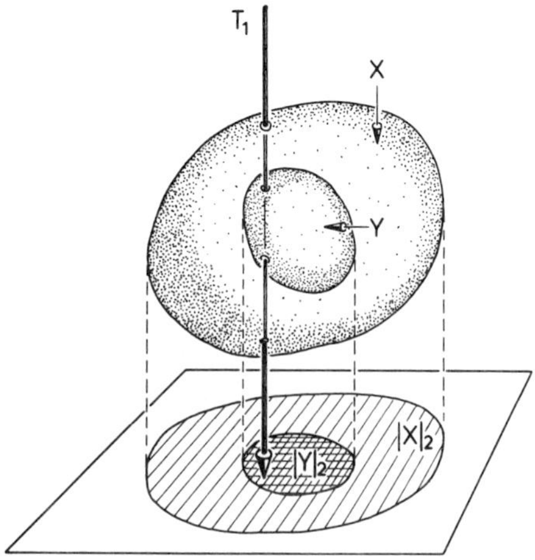

FIG. 2.13: Probability of hitting object within structure with a line probe depends on the ratio of the respective projection areas onto a plane perpendicular to the line probe.

the ratio of the two measures of object Y and structure X, measured on a line perpendicular or orthogonal to the section plane.

In order to derive the probability of cutting the object Y with a random test line T_1 hitting X in a fixed direction we evidently need to determine their "measure" with respect to a plane perpendicular to T_1 (Fig. 2.13); $|Y|_2$ and $|X|_2$ are the projected areas of Y and X respectively onto such a plane. Again following the arguments which lead to eq. (2.34) we find that

$$\Pr\{T_1 \uparrow Y \mid T_1 \uparrow X\} = |Y|_2/|X|_2 \tag{2.46}$$

the ratios of projected areas of object and structure respectively.

Finally, the probability of hitting Y with a point probe T_0, if T_0 lies within X, follows from the arguments which lead to eq. (2.39): the reference set is the volume of X, or its "measure" $|X|_3$, so that

$$\Pr\{T_0 \uparrow Y \mid T_0 \uparrow X\} = |Y|_3/|X|_3 \tag{2.47}$$

where $|Y|_3$ is evidently the volume of Y.

Before we conclude this part it should be pointed out that the three probability statements we have worked out for intersecting planes, lines and points are part of a general scheme worked out by Miles (1969, 1972) for n-dimensional Euclidean space R^n. Without going into details we can formulate his findings in the following way: Consider an n-dimensional space X which contains an n-dimensional subspace Y; $Y \subset X$ hence defines a structure in n-dimensional space. Now cut this with an s-dimensional "flat" space T_s, called s-flat. Note that $s = 1$ defines a line, $s = 2$ a plane, $s = 3$ a solid, and so on. In n-dimensional space each s-flat has an *orthogonal complement* of dimension $s^c = n - s$; the orthogonal complement of a line in 3-dimensional space is a plane: $s^c = 3 - 1 = 2$, and so on (Fig. 2.14). In this general system, the probability that a random s-flat through X (the structure) hits the object $Y \subset X$ is

$$\Pr\{T_s \uparrow Y \mid T_s \uparrow X\} = (|Y|_{s^c})/(|X|_{s^c}) \tag{2.47}$$

where $|Y|_{s^c}$ and $|X|_{s^c}$ are the orthogonal projections of Y and X respectively, i.e. their projections onto the orthogonal space of dimension $s^c = n - s$.

One can easily verify that our three examples comply with this rule. In the case of a line ($s = 1$) intersecting the object the probability was given by the ratio of the (orthogonal) projection areas ($s^c = 2$) of the object and the containing space (Fig. 2.14b); in cutting with a plane ($s = 2$) it was the ratio of tangent plane distances ($s^c = 1$) which defined the probability (Fig. 2.14a); and finally in "cutting" with a random point ($s = 0$) the ratio of volumes ($s^c = 3$) determined the probability.

In this development we have used one important simplification: the structure with the object was mainted in a fixed orientation to the section. If we allow the object to assume all possible orientations the same basic rules will hold, but the general formula must be modified (Miles, 1969, 1972) to read

$$\Pr\{T_s \uparrow Y \mid T_s \uparrow X\} = M_{s^c}\{Y\}/M_{s^c}\{X\} \tag{2.48}$$

where $M_{s^c}\{Y\}$ and $M_{s^c}\{X\}$ are the "mean measures" of Y and X with respect to the section T_s, or the mean projections of Y and X respectively onto the orthogonal complement of the section. We will now have to examine how a "mean projection" can be obtained.

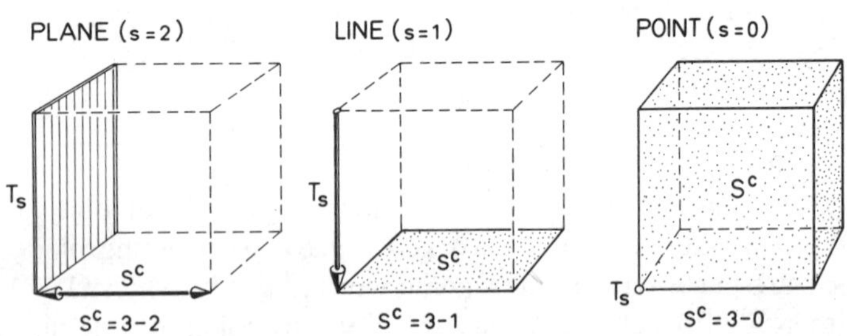

FIG. 2.14: Graphic representation of orthogonal complements, s^c, to plane, line and point in three-dimensional space.

2.3. Averaging with respect to orientation

In the preceding chapter we have seen that the probability that a plane section cuts an object is proportional to the distance H between tangent planes parallel to the section plane. In fact, we have shown that H is really the projection of the object on a line perpendicular to the section plane, its orthogonal complement. Referring to Fig. 2.5 it is quite evident that H depends on the orientation of the object with respect to the section plane, if the object is not a sphere. To consider the probability that a random section—randomly oriented, randomly positioned (see Section 2.4)—cuts the object we must evidently find a way to determine the mean value of H, averaged over all orientations in three-dimensional space. To do this we need a probability measure for orientation.

If we define random orientation to mean that all orientations in space are equally likely we can proceed in the following way. Construct a sphere of radius 1 around the origin of a x, y, z coordinate system and mark the different orientations by rays emanating from the origin (Fig. 2.15). For random orientation we must postulate that the intersection points between the rays and the sphere must be homogeneously dispersed over the sphere surface. We could now subdivide the sphere surface into a large number of very small units of area ds and state that the probability that a ray crosses one of these units is

$$p = ds/(4\pi) \tag{2.49}$$

where 4π is the surface of the sphere of radius 1. This is a straightforward

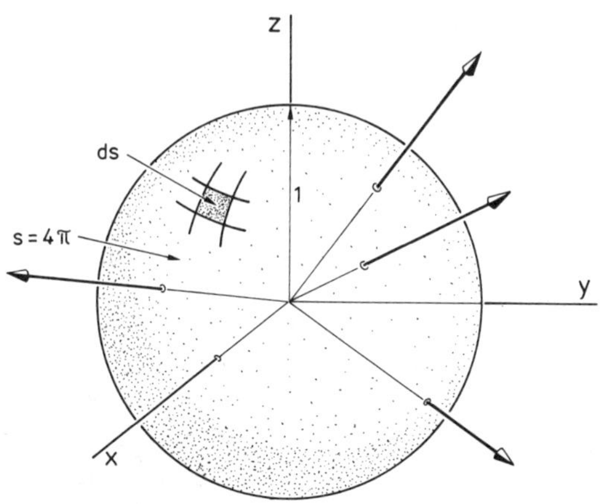

FIG. 2.15: Element of probability measure for orientation is a unit area ds on a sphere of radius 1 and surface area 4π.

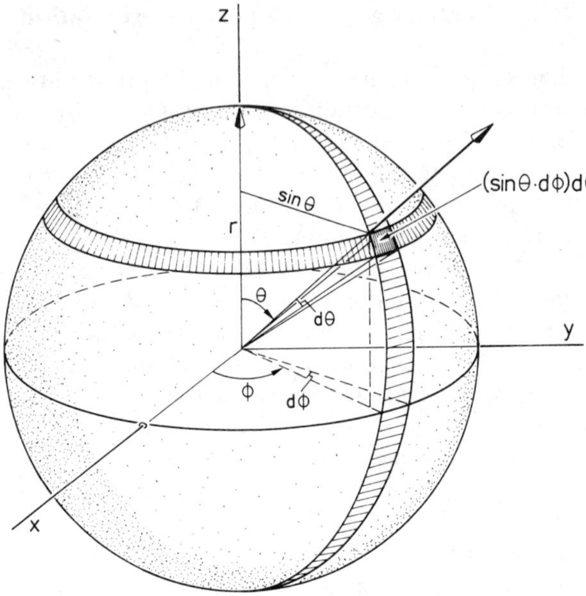

FIG. 2.16: The area of the surface element defined by the polar coordinates θ and ϕ varies in proportion to sin θ.

procedure, identical in concept to those used before: p is the element of probability measure for orientation.

It is most convenient to indicate the orientation of a line by polar coordinates (Fig. 2.16), that is by two angles: θ the angle between the line and the z-axis; ϕ the angle between the x-axis and the vertical projection of the line onto the x–y plane. The angle ϕ can vary around a full circle, i.e. between 0 and 2π (360°), whereas θ can vary between 0 and π, i.e. over a half-circle. (In fact, it would even suffice to vary θ in the range 0 to $\pi/2$, i.e. over a quarter-circle; we would then consider an orientation hemisphere only, because in the other hemisphere each orientation is duplicated.) Now we can ask what is the probability that the angle of orientation of a random line lies in the range $\{\theta \to (\theta + d\theta); \phi \to (\phi + d\phi)\}$ where $d\phi$ and $d\theta$ are small increments in angle; this range is represented by a small pyramid (Fig. 2.16) which, when it intersects the surface of the sphere of radius 1 around the origin, marks off a small quadrangular area on that surface. At first sight one could think that this area would correspond to the unit area ds introduced before; but this is not the case. As shown in Fig. 2.16 the angular increment $d\theta$ defines a zone of equal width which is directly proportional to $d\theta$, if $d\theta$ is small enough; the angular increment $d\phi$, however, defines a wedge-shaped region which tapers to width 0 as it approaches the z-axis, i.e. as $\theta \rightarrow 0$. At each angle θ the width of this wedge is $d\phi \cdot \sin\theta$, since it depends on the radius of the zonal circle at θ, which is $\sin\theta$. Hence the area of this unit at the sphere surface is

$$(\sin\theta \cdot d\phi) \cdot d\theta$$

and the probability of having a line oriented in the range $\{\theta \rightarrow (\theta + d\theta); \phi \rightarrow (\phi + d\phi)\}$ is

$$\Pr\{\theta, \phi\} = \frac{2 \sin\theta \cdot d\phi \cdot d\theta}{4\pi} = \frac{\sin\theta \, d\phi \cdot d\theta}{2\pi} \tag{2.50}$$

obtained by the ratio of the unit orientation area to the total sphere surface area. Note that each orientation is related to two "unit orientation areas", at opposite poles of the orientation sphere, so that the probability is the ratio of the unit orientation area to the surface of the orientation hemisphere. In the following we shall integrate over the hemisphere only. This "elementary" probability measure is not constant but varies with the angle of orientation, in spite of homogeneous distribution of angles, but this is only due to our choice of polar coordinates to describe orientation; it is a property of polar coordinates rather than of the probability model employed! This is in order to eliminate all misunderstanding.

With the probability measure for orientation defined it is now a straight-forward affair to average any parameter $\Pi(\theta, \phi)$ which varies with orientation. By definition, averaging requires integration of the product of parameter times probability. Since orientation is given by two variables, θ and ϕ, a double integration over both angles is required, and this over the ranges $0 < \theta < \pi$, and $0 < \phi < 2\pi$. Hence:

$$\bar{\Pi} = \frac{1}{4\pi} \int_0^{\pi} \int_0^{2\pi} \Pi(\theta, \phi) \cdot \sin\theta \, d\theta \, d\phi \tag{2.51}$$

We might remark parenthetically, that averaging for orientation in a plane, rather than in three-dimensional space, is simpler, but follows the same rationale. Defining orientation by the angle θ to the y-axis (Fig. 2.17) those orientations which lie in the range $\{\theta, (\theta + d\phi)\}$ have probability

$$p(\theta) = d\theta/2\pi \tag{2.52}$$

the circle of reference having radius 1. The mean parameter then is

$$\bar{\Pi} = \frac{1}{2\pi} \int_0^{2\pi} \Pi(\theta) \, d\theta \tag{2.53}$$

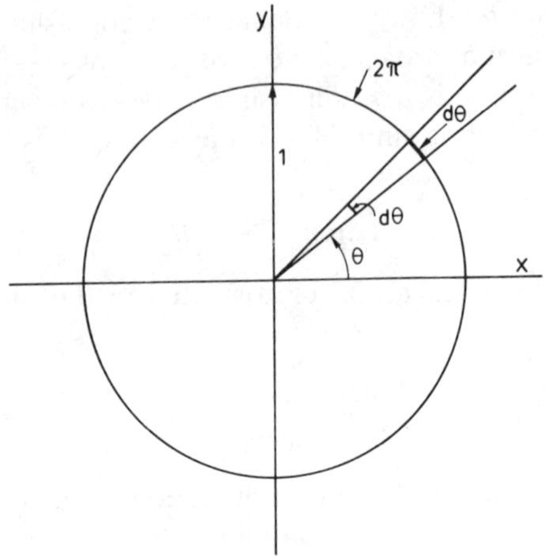

FIG. 2.17: Polar coordinates and averaging for orientation in two-dimensional space.

We may now return to the original problem of averaging the tangent diameter or orthogonal projection H of an object Y cut by a random plane. We must, however, first introduce one restriction on the properties of the object: it must be a convex solid. The definition of convexity demands that the straight line segment connecting any two points of the solid lies entirely within the solid. This restriction is necessary because none of the tangent planes to the solid may intersect the solid. An equivalent definition is that the solid may have no negative curvatures.

Let us now fix the orientation of this solid in space with respect to an x, y, z coordinate system, but allow the orientation of the section plane to vary. The orientation of a plane is determined by the direction of its normal, i.e. a line perpendicular to the plane (Fig. 2.18). The normal is evidently also the orthogonal complement to the section plane, on which we had measured the tangent diameter H. The direction of the normal is given by the angles of orientation θ and ϕ. At each angle one can now determine a projection length of the object on to the orthogonal complement to the section plane (or the tangent diameter) $H(\theta, \phi)$, which is the variable parameter (Fig. 2.18). Substituting $H(\theta, \phi)$ for $\Pi(\theta, \phi)$ in eq. (2.51) we obtain the average projection length

$$\bar{H} = \frac{1}{4\pi} \int_0^{\pi} \int_0^{2\pi} H(\theta, \phi) \cdot \sin \theta \, d\theta \, d\phi \tag{2.54}$$

which corresponds to the parameter $M_1\{Y\}$ in eq. (2.48). To calculate the probability of cutting the object on the basis of eq. (2.48) we would also

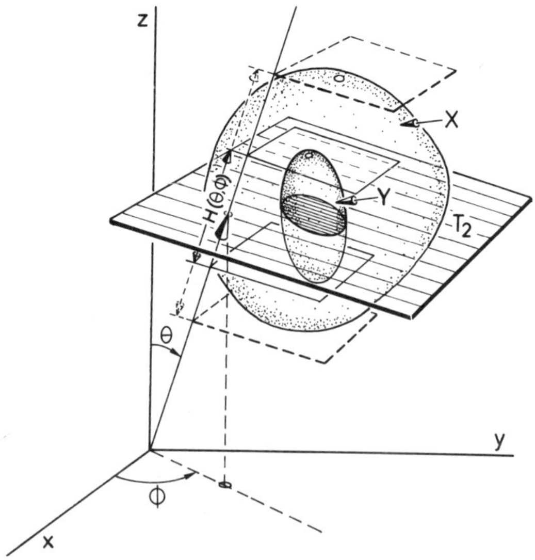

FIG. 2.18: Averaging the tangent diameter of an object for orientation: at each orientation (θ, ϕ) the projection of the object Y (and of the structure X) onto the orthogonal complement to the section, i.e. the normal to the section plane, is determined.

have to determine the mean projection length of the structure, i.e. of the containing space X, or, in the notation of Miles $M_1\{X\}$. This follows the same procedure (Fig. 2.18). Designating with $H(Y)$ and $H(X)$ the tangent diameter of object and containing space respectively, we find

$$\Pr\{T_2 \uparrow Y_3 \mid T_2 \uparrow X_3\} = \frac{\overline{H(Y)}}{\overline{H(X)}}$$

$$= \frac{\dfrac{1}{4\pi}\displaystyle\int_0^{\pi}\int_0^{2\pi} H(Y \mid \theta, \phi) \cdot \sin\theta \, d\phi \, d\theta}{\dfrac{1}{4\pi}\displaystyle\int_0^{\pi}\int_0^{2\pi} H(X \mid \theta, \phi) \cdot \sin\theta \, d\phi \, d\theta} = \frac{M_1\{Y\}}{M_1\{X\}} \tag{2.55}$$

The important result of this derivation is that the mean measures of Y and of X are determined independently; this allows for complete independence of the orientations of Y with respect to X (see Section 2.4).

We can now also derive the probability that a random line intersects the object. We again fix the same convex object with respect to the coordinate system (Fig. 2.19). The direction (θ, ϕ) now marks the orientation of the test line with respect to the coordinates. It also marks, as its

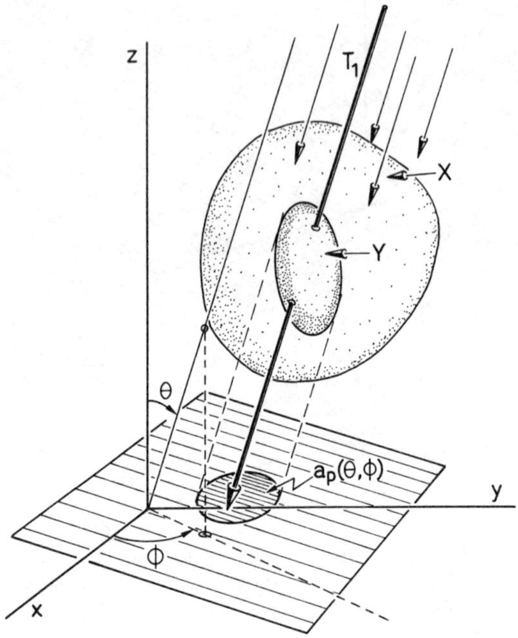

FIG. 2.19: The probability of a line intersecting the object is averaged for orientation by determining the orthogonal projection area of the object Y (and of the structure X) at each orientation.

normal, the orientation of the orthogonal projection plane ($s^c = n - s = 2$ since in this case $s = 1$). At each orientation the projection area of the object on to the orthogonal complement to the test lines, the projection plane, is $a_p(\theta, \phi)$. The mean projection area, averaged over all orientations is

$$\bar{a}_p = \frac{1}{4\pi} \int_0^{\pi} \int_0^{2\pi} a_p(\theta, \phi) \cdot \sin\theta \, d\phi \, d\theta = M_2\{Y\} \tag{2.56}$$

which is the mean measure of Y on the orthogonal complement to the test lines.

All we need to do is to estimate $M_2\{X\}$ which is the mean projection area of the containing space in order to estimate the average probability for a line T_1 to hit a spatial object Y_3 in the structure X_3:

$$\Pr\{T_1 \uparrow Y_3 \mid T_1 \uparrow X_3\} = \frac{M_2\{Y\}}{M_2\{X\}} = \frac{\int_0^{\pi} \int_0^{2\pi} a_p(Y \mid \theta, \phi) \cdot \sin\theta \, d\phi \, d\theta}{\int_0^{\pi} \int_0^{2\pi} a_p(X \mid \theta, \phi) \cdot \sin\theta \, d\phi \, d\theta} \tag{2.57}$$

Finally we had calculated the probability that a random point intersects the object. This is not affected by orientation of the object: a point is dimensionless, its "orientation" cannot be defined, nor that of its orthogonal complement.

For these considerations to be correct we have assumed the object to be in fixed position within a containing space of limited (measurable) size; we have allowed the sections or line probes to take any orientation with respect to both the object and the containing structure. In practice one is often dealing with containing spaces that are very large compared with the object. One is then keeping the size of the test plane l^2 or the length of the test lines l constant and is considering a cube of tissue which has its faces and edges parallel to test plane or test line (Fig. 2.20); in other words the section is kept in fixed position and the object is allowed to take any orientation. In this case one may derive simplified versions of eqs (2.55) and (2.57):

$$\Pr\{T_2 \uparrow Y \mid Y \subset X\} = \bar{H}/l \tag{2.58}$$

$$\Pr\{T_1 \uparrow Y \mid Y \subset X\} = \overline{a_p}/l^2 \tag{2.59}$$

where l and l^2 are the projections of the containing tissue cube parallel to one of its edges (Fig. 2.20).

The purpose of this chapter was to introduce the methodology of averaging over orientation. We are going to make extensive use of this in subsequent chapters where the parameters $\Pi(\theta, \phi)$ can be defined on the basis of geometric properties of the objects.

2.4 Random sectioning and general conditions for the validity of stereological principles

We have so far rather loosely used the term "random section". We now need to specify the conditions of random sectioning as a basis for the development of stereological principles; we shall borrow mainly from Miles and Davy (1976a, b) and Miles (1978), where the reader can find further details.

Consider a specimen X fixed with respect to an x, y, z coordinate system and cut by a plane T, the section (Fig. 2.21). The section plane is specified by the spherical polar coordinates (θ, ϕ) with range $\{0 \leqslant \theta \leqslant \pi/2;$

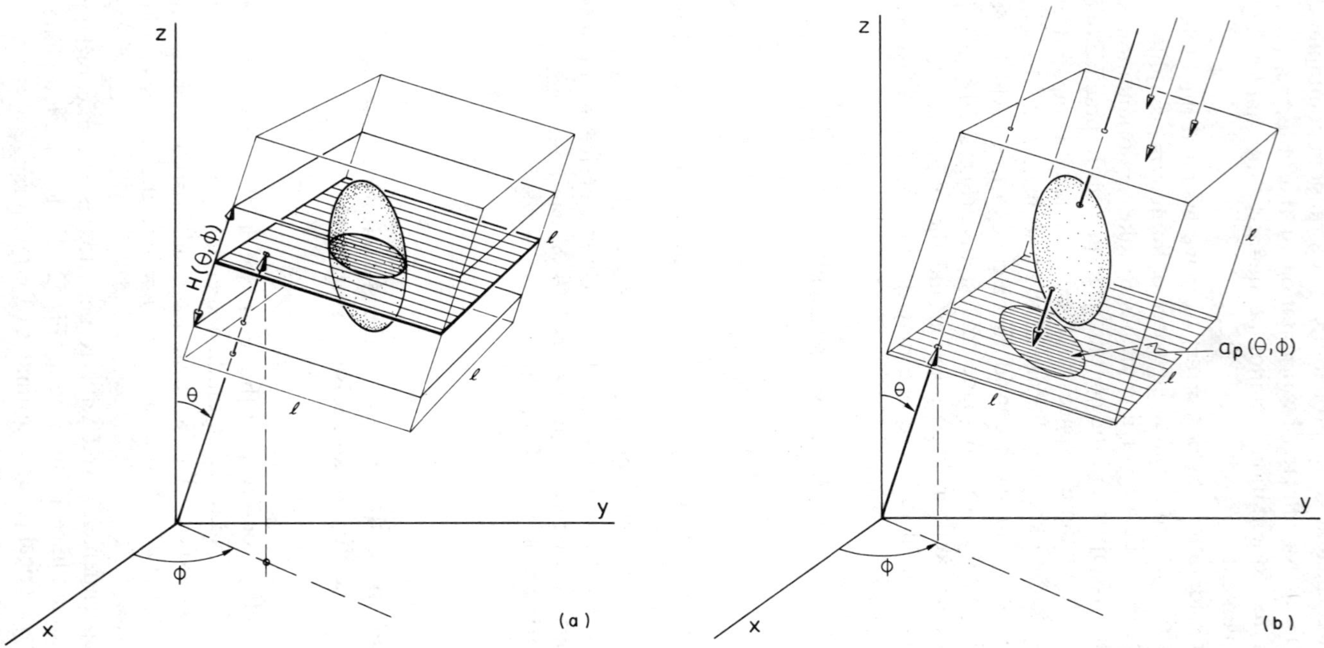

FIG. 2.20: If the object Y is contained in a very large structure W (see Fig. 2.24) it suffices to average for orientation the measure of the object with respect to the section plane (a) or line probe (b), considering the projection of the sample X to be constant at all orientations.

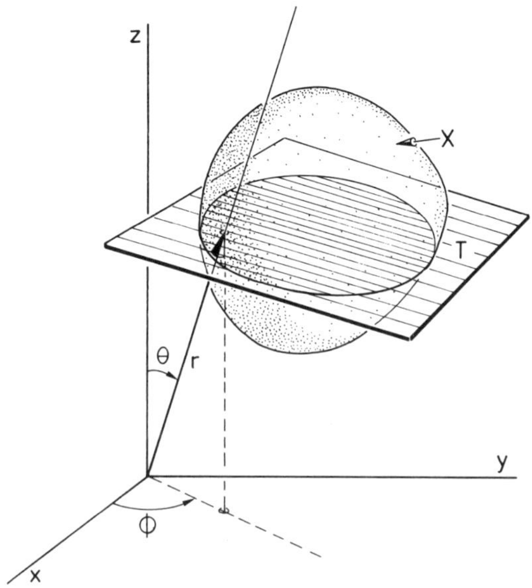

FIG. 2.21: The position of the section T with respect to a structure X is determined by the polar coordinates $(\theta, \phi; r)$.

$0 \leqslant \phi \leqslant 2\pi\}$ for orientation determined by the direction of the normal to T, and by the distance r from the origin in the direction (θ, ϕ) with range $\{-\infty < r < \infty\}$ thus $T = T(\theta, \phi; r)$. Projecting all points of the specimen X orthogonally on to a line at (θ, ϕ) which passes through the origin yields a linear set $X_1(\theta, \phi)$ whose length $H(X_1 \mid \theta, \phi)$ specifies the range for r over which T hits X if the orientation (θ, ϕ) is maintained. In order to specify a *random* plane through X we now need to consider the set of values $(\theta, \phi; r)$ which satisfies certain conditions of randomness, and these will have to account for orientation and position. There is an infinite number of distributions of $(\theta, \phi; r)$, but one possesses certain natural properties, that specifying *isotropic uniform random* (IUR) planes.

We have seen above that "isotropic random orientation" means uniform distribution over the orientation hemisphere with probability density

$$p(\theta, \phi) = (\sin\theta)/(2\pi) \tag{2.60}$$

A random plane through an arbitrary point in X with such an orientation distribution is called an isotropic random plane through X.

For such a plane to be an IUR plane section its position must be uniformly distributed over the orthogonal projection of X at the given orientation, i.e. uniformly distributed over the range $X_1(\theta, \phi)$ of length

$H(X_1 \mid \theta, \phi)$. Assuming isotropic orientation distribution, the mean linear projection of X is

$$H(X) = (2\pi)^{-1} \int_0^{2\pi} \int_0^{\pi/2} H(X_1 \mid \theta, \phi) \cdot \sin\theta \, d\theta \, d\phi \tag{2.61}$$

Now consider the probability that an IUR plane through X hits a given subdomain Y of X, i.e. an object contained in the specimen. It is natural to suppose that for every orientation (θ, ϕ) the conditional distribution of r is uniform over the linear projection of the specimen $X_1(\theta, \phi)$, i.e. it has constant probability density $1/H(X_1 \mid \theta, \phi)$. Then the probability that T hits Y, given it has orientation (θ, ϕ), is

$$\Pr\{T \uparrow Y \mid \theta, \phi) = H(Y_1 \mid \theta, \phi)/H(X_1 \mid \theta, \phi) \tag{2.62}$$

Using the probability density for orientation $p(\theta, \phi)$ (eq. 2.60) and knowing that the probability to cut the specimen is proportional to $H(X_1 \mid \theta, \phi)$, we derive the probability density for an IUR plane in the specimen X to be

$$f(\theta, \phi) = \frac{H(X_1 \mid \theta, \phi) \cdot \sin\theta}{\iint H(X_1 \mid \theta, \phi) \cdot \sin\theta \, d\phi \, d\theta} \tag{2.63}$$

We now find the probability that an IUR plane hits Y as

$$\Pr(T \uparrow Y \mid T \uparrow X) = \iint \Pr(T \uparrow Y \mid \theta, \phi) \cdot f(\theta, \phi) \, d\phi \, d\theta \tag{2.64}$$

and this turns out to be

$$\Pr(T \uparrow Y \mid T \uparrow X) = H(Y)/H(X) \tag{2.65}$$

the ratio of the mean linear projections (or caliper diameters for single objects) of the object and the specimen.

Two important aspects of this result need to be stressed: (1) for all subdomains Y of X, the probability that T hits Y is independent of the location and orientation of Y within X. This is due to the fact that we have used an IUR plane as T; (2) if an IUR plane through X hits Y, then it is also an IUR plane through Y (Miles and Davy, 1976b).

IUR sections so defined have hitherto been considered to be the proper choice as "random section" for stereology because they hit every locality of the specimen equivalently (Miles, 1972). Miles and Davy (1976b) have however pointed out that they may lead to biased estimates of stereological parameters. They offer as an alternative which yields unbiased stereological ratio estimators the "A-weighted random plane section of X". What they mean is that the distribution of the isotropic random plane

T is weighted by the area of the specimen section obtained, i.e. by $A(T \cap X)$; large sections have more weight than smaller ones. The probability density for such an A-weighted section is

$$p_a(\theta, \phi; r) = \frac{A(T \cap X)\sin\theta}{\int A(T \cap X)\sin\theta\, d\theta} \tag{2.66}$$

where the subscript a stands for "A-weighted". Miles and Davy (1976b) also show that results obtained with A-weighted sections are not only unbiased but that they also are likely to have a smaller mean square error. They require no restriction on the properties of the specimen: the subdomain Y can be inhomogeneously distributed; it can even be anisotropic.

How are such random planes generated? We follow the instructions of Miles and Davy (1976b).

IUR plane (Fig. 2.22). (1) Measure the linear projection of the specimen $H(X_1)$ in a sufficient number of orientations; (2) select a random orientation having an isotropic distribution weighted by $H(X_1)$, eq. (2.63); (3) select a random plane with this orientation and with location uniformly distributed over the "hitting positions" that are in the range $\{H(X_1)\}$ between the two tangent planes to the specimen. That is not a

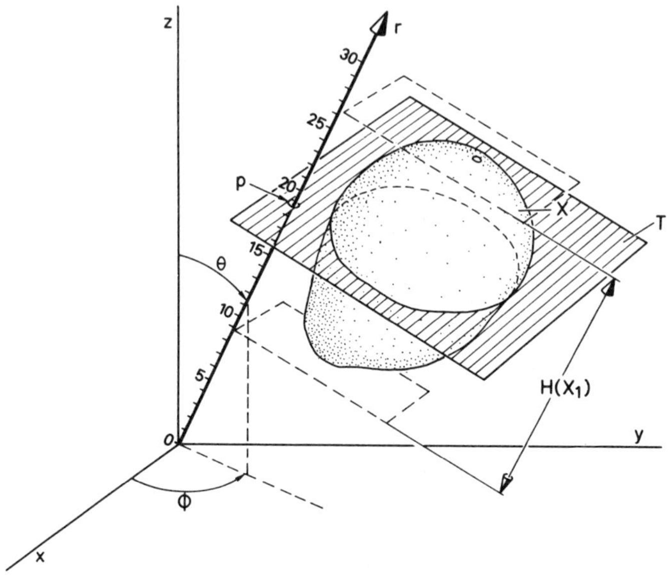

FIG. 2.22: Construction of IUR section through specimen X.

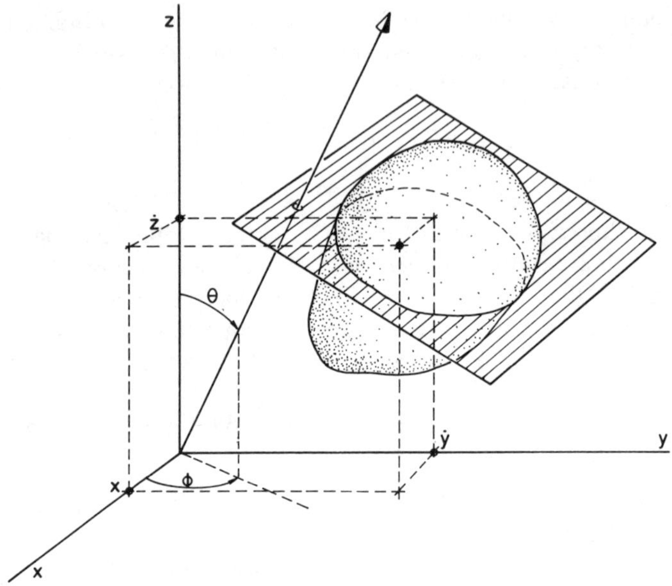

FIG. 2.23: Construction of A-weighted section through specimen X.

very easy procedure; to put it into practice requires some ingenuity (see Volume 1, Section 3.3).

A-weighted section (Fig. 2.23). The method for generating such a section follows from a theorem (2a in Miles and Davy, 1976b) which says that *an isotropic random plane through a uniform random point in X is equivalently an A-weighted random plane through X*. The first step is to "geometrically map" the specimen X so as to identify a sufficiently large number of uniformly distributed points in X, then to identify one of these points, e.g. by means of random numbers. Then, select the orientation of the isotropic random plane and let it pass through the random point previously identified. With some ingenuity it should be possible to put this procedure into practice (see Volume 1, Section 3.3).

This description of the sectioning procedure reveals that for A-weighted sections to be produced we need to be able to encompass the entire specimen. That is to say that the specimen must be relatively small and that we want to examine sections that cut across the entire specimen; in such a case edge effects are an important consideration, and the bias resulting from them can be eliminated by A-weighted sectioning. Miles calls this the "*restricted case*". Such cases occur, of course, if small organs (e.g. the rat's thyroid) are to be sectioned for light microscopy.

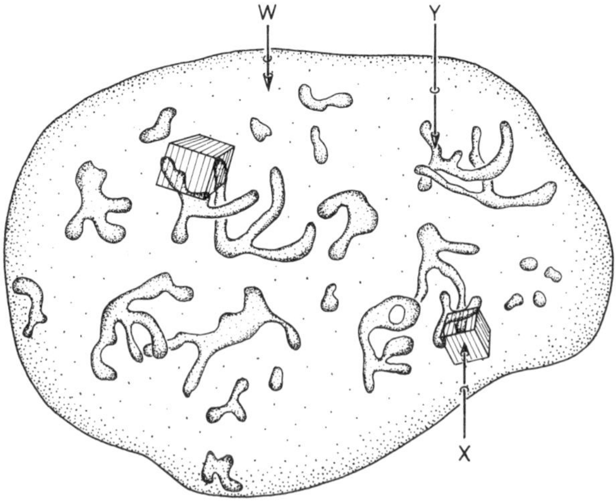

FIG. 2.24: The objects Y are contained in a very large specimen W (extended case); X is an IUR sample of W.

Very often, however, the specimen X is itself a sample from a much larger body W, and the area of the section $A(T \cap X)$ used for microscopic study is very small compared to the size of W; Miles calls this the "*extended case*" (Fig. 2.24). In this case Y is really a subset of W rather than X and we want to estimate some stereological parameters of Y with respect to W. Because W is large we need to sample it in two stages: (1) we cut out a little "specimen" X (or several of them); (2) we section this specimen with a plane T. It now turns out that the critical step is really the first one: cutting X out of W. If X is obtained by an IUR procedure—e.g. locating by random numbers its position and orientation within W—then X can itself be regarded as a "volume weighted" sample and will thus necessarily yield an A-weighted section, regardless of how the section is now cut. Indeed, an arbitrary section of the specimen block X thus obtained is adequate. It is however essential that true randomness enters at the first step of the sampling procedure. It can be shown that such sections yield unbiased ratio estimators for all basic stereological formulae in which the denominator of the ratio is the volume of the containing space W.

This consideration of extended cases justifies the simple models used for deriving stereological formulae in an easy fashion: one often considers cubic specimens and keeps the section plane parallel to one of the cube's faces (Weibel, 1963; Underwood, 1970; De Hoff and Rhines, 1968; Giger, 1970). Such a section is evidently non-random, it is arbitrary; this is, however, an acceptable procedure provided we consider the cube to be an IUR "volume weighted" sample of the true, much larger specimen.

In conclusion, definition of what is meant by "random sectioning" is fundamental in deriving valid stereological methods, i.e. methods that yield unbiased ratio estimators. For the general case, A-weighted sections, favouring large sections of the specimen, are the correct type of random section. Their use is compulsory for relatively small specimens (restricted case). For the extended case, where the section is cut from an IUR random sample of the very large specimen, arbitrary sections yield unbiased ratio estimators.

We must finally mention that analogous conditions prevail with respect to "cutting" the specimen with a test line. For the restricted case unbiased ratio estimators are only obtained if the test line is an L-weighted one-dimensional section of the three-dimensional specimen X; it is represented by an isotropic line through a uniform random point in X. For practical purposes it is important to note that an L-weighted line section of an A-weighted plane section of X is equivalently an L-weighted line section of X (Miles and Davy, 1976b). Again, in the extended case arbitrary line sections of the sample X out of the specimen W yield unbiased ratio estimators, provided X is an IUR sample of W.

In Volume 1, Section 3.3, some practical methods are discussed which allow these sampling rules to be implemented. For further reading on geometric probability the reader is directed particularly to Kendall and Moran (1963) and Santaló (1976).

CHAPTER 3

Estimation of basic stereological parameters

In this chapter we shall present the mathematical justification for what are often called the basic stereological methods, namely those which allow an estimate of the basic descriptors of structure such as volume density, V_V, surface density, S_V, length density, J_V, and mean curvature density, K_V, from a measurement done on infinitesimally thin sections: true planes, true lines, true points.

In order to make the necessary mathematical developments accessible also to stereologists with limited mathematical experience we shall proceed as follows. For each principle to be derived we shall first present a definition of the structure, of the section and of the resulting "profile" in terms of set theory (see Sections 1.6 and 2.1). Secondly, we shall go through a derivation which makes use of elementary mathematical tools, which will require some limitations on the general validity of the result. Thirdly, we shall start afresh and proceed through a rigorous derivation using the notions of geometric probability developed in Sections 2.2 and 2.4, following essentially the methods of Miles and Davy (1976b). The use of these two approaches will necessitate some modifications in the notation used. In the first elementary approach we shall maintain the description of the structure as being made up of phases (labelled *a*, *b* etc.), in analogy with the notation used in presenting the practical methods in Volume 1. In the second approach we shall need to specify that both the specimen and the objects it contains are finite pieces of material, and we shall call them *X* and *Y* respectively, in analogy with the notation used by Miles and Davy (1976b).

3.1. Volume density measurement

We have shown in Section 2.1 that sectioning a three-dimensional object Y_3 with sections T_2, T_1 and T_0 leaves a trace on the section. We can therefore expect three methods to be available for estimating the volume density V_V of an object within the structure. In the following we shall first consider the case of a very large structure containing a very large number of objects of completely arbitrary shape (extended case). A small cube of tissue is cut out of this structure randomly, i.e. following an IUR sampling procedure (Fig. 2.25); an arbitrary section of this cube should therefore yield an unbiased ratio estimate for V_V (see Section 2.4).

The model structure on which we shall base the following developments hence consists of a representative cube of tissue, c, containing some objects which we shall all assign to one subset $a \subset c$, regardless of whether the objects are continuous or discrete; $a \subset c$ is conceived as a phase of the structure. These objects are embedded in a second phase $b \subset c$, the matrix. The definition of this two-phase structure as a set is

$$a \cup b = c \tag{3.1}$$

where c symbolizes the structure or containing space; a and b are space-filling and not overlapping:

$$a \cap b = \emptyset$$

If $V(a)$ and $V(b)$ denote the volumes of subsets a and b, respectively, then

$$V(a) + V(b) = V(c) = V(a \cup b) \tag{3.2}$$

From this we derive the definition of volume density:

$$\begin{aligned} V_{Va} &= V(a)/V(c) \\ V_{Vb} &= V(b)/V(c) \end{aligned} \tag{3.3}$$

$$V_{Va} + V_{Vb} = 1 \tag{3.4}$$

3.1.1. THE PRINCIPLE OF DELESSE: $V_V = A_A$

The first fundamental principle of stereology was formulated in 1847 by A. Delesse as follows: if a structure containing some objects of phase a is sectioned randomly, then the areal density A_A of profiles of a on the section is equal to the volume density V_V of the objects in the structure:

$$V_{Va} = A_{Aa} \tag{3.5}$$

where

$$A_{Aa} = A(a)/A(c) \qquad (3.6)$$

is the ratio of the area of all a-profiles to the area of the section through the structure c.

We shall now prove this principle by taking two different approaches. The first uses a procedure of slicing and integration (Weibel, 1963), the second employs the notions of geometric probability theory.

First proof

Consider a model structure in form of a cube of volume

$$V(c) = l^3 \qquad (3.7)$$

to be placed in an x, y, z coordinate system (Fig. 3.1). It contains granules of any shape and size belonging to phase a which together have a volume

$$V(a) = \rho \cdot V(c) \qquad (3.8)$$

Consider, now, a thin slice of this cube of thickness dy parallel to the x–z plane having a volume

$$dV(c) = l^2 \cdot dy \qquad (3.9)$$

In this slice a volume

$$dV(a) = \eta(y) \cdot dV(c) = \eta(y) \cdot l^2 \cdot dy \qquad (3.10)$$

will contain slices of the granules. l^2 and dy are constant whenever we

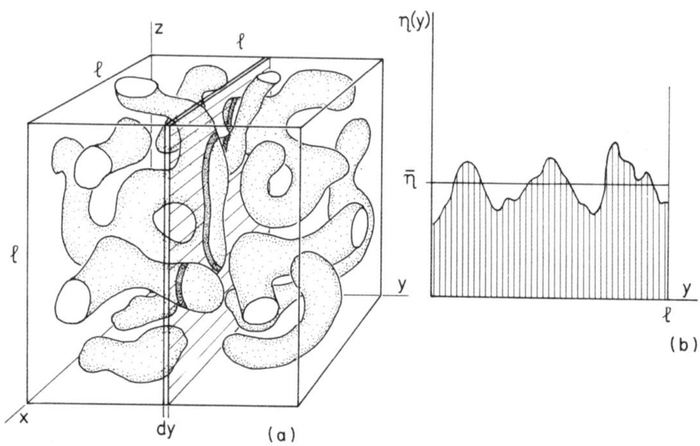

FIG. 3.1: Demonstration of Delesse principle.

place the slice but $\eta(y)$ will vary with y, as is indicated in Fig. 3.lb. If $dy \to 0$ the total volume of the granules $V(a)$ is

$$V(a)=\int_0^l dV(a)=l^2\cdot\int_0^l \eta(y)\,dy=\rho\cdot l^3 \tag{3.11}$$

But

$$\frac{1}{l}\int_0^l \eta(y)\,dy=\bar{\eta} \tag{3.12}$$

is the average value of the coefficient $\eta(y)$ between 0 and l so that it follows from eqs (3.11) and (3.12) that

$$\bar{\eta}\cdot l^3=\rho\cdot l^3 \tag{3.13}$$

or

$$\bar{\eta}=\rho \tag{3.14}$$

It is evident from eq. (3.8) that

$$\rho=V(a)/V(c)=V_{Va} \tag{3.15}$$

is the volume density of the structures in the tissue. Similarly it follows from eq. (3.10) that

$$\eta(y)\cdot l^2=A[a](y) \tag{3.16}$$

is the area of all profiles of a on the section at y, so that with

$$l^2=A(c) \tag{3.17}$$

as the area of the test field on the structure we find that

$$\eta=A(a)/A(c)=A_{Aa} \tag{3.18}$$

is the areal or planimetric density of profiles on the sections. Hence we may rewrite eq. (3.14) as

$$\overline{A_{Aa}}=V_{Va} \tag{3.19}$$

Our conclusion is that the volume density is represented on sections by the areal density of profiles. More exactly: *the areal density of profiles on sections is an unbiased estimator of the volume density of structures.*

Second Proof

Consider a more general specimen X of volume $V(X)$ with some components Y belonging to phase a embedded (Fig. 3.2), their combined volume being $V(Y)$. (Note that $X=c$, $Y=a$ of the first proof.) We shall

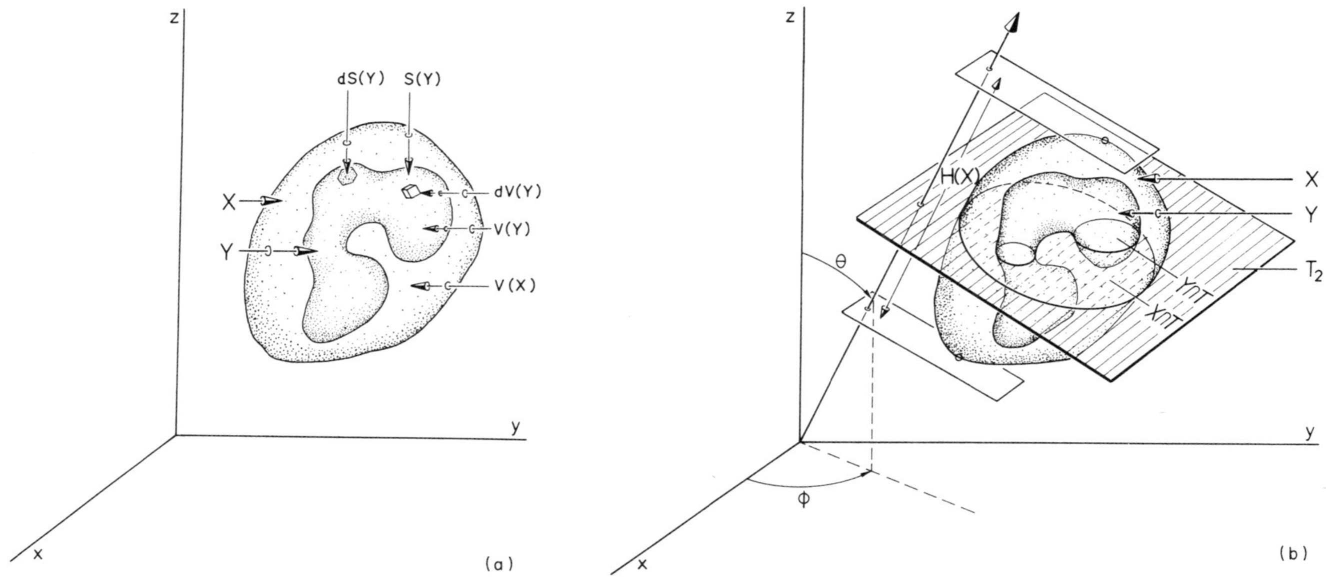

FIG. 3.2: (a) Structure X containing object Y, with elements of volume and surface of Y. (b) Sectioning this structure with a plane T_2 depends on caliper diameter $H(X)$.

now derive the formula of Delesse using results of geometrical probability theory.

Suppose dV is an infinitesimal volume element of Y. If dV is intersected by a section plane $T = T_2$ a profile is formed whose total infinitesimal area is $A(dV \cap T)$. Now it is evident that the total profile area in Y is

$$A(Y \cap T) = \int_Y A(dV \cap T) \tag{3.20}$$

the integral of all infinitesimal profile areas of the volume formed in Y.

The average or expected profile area of Y is

$$E\{A(Y \cap T)\} = E\{\int_Y A(dV \cap T)\} \tag{3.21}$$

The expected value of an integral being equal to the integral of the expected value we write

$$\begin{aligned} E\{A(Y \cap T)\} &= \int_Y E\{A(dV \cap T)\} \\ &= \int_Y E\{A(dV \cap T) \mid T \uparrow dV\} \Pr(T \uparrow dV) \end{aligned} \tag{3.22}$$

considering the probability that T hits dV. This probability is

$$\Pr(T \uparrow dV) = H(dV)/H(X) \tag{3.23}$$

the ratio of the mean linear projection lengths of dV and X. Furthermore we know that the mean profile area of dV, provided T hits dV, is

$$E\{A(dV \cap T) \mid T \uparrow dV\} = V(dV)/H(dV) \tag{3.24}$$

the volume of dV divided by its mean caliper diameter. Hence

$$E\{A(Y \cap T)\} = \int_Y \frac{H(dV)}{H(X)} \cdot \frac{V(dV)}{H(dV)} = \frac{1}{H(X)} \cdot \int_Y V(dV) \tag{3.25}$$

But since

$$\int_Y V(dV) = V(Y) \qquad 3.26)$$

we find

$$E\{A(Y \cap T)\} = V(Y)/H(X) \tag{3.27}$$

The mean profile area of $X \cap T$ is

$$E\{A(X \cap T)\} = V(X)/H(X) \tag{3.28}$$

Dividing eq. (3.27) by eq. (3.28) we find eventually

$$\frac{E\{A(Y \cap T)\}}{E\{A(X \cap T)\}} = \frac{V(Y)}{V(X)} \tag{3.29}$$

or in words: the ratio of the mean profile areas of Y and X is an estimator of the volume ratio of Y and X.

Note the difference to the first proof: we did not restrict the shape of the specimen X and we did not arbitrarily fix the direction of the sectioning plane T. Indeed we have supposed that T was an IUR plane through X. Note, however, that for a "restricted specimen" only an A-weighted random section will make eq. (3.29) an unbiased estimator of volume density (see Section 2.4).

3.1.2. THE LINEAR INTEGRATION PRINCIPLE OF ROSIWAL: $V_V = L_L$

Extending the principle of Delesse, the geologist Rosiwal demonstrated in 1898 that the fraction of a line traversing a tissue containing a randomly distributed component a is equal to the volume density of the component in the tissue:

$$V_{Va} = L_{La} \tag{3.30}$$

In proving this principle we shall take the same approach as with the Delesse principle. In a first proof we shall proceed using an integration, in a second proof use considerations of geometric probability.

First proof

Consider the model of Fig. 3.1 in an x, y, z coordinate system. We have cut sections and have shown that on each section of area $A(c) = l^2$, a fraction η will be covered by profiles of a, so that the area covered by a is

$$A(a) = \eta \cdot A(c) \tag{3.31}$$

Figure 3.3 shows a representative section of this kind with the coordinates of the x–z plane indicated. We now assume that the test line is deposited at random on this section, but parallel to the z-axis; the length of line included in the section is l. Consider a strip of width dx and area

$$dA = l \cdot dx \tag{3.32}$$

A fraction

$$dA(a) = \lambda \cdot dA = \lambda \cdot l\,dx \tag{3.33}$$

of this strip will be covered by profiles of a. We now find that $\lambda(x)$ again

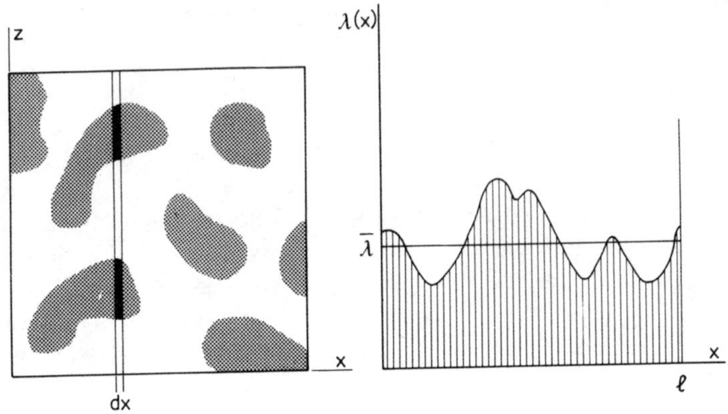

FIG. 3.3: Demonstration of linear integration on section.

varies with x (Fig. 3.3). If $dx \to 0$

$$A(a) = l \int_0^l \lambda(x) \cdot dx = \eta \cdot l^2 \tag{3.34}$$

But since

$$\frac{1}{l} \int_0^l \lambda(x)\, dx = \bar{\lambda} \tag{3.35}$$

is the average value of $\lambda(x)$ between 0 and l, it follows that

$$\bar{\lambda} = \eta \tag{3.36}$$

As seen from eq. (3.31), η is the areal fraction of the test area occupied by profiles, or A_{Aa}, whereas λ is the fraction L_{La} of test lines enclosed in profiles. We can therefore rewrite eq. (3.36) as

$$A_{Aa} = \overline{L_{La}} \tag{3.37}$$

which says that the mean of the linear fraction L_{La} of test lines enclosed in profiles is an unbiased estimate of A_{Aa}.

In deriving Delesse's principle we had shown that the average of A_{Aa}, is an unbiased estimate of the volume fraction V_{Va}. We therefore find that

$$V_{Va} = \overline{A_{Aa}} = \overline{\overline{L_{La}}} \tag{3.38}$$

Second proof
Consider a more general specimen X, as for the second proof of the Delesse principle (Fig. 3.4); its volume is $V(X)$ and the volume of the

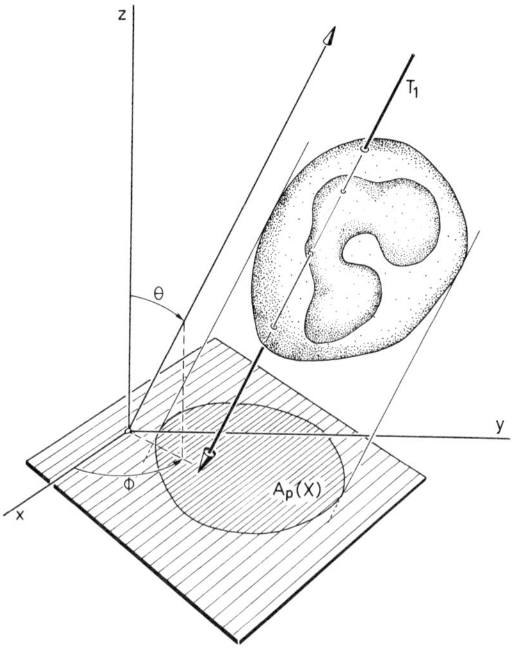

FIG. 3.4: Sectioning structure with random test line T_1 depends on projected area $A_p(X)$.

phase Y it contains is $V(Y)$. This specimen is now cut by a random *line* $T = T_1$, i.e. by a one-dimensional section. The proof follows along the same line as that for the Delesse principle.

Suppose dV is an infinitesimal volume element of Y (Fig. 3.2); if it is intercepted by the line T the total infinitesimal length of the intercept is $L(dV \cap T)$. The total intercept length in Y is

$$L(Y \cap T) = \int_Y L(dV \cap T) \tag{3.39}$$

The mean total intercept length of Y on any random line traversing X is

$$\begin{aligned} E\{L(Y \cap T)\} &= E\left\{\int_Y L(dV \cap T)\right\} \\ &= \int_Y E\{L(dV \cap T)\} \\ &= \int_Y E\{L(dV \cap T) \mid T \uparrow dV\} \cdot \Pr(T \uparrow dV) \end{aligned} \tag{3.40}$$

The probability that T hits dV is

$$\Pr(T\uparrow dV) = A_p(dV)/A_p(X) \tag{3.41}$$

the ratio of the projected areas of dV and X. The mean intercept length in dV is

$$E\{L(dV\cap T)\mid T\uparrow dV\} = V(dV)/A_p(dV) \tag{3.42}$$

the ratio of volume to projected area of dV. Therefore

$$\begin{aligned} E\{L(Y\cap T)\} &= \int_Y \frac{V(dV)}{A_p(dV)}\cdot\frac{A_p(dV)}{A_p(X)} \\ &= \frac{1}{A_p(X)}\cdot\int_Y V(dV) \\ &= V(Y)/A_p(X) \end{aligned} \tag{3.43}$$

Since the mean intercept of the random line in X is

$$E\{L(X\cap T)\} = V(X)/A_p(X) \tag{3.44}$$

we obtain by dividing eq. (3.43) by eq. (3.44)

$$E\{L(Y\cap T)\}/E\{L(X\cap T)\} = V(Y)/V(X) \tag{3.45}$$

The ratio of the mean intercept lengths in Y and X is an estimator of the volume ratio of Y and X. As in the second proof of the Delesse principle we assumed T to be an IUR line through X. For a restricted specimen only an L-weighted random line section will make eq. (3.45) an unbiased estimator of volume density. Although this proof has used random lines shot directly into the specimen it is equally valid for a two-stage procedure where lines are deposited on plane sections of the specimen; according to Miles and Davy (1976b) an L-weighted line section of an A-weighted plane section of a spatial domain X is equivalently an L-weighted line section of X (see Section 2.4).

3.1.3. THE POINT COUNTING PRINCIPLE: $V_V = P_P$

In describing the process of sectioning we have said that the intersection between a 0-section (a set of test points T_0) and an object Y_3, $T_0\cap Y_3$, is a set of object points which can be counted.

Thomson (1930) and Glagolev (1933) were apparently first to recognize that this could be used to estimate the volume density of a phase $a\subset c$, their statement being

$$V_{Va} = P_{Pa} = P(a)/P(c) \tag{3.46}$$

The most direct proof of this principle has already been given in Section 2.3 in relation to estimating the probability that a random test point T_0 in space lies within an object $Y \subset X$, where we found

$$\Pr(T_0 \uparrow Y \mid Y \subset X\} = V(Y)/V(X) = V_V(Y) \quad (3.47)$$

Testing experimentally for this probability amounts to randomly spreading a certain number of points $P(c)$ in the structure $X = c$ and counting those points $P(a)$ which are enclosed within $Y = a$, so that

$$\Pr\{T_0 \uparrow Y\} = P(a)/P(c) = P_{Pa} = V_{Va} \quad (3.48)$$

This demonstration corresponds in approach to the second proofs brought to the foregoing principles. To make our demonstrations consistent we should also prove the validity of this principle for a point grid deposited on a section whose area $A(c)$ is covered over an area $A(a)$ with profiles of phase a; in other words, that

$$A_{Aa} = P_{Pa} \quad (3.49)$$

We may take again the approach of geometric probability, but this time in two-dimensional space (Fig. 3.5). The section has area $A(X)$ and is covered over an area $A(Y)$ with profiles of phase $Y = a$.

Suppose dA is an infinitesimal area element of area $A(dA)$. The probability that a random point T_0 on X hits dA is

$$\Pr(T_0 \uparrow dA \mid T_0 \uparrow X) = A(dA)/A(X) \quad (3.50)$$

The average probability for the random point to lie in Y is

$$E\{\Pr(Y \uparrow T_0)\} = \int_Y E\{\Pr(dA \uparrow T_0 \mid T_0 \uparrow dA)\}\Pr(T_0 \uparrow dA \mid T_0 \uparrow X) \quad (3.51)$$

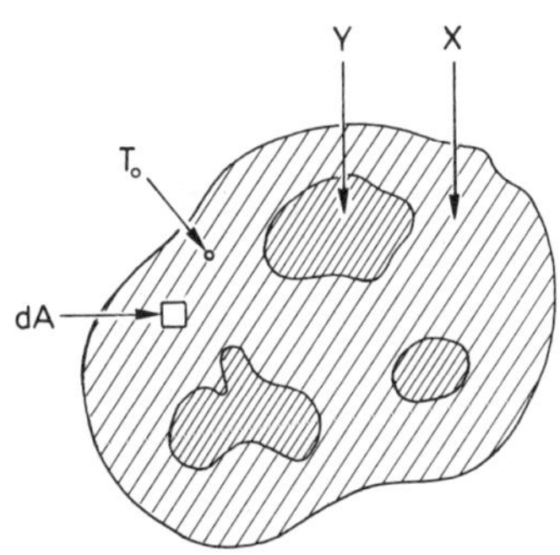

FIG. 3.5: Demonstration of point counting principle on section.

but evidently

$$\Pr(dA \uparrow T_0 \mid T_0 \uparrow dA) = 1$$

so that

$$\begin{aligned} E\{\Pr(Y \uparrow T_0)\} &= \int_Y \Pr(T_0 \uparrow dA \mid T_0 \uparrow X) \\ &= \int_Y \frac{A(dA)}{A(X)} \\ &= A(Y)/A(X) \end{aligned} \tag{3.52}$$

Applying $P(X)$ test points to the section we find, experimentally,

$$\Pr(Y \uparrow T_0) = P(Y)/P(X) \tag{3.53}$$

so that

$$E\{P(Y)/P(X)\} = A(Y)/A(X) \tag{3.54}$$

or

$$\overline{P_{Pa}} = A_{Aa} \tag{3.55}$$

Hence the principle also holds for a set of test points spread onto a section, provided this section is a random section. In this case, however, we have gone through a two-stage sampling procedure: (1) random sectioning; (2) testing the section with random points. If the section has been obtained by an A-weighted random sampling procedure then eq. (3.55) also yields an unbiased estimate of V_{Va} (see Section 2.4).

In practice (see Volume 1, Chapter 4) one tends to estimate V_V by applying regular test point lattices to several sections. Considering that this amounts to maintaining $P(X)$ proportional to $A(X)$, such that

$$P(X) = \gamma \cdot A(X) \tag{3.56}$$

where γ is the test point density in the area, we can rewrite eq. (3.54) as

$$\frac{A(Y)}{A(X)} = E\left\{\frac{P(Y)}{P(X)}\right\} = \frac{E\{P(Y)\}}{\gamma \cdot A(X)} \tag{3.57}$$

because $\gamma \cdot A(X)$ is constant. If we now test many sections of variable area $A(X)$ with the same test point lattice of density γ we obtain

$$E\{A(X)\} = (1/\gamma) \cdot E\{P(X)\} \tag{3.58}$$

$$E\{A(Y)\} = (1/\gamma) \cdot E\{P(Y)\} \tag{3.59}$$

Using eq. (3.29), the formulation of the Delesse principle, we find the estimator for volume density to be

$$\frac{V(Y)}{V(X)}=\frac{E\{A(Y\cap T)\}}{E\{A(X\cap T)\}}=\frac{E\{P(Y)\}}{E\{P(X)\}} \tag{3.60}$$

As discussed in Volume 1, Section 4.3.1, this amounts to considering the "weight" of each section, i.e. its area $A(X\cap T)$ contributed to the total sample. It should be stressed that eq. (3.55) is correct only if A_A is to be determined on a given section, whereas eq. (3.60) is more appropriate when the sample consists of several sections of varying area $A(X\cap T)$.

It would be easy to extend this type of reasoning to show that also

$$L_{La}=\overline{P_{Pa}}$$

The probability considerations are here restricted to one-dimensional space, a line, and the element of probability measure is a short unit segment, say dz. The reader may want to execute this proof as an exercise.

3.2. Surface density measurement

In Section 2.1 we have shown that sectioning an object whose set of surface points was called Y_2 will leave a trace only if the section is a plane (T_2) or a line (T_1). We must therefore expect two possibilities for estimating the surface density. For $Y_2\cap T_2$ the trace is a line (Z_1) and for $Y_2\cap T_1$ the trace is an intersection point (Z_0). In one principle, therefore, the surface density must be proportional to the boundary length of the profiles, in the other to the number of intersection points with test lines.

In terms of set theory the structure c is the union of two sets a and b

$$c=a\cup b$$

The "surface" is the set of points Y_2 at the interface between a and b, which we shall assign to the "object" set a, which is hence considered as a closed set (see Section 2.1). The extent of the surface subset Y_2 is measured by the surface area of phase a, $S(a)$, that of the structure by its volume, $V(c)$. The surface density of a in c is defined as

$$S_{Va}=S(a)/V(c) \tag{3.61}$$

or, in terms of the specimen shown in Fig. 3.2

$$S_V(Y_2)=S(Y_2)/V(X) \tag{3.62}$$

3.2.1. SURFACE DENSITY S_V AND DENSITY OF PROFILE BOUNDARIES ON SECTION B_A

If a 2-section T_2 intersects the structure $X \supset Y_2$ the trace of X will be an area Z_2, that of Y_2 will be a line Z_1:

$$X \cap T_2 = Z_2(X) \tag{3.63}$$

$$Y_2 \cap T_2 = Z_1(Y) \tag{3.64}$$

If the extent of $Z_2(X)$ is measured by the section area through X, $A(X \cap T) = A(c)$, and the extent of $Z_1(Y)$ by the length of the boundaries of the profiles of a, $B(Y \cap T) = B(a)$, then we define a boundary length density on the section area as

$$B_{Aa} = \frac{B(Y \cap T)}{A(X \cap T)} = \frac{B(a)}{A(c)} \tag{3.65}$$

It is intuitively plausible that B_A should be proportional to S_V. It has been shown by Saltykov (1945) and, independently, by Smith and Guttmann (1953) that

$$S_V = (4/\pi)B_A \tag{3.66}$$

We shall demonstrate this principle using results of geometric probability. Consider the specimen X with volume $V(X)$ containing components Y with smooth surface Y_2 of area $S(Y)$ (Fig. 3.2a); since we are only interested in the surface set of the components we shall use $Y = Y_2$ for simplicitly in this demonstration. If a random plane section $T = T_2$ hits Y, then $Y \cap T$ is a random curve of length $B(Y \cap T)$.

Suppose dS is an infinitesimal surface element (Fig. 3.2a); since Y is smooth dS may be regarded as an essentially planar disc. If dS is hit by T a line segment is formed whose total infinitesimal length we may designate $B(dS \cap T)$. The length of the curve $Y \cap T$ is the integral of all $B(dS \cap T)$ that are hit by T:

$$B(Y \cap T) = \int_Y B(dS \cap T) \tag{3.67}$$

The mean curve length for any random section through X is found, in analogy with the proof offered for the Delesse principle, by

$$\begin{aligned} E\{B(Y \cap T)\} &= E\left\{\int_Y B(dS \cap T)\right\} \\ &= \int_Y E\{B(dS \cap T)\} \end{aligned} \tag{3.68}$$

Considering the probability for T hitting dS this becomes

$$E\{B(Y \cap T)\} = \int_Y E\{B(dS \cap T) \mid T \uparrow dS\} \cdot \Pr(T \uparrow dS) \qquad (3.69)$$

Now the probability that a section through X cuts one infinitesimal surface element is (in analogy with eq. 2.55)

$$\Pr(T \uparrow dS) = H(dS)/H(X) \qquad (3.70)$$

where H is the mean caliper diameter of dS (Fig. 3.6) and X (Fig. 3.2b). The mean intercept length $B(dS \cap T)$, provided it is hit by T, is

$$E\{B(dS \cap T) \mid T \uparrow dS\} = S(dS)/h(dS) \qquad (3.71)$$

where $S(dS)$ is the surface area of the element, and $h(dS)$ is the mean projection of dS in the plane of the element (Fig. 3.6).

$H(dS)$ and $h(dS)$ are both linear characteristics of the element dS; what is their relationship? Let us first look at the simple case of a circular disc, where $h(dS) = 2R$, so that the mean linear intercept becomes $(\pi/2) \cdot R$. We find $H(dS)$ by integrating the caliper diameter of dS for all orientations (θ, ϕ) in space (Fig. 3.6); for a circular dS it comes out as $H(dS) = (\pi/2) \cdot R$. We therefore find that

$$H(dS)/h(dS) = \pi/4 \qquad (3.72)$$

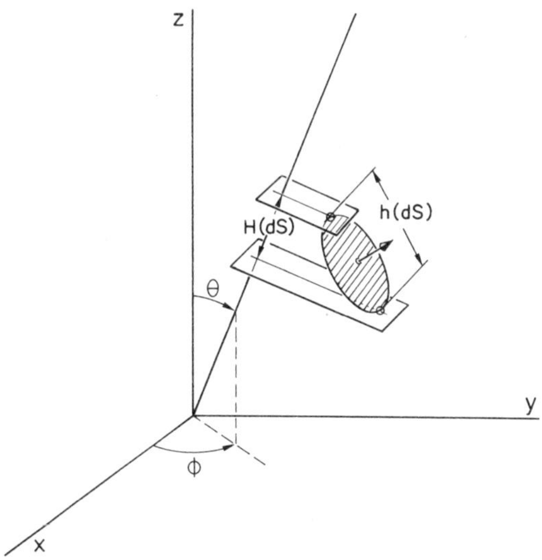

FIG. 3.6: Measures of surface element dS.

This can be shown to hold for any dS with a convex contour; for the general case one finds that $h(dS) = b(dS)/\pi$ and $H(dS) = b(dS)/4$, where $b(dS)$ is the perimeter of dS.

With these bits of information we can now return to the estimation of the mean curve length of $Y \cap T$; substituting eqs (3.70) to (3.72) into eq. (3.69) we find

$$E\{B(Y \cap T)\} = \int_Y \frac{S(dS)}{h(dS)} \cdot \frac{H(dS)}{H(X)}$$

$$= \frac{\pi}{4} \cdot \frac{1}{H(X)} \int_Y S(dS) \tag{3.73}$$

The integral is evidently $S(Y)$ so that

$$E\{B(Y \cap T)\} = \frac{\pi}{4} \cdot \frac{S(Y)}{H(X)} \tag{3.74}$$

Now the mean area of the section of $X \cap T$ is

$$E\{A(X \cap T)\} = V(X)/H(X) \tag{3.75}$$

so that, dividing eq. (3.74) by eq. (3.75)

$$B_A = \frac{E\{B(Y \cap T)\}}{E\{A(X \cap T)\}} = \frac{\pi}{4} \cdot \frac{S(Y)}{V(X)} = \frac{\pi}{4} S_V \tag{3.76}$$

In words, the ratio of the mean profile boundary length to the mean section area of the specimen multiplied with $4/\pi$ is an estimator of S_V. According to Miles and Davy (1976b) this estimator is only unbiased if A-weighted random sections are used, at least in the "restricted case" of a small specimen. For the "extended case", where the section is obtained from an IUR sample of the large specimen, an arbitrary section plane yields an unbiased result (see Section 2.4).

3.2.2. SURFACE DENSITY S_V AND DENSITY OF INTERSECTIONS ON TEST LINES I_L

The intersection of a 1-section or line T_1 with the structure $X \supset Y_2$ leads to the following traces:

$$X \cap T_1 = Z_1(X) \tag{3.77}$$

$$Y_2 \cap T_1 = Z_0(Y_2) \tag{3.78}$$

If the extent of $Z_1(X)$ is measured by the length of the test line enclosed in X, $L(X \cap T_1) = L(c)$, and $Z_0(Y_2)$ by the number of intersections between the line and the object surface $I(Y_2 \cap T_1) = I(a)$, then we define the intersection density on test lines as

$$I_{La} = \frac{I(a)}{L(c)} = \frac{I(Y_2 \cap T_1)}{L(X \cap T_1)} = \frac{I(Y \cap T)}{L(X \cap T)} \tag{3.79}$$

It is again intuitively plausible that I_L should be proportional to S_V. The fundamental relationship

$$S_V = 2I_L \tag{3.80}$$

has been derived independently at least seven times since the first independent derivations by Tomkeieff and by Saltykov in 1945 (Smith and Guttman, 1953; Duffin *et al.*, 1953; Horikawa, 1954; Corrsin, 1954; Hennig, 1956). One keeps finding "new" independent developments, which points to the fact that interdisciplinary communication on stereological problems was very poor until 1961, the founding year of the International Society for Stereology.

We shall prove this principle by two approaches, the first (Weibel, 1963) being more elementary and restrictive than the second which follows Miles and Davy (1976b).

First proof

Consider a cubic specimen of volume $V(c) = l^3$ which contains objects of phase a (Fig. 3.7). Suppose the surface of set a to be represented by a polyhedral surface formed of very small polygons of equal area dS (Fig. 3.8). The surface area $S(a)$ is evidently $S(dS)$ multiplied with the number N

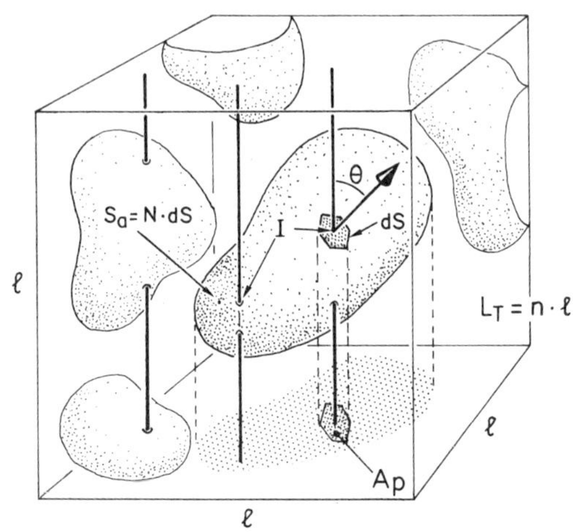

FIG. 3.7: Demonstration of surface density estimation from intersection count on test lines.

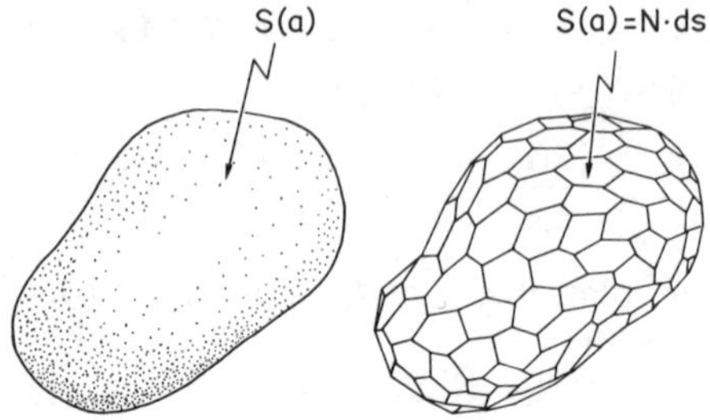

FIG. 3.8: Polyhedral approximation of smooth surface to form planar surface elements.

of these elementary polygons:

$$S(a) = N \cdot S(dS) \tag{3.81}$$

We now postulate that the surface $S(a)$ is isotropic, i.e. that the surface elements dS take all orientations in space. Single out one of these surface elements and suspend it at random in the cube (Fig. 3.9); the structure is now $dS \subset c$. Section this structure with a test line parallel to the z-axis; it is required to find the probability that the line intersects dS.

From what we have learned in Section 2.3 this probability will be given by the ratio of the orthogonal projection of dS, which we will call A_p, to the orthogonal projection of c, which is l^2 (Fig. 3.9):

$$p(I) = A_p(\theta, \phi)/l^2 \tag{3.82}$$

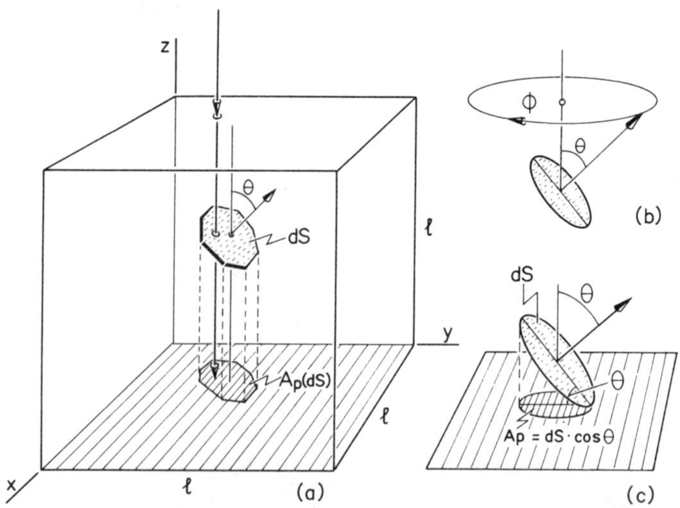

FIG. 3.9: Probability of hitting surface element depends on its projected area (a); this depends on angle θ (c), but is unaffected by ϕ (b).

The projection area A_p evidently depends on the orientation of dS with respect to the test lines, measured, as usual, by the angles θ and ϕ formed between the normal to dS and the coordinate system; we must therefore average for orientation to obtain the average probability

$$\Pr(I) = \bar{A}_p / l^2 \tag{3.83}$$

Fortunately it happens that A_p is insensitive to changes in the angle ϕ; one can easily verify that (Fig. 3.9b): if one rotates dS at a fixed angle θ, thus varying ϕ, the projection area does not change. Hence we find (Fig. 3.9c)

$$A_p(\theta) = S(dS) \cdot \cos\theta \tag{3.84}$$

and, since dS is constant,

$$\overline{A_p} = S(dS) \cdot \overline{\cos\theta} \tag{3.85}$$

Because of the exclusive dependence on θ the procedure of integration over all angles becomes simpler. We must now only consider the probability densities with respect to θ. The set of all orientations $\{\theta \rightarrow (\theta + d\theta)\}$ lies within a zone of width $d\theta$ and circumferential length $2\pi \sin\theta$ on the orientation sphere of radius 1 around the origin (Fig. 2.16) as demonstrated in Section 2.3; the area of this zone is $2\pi \sin\theta\, d\theta$. If we allow θ to vary from 0 to $\pi/2$ (since from $\pi/2$ to π the values of $\cos\theta$ are simply duplicated) the probability density of θ is

$$p(\theta) = \frac{2\pi \sin\theta\, d\theta}{2\pi} = \sin\theta\, d\theta \tag{3.86}$$

where 2π in the denominator is the surface of the hemisphere, the orientation reference set (Section 2.3).

The average value of $\cos\theta$ is obtained by integrating the product of probability θ with $\cos\theta$ over the limits $0 < \theta < \pi/2$, the range covering the hemisphere:

$$\overline{\cos\theta} = \int_0^{\pi/2} \cos\theta \cdot \sin\theta\, d\theta \tag{3.87}$$

Since

$$\cos\theta\, d\theta = d \sin\theta \tag{3.88}$$

we immediately obtain:

$$\begin{aligned} \overline{\cos\theta} &= \int_0^{\pi/2} \sin\theta \cdot d\sin\theta \\ &= \tfrac{1}{2}\Big[\sin^2\theta\Big]_0^{\pi/2} = \tfrac{1}{2} \end{aligned} \tag{3.89}$$

The mean projection area of the surface element dS is therefore

$$\overline{A_p} = \tfrac{1}{2} S(dS) \tag{3.90}$$

and the probability of having an intersection between the surface element and one of the test lines is:

$$\Pr\{I\} = \tfrac{1}{2} \frac{S(dS)}{l^2} \tag{3.91}$$

Let us now return to the structure itself. The surface of its objects $S(a)$ had been subdivided into N polygons of area $S(dS)$ so that

$$S(a) = N \cdot S(dS) \tag{3.92}$$

Furthermore, assume that the structure is probed with n vertical test lines of length l making a total test line length in c of

$$L(c) = n \cdot l \tag{3.93}$$

It is now evident that the expected number of intersections, $E\{I\}$, between test lines and object surface must be proportional to both the number of surface elements and the number of test lines, as well as to the probability of obtaining an intersection with the surface element:

$$E\{I\} = \Pr\{I\} \cdot N \cdot n \tag{3.94}$$

Appropriately substituting eqs (3.91) to (3.93) into eq. (3.94) yields:

$$E\{I\} = \tfrac{1}{2} S(a) \cdot L(c) l^3$$

or

$$E\{I\}/L(c) = \tfrac{1}{2} \cdot S(a)/V(c) \tag{3.95}$$

It is easily seen that this is the relationship we sought:

$$S_{Va} = 2 \cdot I_{La}$$

whereby $E\{I\}$ signifies that the number of intersections truly observed will be an estimate of a theoretically expected value.

Second proof

Consider now the more general specimen X of volume $V(X)$ which contains Y with a surface area $S(Y) = S(Y_2)$ and allow it to be sectioned with a random line T_1 (Figs 3.2 and 3.4). If T_1 hits Y_2, then $Y_2 \cap T_1$ is an intersection point; their number is $I(Y \cap T)$.

Suppose dS is again an infinitesimal surface element of Y which we may assume to be planar; $dS \cap T$ is again an intersection point. Note that

$I(dS \cap T)$ is 1 if T hits dS and 0 otherwise. We have

$$I(Y \cap T) = \int_Y I(dS \cap T) \tag{3.96}$$

for the number of intersections formed with one line section, and on the average

$$\begin{aligned} E\{I(Y \cap T)\} &= E\left\{\int_Y I(dS \cap T)\right\} \\ &= \int_Y E\{I(dS \cap T)\} \\ &= \int_Y E\{I(dS \cap T) \mid T \uparrow dS\} \cdot \Pr(T \uparrow dS) \end{aligned} \tag{3.97}$$

We have already said that

$$E\{I(dS \cap T \mid T \uparrow dS)\} = 1 \tag{3.98}$$

The probability that T hits dS is

$$\Pr(T \uparrow dS) = A_p(dS)/A_p(X) \tag{3.99}$$

the ratio of projected areas of dS and X. The projected area of a planar element of surface $S(dS)$ is—we have derived it in the first proof—

$$A_p(dS) = S(dS)/2 \tag{3.100}$$

so that

$$\begin{aligned} E\{I(Y \cap T)\} &= \int_Y 1 \cdot S(dS)/[2A_p(X)] \\ &= \frac{1}{2A_p(X)} \int_Y S(dS) \\ &= \tfrac{1}{2} \cdot S(Y)/A_p(X) \end{aligned} \tag{3.101}$$

since the integral of $S(dS)$ over the objects Y is the object surface.

The mean length of test line enclosed in X is

$$E\{L(X \cap T)\} = V(X)/A_p(X) \tag{3.102}$$

as can easily be seen (Fig. 3.4). Dividing eq. (3.101) by eq. (3.102) we obtain

$$I_L = \frac{E\{I(Y \cap T)\}}{E\{L(X \cap T)\}} = \frac{1}{2} \cdot \frac{S(Y)}{V(X)} = \frac{1}{2} S_V \tag{3.103}$$

In this second proof we have not imposed any conditions on the arrangement of the surface Y_2, in particular it did not have to be isotropic. However we have assumed that $T \cap X$ was a random line section of X. Again, if X is small the sampling conditions of the "restricted case" apply, and only an L-weighted line section will yield an unbiased estimator of S_V (Miles and Davy, 1976b). An L-weighted section is an isotropic random line through a uniform random point in X (see Section 2.4). If X is an IUR sample from a large body then an arbitrary line section yields an unbiased estimate.

3.2.3. BOUNDARY DENSITY B_A AND INTERSECTION DENSITY I_L: BUFFON'S NEEDLE PROBLEM

The practical procedure for estimating S_V by the intersection method involves two-stage sampling: (1) a section is cut, exposing profile boundaries, B_A; (2) test lines are deposited on the section to determine I_L. The problem therefore is to find whether the formula $S_V = 2I_L$ still holds for this procedure. It is first necessary to find the relationship between B_A and I_L.

It can easily be shown that B_A can be estimated by placing a set of test lines of known length onto the section and determining the number I of intersections with these lines: the formula

$$B_A = \frac{\pi}{2} \cdot I_L \tag{3.104}$$

is derived from Buffon's needle problem (Buffon, 1777; see Miles and Serra, 1978).

The profile boundaries of $(Y \cap T_2)$ form a curve of length $B(Y \cap T_2)$ (see Section 3.2.1); if the curve is smooth it can be cut into (infinitesimal) short straight line segments dB of equal length $l(dB)$ (Fig. 3.10) such that

$$B = \int_{Y \cap T} l\,(dB) = n \cdot l(dB) \tag{3.105}$$

We assume the curve to be isotropic in the plane, i.e. all orientations of the segments dB are equally likely. We now place a grid of parallel equidistant test lines, T_1, onto the section plane (Fig. 3.10); the distance of the test lines is $d > l(dB)$. The problem is to find the number I of the n segments dB which will intersect the test lines. Evidently this amounts to finding the probability

$$\Pr\{dB \uparrow T_1\} = I/n \tag{3.106}$$

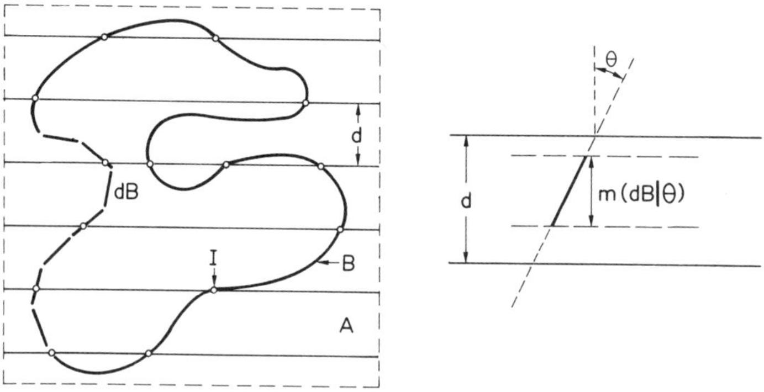

FIG. 3.10: Buffon principle: probability that line segment ("needle") dB intersects grid of lines as basis for estimating length of curve on plane.

in terms of d and $l(dB)$, allowing for all orientations of dB, measured by the angle θ between dB and the test lines (Fig. 3.10).

The measure of $dB(\theta)$ is the "height" of dB or its projection onto the orthogonal complement to the grid T_1:

$$m(dB \mid \theta) = l(dB) \cdot \cos \theta \tag{3.107}$$

The probability that $dB(\theta)$ intersects T_1 is

$$\begin{aligned} \Pr\{dB \uparrow T_1 \mid \theta\} &= m(dB \mid \theta)/d \\ &= l(dB) \cdot \cos \theta/d \end{aligned} \tag{3.108}$$

For isotropic orientation of dB we must allow θ to vary uniformly over the range $\{0 < \theta < \pi/2\}$, and average their measure over this range, so that

$$\begin{aligned} \Pr\{dB \uparrow T_1\} &= \frac{2}{\pi} \int_0^{\pi/2} \Pr\{dB \uparrow T_1 \mid \theta\}\, d\theta \\ &= \frac{2 \cdot l(dB)}{\pi \cdot d} \int_0^{\pi/2} \cos \theta\, d\theta \end{aligned} \tag{3.109}$$

Since $\int_0^{\pi/2} \cos \theta\, d\theta = 1$ we obtain, by eqs (3.106) and (3.109)

$$\Pr\{dB \uparrow T_1\} = \frac{2 \cdot l(dB)}{\pi \cdot d} = I/n \tag{3.110}$$

This is the result obtained by Buffon in 1777, published in his "Essai d'Arithmétique Morale", which is considered the first problem in geometric probability. Discussions of this problem are found in Kendall and Moran (1963), Santaló (1976) and in Miles and Serra (1978) which compiles the proceedings of a Symposium on Geometrical Probability and

Biological Structure held in Paris in 1977 to commemorate the 200th anniversary of this fundamental work of Buffon, in fact the first step that eventually led to the mathematical foundations for stereology. The historical background of Buffon's involvement in this kind of work—he was one of the great naturalists and universalists of the eighteenth century—is discussed by Binet and Roger (1977).

If we now rearrange eq. (3.110) and substitute into eq. (3.105) we find the relationship between the number of intersections, *I*, and the length of the curve as

$$B = \tfrac{1}{2}\pi \cdot I \cdot d \tag{3.111}$$

If both the curve B and the test grid T_1 are enclosed in an area A (Fig. 3.10), the length of the test line is $L = A/d$. Substituting this for d in eq. (3.111) yields the relationship between the curve length density B_A and the number of intersections per unit test line length:

$$B_A = \frac{B}{A} = \frac{\pi}{2} \cdot \frac{I}{L} = \frac{\pi}{2} \cdot I_L \tag{3.112}$$

The preceding development assumed a grid of parallel lines to be deposited on the test area, and the curve B was assumed to be isotropic. But the same result can be obtained using a single test line and following the reasoning repeatedly used in preceding sections. Consider a specimen X, this time a planar surface of area $A(X)$, which contains an "object" curve Y of length $B(Y)$ (Fig. 3.11); imagine that X is a profile of a three-dimensional specimen and Y the trace of an enclosed object surface, obtained by a random section. Deposit a random test line T_1 on this

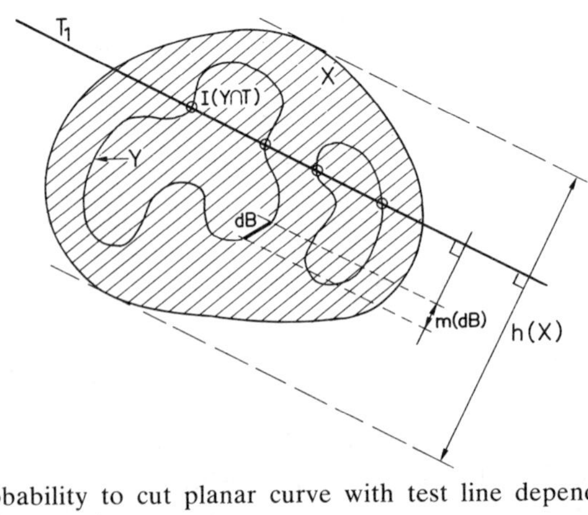

FIG. 3.11: Probability to cut planar curve with test line depends on projected width of curve segment.

planar specimen. We would like to find the expected number of intersections $I(Y \cap T_1)$.

Breaking up the curve into infinitesimal segments dB we have

$$I(Y \cap T) = \int_Y I(dB \cap T) \tag{3.113}$$

and, following the development given in relation to eq. (3.97) on the average,

$$\begin{aligned} E\{I(Y \cap T)\} &= E\left\{\int_Y I(dB \cap T)\right\} \\ &= \int_Y E\{I(dB \cap T)\} \\ &= \int_Y E\{I(dB \cap T) \mid T\uparrow dB\} \cdot \Pr(T\uparrow dB) \end{aligned} \tag{3.114}$$

Evidently,

$$E\{I(dB \cap T) \mid T\uparrow dB\} = 1 \tag{3.115}$$

The probability that T hits dB is

$$\Pr\{T\uparrow dB\} = m(dB)/h(X) \tag{3.116}$$

the ratio of the projections of dB and X on to the orthogonal complement to the test line (Fig. 3.11). If dB is allowed to assume all orientations θ with respect to the test line

$$\Pr\{T\uparrow dB\} = \bar{m}(dB)/h(X) \tag{3.117}$$

and, since by eq. (3.109), $\bar{m}(dB) = (2/\pi) \cdot l(dB)$,

$$Pr\{T\uparrow dB\} = \frac{2}{\pi} l(dB)/h(X) \tag{3.118}$$

Consequently,

$$\begin{aligned} E\{I \cap T\} &= \frac{2}{\pi} \int_Y l(dB)/h(X) \\ &= \frac{2}{\pi} \frac{B(Y)}{h(X)} \end{aligned} \tag{3.119}$$

If T_1 is a random line through X we find the average length of the test line included in X as

$$E\{L(X \cap T)\} = A(X)/h(X) \tag{3.120}$$

From eqs (3.119) and (3.120) we find

$$\frac{E\{I(Y \cap T\}}{E\{L(X \cap T\}} = \frac{2}{\pi} \cdot \frac{B(Y)}{A(X)} \tag{3.121}$$

or

$$B_A = \frac{\pi}{2} \cdot \frac{E\{I(Y \cap T)\}}{E\{L(X \cap T)\}} \tag{3.122}$$

which is equivalent to eq. (3.112).

In this second derivation we have not imposed any conditions on the structure; Y_2 did not have to be isotropic. However, we assumed that T_1 was a random line section of the planar specimen X. If X is small the sampling conditions of the restricted case apply (see Section 2.4), so that only an L-weighted line through X, i.e. an isotropic random line through a uniform random point in X, will yield an unbiased estimate (Miles and Davy, 1976b). If X is a planar IUR sample of a larger body then an arbitrary line section yields an unbiased estimate.

3.2.4. ESTIMATING SURFACE DENSITY FROM LINE INTERSECTION COUNTS ON RANDOM SECTIONS

It now remains to show that an intersection count performed on a random section allows estimation of surface density. A first result is evident: substituting eq. (3.112) into eq. (3.66) yields eq. (3.80):

$$S_V = \frac{4}{\pi} B_A$$

$$B_A = \frac{\pi}{2} \cdot I_L$$

$$\overline{S_V = 2 \cdot I_L}$$

To do this is valid only if the two-stage sampling procedure—sectioning and testing the section with a line grid—does not introduce bias (Fig. 3.12).

The relevant result in this respect is the theorem of Miles and Davy (1976b), whereby "*an L-weighted line section of an A-weighted plane section of a spatial domain X is equivalently an L-weighted line section of X*". Observing these sampling conditions we had derived the following

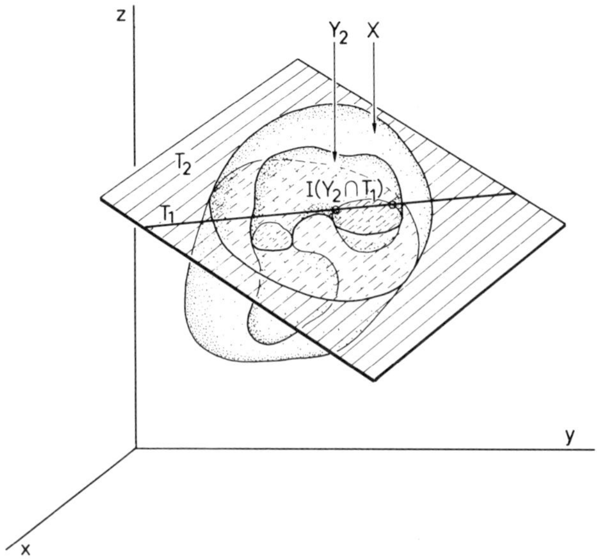

FIG. 3.12: Model for estimating surface density of spatial structure from intersection count on test line T_1 on section T_2 by two-stage sampling procedure.

set of relations (the subscripts specify the dimension s of structure, object and section):

$$\frac{E\{B(Y_2 \cap T_2)\}}{E\{A(X_3 \cap T_2)\}} = \frac{\pi}{4} \cdot S_V \tag{3.76}$$

$$\frac{E\{I(Y_2 \cap T_1)\}}{E\{L(X_3 \cap T_1)\}} = \tfrac{1}{2} \cdot S_V \tag{3.103}$$

$$\frac{E\{I(Y_1 \cap T_1)\}}{E\{L(X_2 \cap T_1)\}} = \frac{2}{\pi} \cdot B_A \tag{3.121}$$

If Y_1, X_2 are obtained as an A-weighted random section of Y_2, X_3, then the theorem of Miles and Davy (1976b) leads to the conclusion that

$$E\{L(X_2 \cap T_1)\} = E\{L(X_3 \cap T_1)\} \tag{3.123}$$

$$E\{I(Y_1 \cap T_1)\} = E\{I(Y_2 \cap T_1)\} \tag{3.124}$$

and, consequently,

$$\frac{E\{I(Y_1 \cap T_1)\}}{E\{L(X_2 \cap T_1)\}} = \frac{E\{I(Y_2 \cap T_1)\}}{E\{L(X_3 \cap T_1)\}} = \frac{1}{2} S_V \tag{3.125}$$

Observing the sampling conditions outlined in Section 2.4 an intersection count performed on test lines deposited on a section of the specimen yields an unbiased estimate of the surface density within the specimen.

3.3. Length density measurement

Consider a model structure X containing an object $Y_1 = a$ in the form of a long thin filament (Fig. 3.13). Alternatively Y_1 could be the set of edges of polyhedral cells contained in X (Fig. 2.3). We would like to determine the length density of these filaments or edges related to the volume of the structure

$$J_V = J(Y_1)/V(X) \tag{3.126}$$

If a plane T_2 intersects $X \supset Y_1$ the traces are

$$X \cap T_2 = Z_2$$
$$Y_1 \cap T_2 = Z_0$$

The extent of Z_2 is measured by the section area $A(X)$, that of Z_0 by the number of transections of Y_1, $Q(Y_1)$. Hence, the density of transections on the section area is

$$Q_A = Q(Y_1)/A(X) \tag{3.127}$$

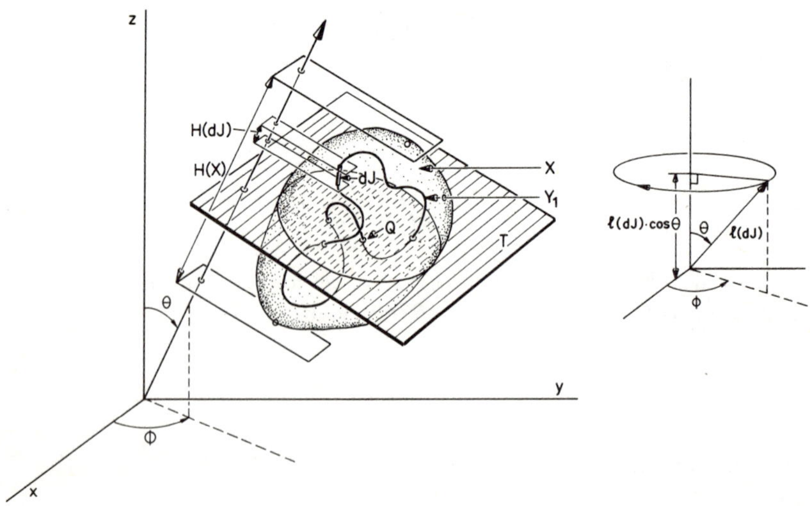

FIG. 3.13: Model for estimating curve length density in spatial structure from transections with test plane.

It has been shown by Saltykov (1945), Smith and Guttman (1953) and Hennig (1963a, b) that

$$J_V = 2 \cdot Q_A \tag{3.128}$$

a formula identical to that for the relationship between surface density and intersection density; in fact the two formulae are identical as can easily be demonstrated by setting $S = A$, $J = L$ and $Q = I$ and slightly rearranging.

A formal demonstration of this equation must follow the same arguments as those exposed in deriving $S_V = 2I_L$. There we had broken the surface into small unit areas which were allowed to assume any orientation and we kept the direction of test lines constant. Here we would keep a flat test area constant and would break up the lines into small segments allowing them to assume any direction in space. The probability of orientation of lines to the plane would be the same, i.e. on the average

$$\overline{\cos\theta} = \tfrac{1}{2} \tag{3.129}$$

Using equivalent notation, but appropriately changed, eq. (3.80) would read

$$A/V = 2 \cdot Q/J \tag{3.130}$$

or, by rearranging and specifying the parameters

$$J(Y_1)/V(X) = 2 \cdot Q(Y_1)/A(X) \tag{3.131}$$

which is equivalent to eq. (3.128).

We can, however, go through a rigorous proof using the notions of geometric probability and following essentially the same reasoning as in the "second proofs" presented in preceding sections, e.g. Section 3.2.2. The structure we consider is a specimen X of volume $V(X)$ containing a smooth spatial curve $Y_1 = Y$ of length $J(Y)$. This specimen is cut by a random plane section $T_2 = T$ (Fig. 3.13).

Let dJ be an infinitesimal line element of Y which we can assume to be a straight line segment because Y is a smooth curve; its total infinitesimal length is $l(dJ)$. When dJ is cut by T intersection points are formed, which we shall call $Q(dJ \cap T)$; note that $Q(dJ \cap T)$ is 1 if T hits dJ and 0 otherwise.

The number of intersections formed with Y is

$$Q(Y \cap T) = \int_Y Q(dJ \cap T) \tag{3.132}$$

and the mean for all possible sections is

$$E\{Q(Y \cap T)\} = E\left\{\int_Y Q(dJ \cap T)\right\}$$
$$= \int_Y E\{Q(dJ \cap T) \mid T \uparrow dJ\} \cdot \Pr(T \uparrow dJ) \quad (3.133)$$

a general result we have now repeatedly seen. Since

$$E\{Q(dJ \cap T \mid T \uparrow dJ) = 1 \quad (3.134)$$

as we have said above, all that remains is to determine the probability that T hits dJ. Since we are cutting with a plane section the probability is estimated from the orthogonal projections of dJ and X on to the normal to T:

$$\Pr(T \uparrow dJ) = H(dJ)/H(X) \quad (3.135)$$

The main question is: what is $H(dJ)$? Figure 3.13 shows that the projection of dJ on to the orthogonal complement to T (its normal) is

$$H(dJ \mid \theta, \phi) = l(dJ)\cos\theta \quad (3.136)$$

it is invariant with respect to ϕ. The probability density for θ being $\sin\theta$, as shown in Section 2.3, the average projection is obtained by

$$H(dJ) = l(dJ)\int_0^{\pi/2} \cos\theta \cdot \sin\theta \, d\theta \quad (3.137)$$

since $\cos\theta \, d\theta = d\sin\theta$ we obtain

$$H(dJ) = l(dJ)\int_0^{\pi/2} (\sin\theta) \cdot d(\sin\theta)$$
$$= l(dJ)\tfrac{1}{2}[\sin^2\theta]_0^{\pi/2}$$
$$= \tfrac{1}{2}l(dJ) \quad (3.138)$$

so that

$$\Pr(T \uparrow dJ) = \frac{1}{2} \cdot \frac{l(dJ)}{H(X)} \quad (3.139)$$

Therefore

$$E\{Q(Y \cap T)\} = \int_Y 1 \cdot \tfrac{1}{2}l(dJ)/H(X)$$
$$= \frac{1}{2H(X)} \cdot \int_Y l(dJ)$$
$$= \tfrac{1}{2} \cdot J(Y)/H(X) \quad (3.140)$$

The mean section area of X is

$$E\{A(X \cap T)\} = V(X)/H(X) \tag{3.141}$$

so that, by dividing eq. (3.140) by eq. (3.141), we obtain

$$Q_A = \frac{E\{Q(Y \cap T)\}}{E\{A(X \cap T)\}} = \frac{1}{2} \cdot \frac{J(Y)}{V(X)} = \frac{1}{2} J_V \tag{3.142}$$

which is again equivalent to eq. (3.128).

The same holds as before: the ratio of mean intersections to mean section area yields an unbiased estimator of J_V if proper sampling procedures are observed. In particular, for a small specimen, the "restricted case", T must be an A-weighted random plane section (Miles and Davy, 1976b).

3.4. Measurement of mean curvature

Any curve which is not a straight line, or any surface which is not a plane, has curvature. The local curvature κ may vary from point to point along the curve or over the surface; one measure of κ is the reciprocal of the radius ρ of a circle which can be fitted to three closely spaced points.‡

3.4.1. THE CURVATURE OF PLANAR CURVES

Let us first look at a planar curve Y_1 of length B (Fig. 3.14), and let us determine the local curvature $\kappa(u_1)$ at point u_1. We look at a small segment dB around u_1 and note that, because dB is curved the direction of the normal to dB varies from one end of dB to the other; it is said that dB subtends as an arc an angle $d\psi$ between the normals at the ends of dB. The local curvature at point u_1 is

$$\kappa(u_1) = d\psi(u_1)/dB \tag{3.143}$$

The angular change $d\psi$ can also be defined as the change in the direction of tangents as we move along the curve by a small length dB (Fig. 3.14a). The change in angle, $d\psi$, can be either positive or negative. By convention one calls a counterclockwise motion along the curve positive; a clockwise change is negative (Fig. 3.14a), and, according to eq.

‡ The dimensionality of curvature is not immediately apparent. The integral of mean curvature is $K = (1/r) \cdot S$, and its dimension $d(K)$ is obtained from the logarithm of $L^d = L^{-1} \cdot L^2$; $d(K) = 1$. From eq. (2.12) if follows that a two-dimensional section is required to test it. On the plane, $C = (1/r) \cdot B$; $L^d = L^{-1} \cdot L$; $d(C) = 0$, so that C again needs to be measured in relation to a plane. K_V and C_A both have dimension $d = -2$, as has T_A. Mean curvature has dimension $d = -1$.

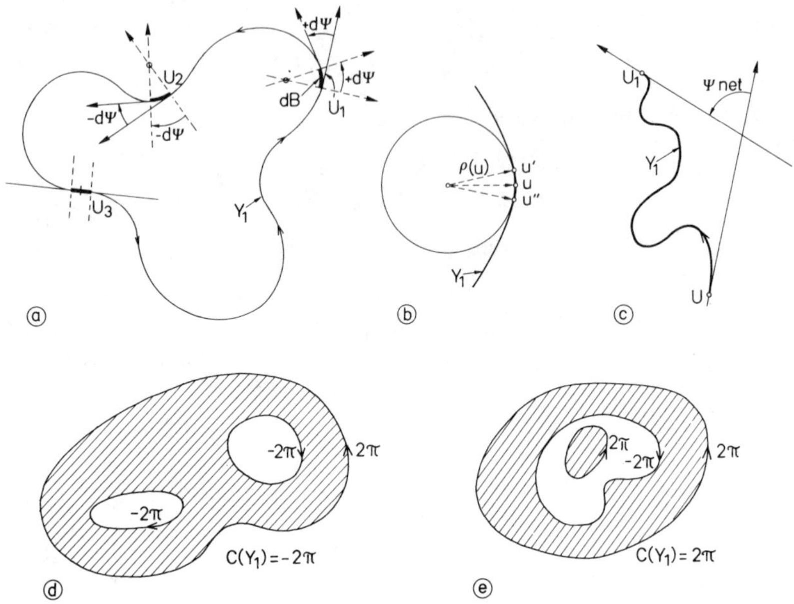

FIG. 3.14: Curvature of planar curves: (a) closed loop; (b) radius of curvature at point u; (c) total curvature of open curve; (d) and (e) total curvature of planar domains with holes.

(3.143), the local curvature $\kappa(u)$ will be positive or negative respectively in these two instances. Note that the normals to the curve at the end points of dB intersect to the left of the curve if $d\psi$ is positive, and to the right if $d\psi$ is negative. Note that we move along the curve in a counterclockwise direction (Fig. 3.14a).

Alternatively the local curvature is defined as the reciprocal of the radius, $\rho(u)$, of a circle that can be fitted through three closely spaced points u, u' and u'' on the element of curve dB (Fig. 3.14b):

$$\kappa(u) = 1/\rho(u) \tag{3.144}$$

A closed curve is called simple if it has no double points; in this case it bounds a domain Y_2. The convention about the orientation of the boundary Y_1 of Y_2 is that the domain is to the left of the boundary: the outer boundary runs counterclockwise, whereas inner boundaries run clockwise (Fig. 3.14d and e). If the local curvature $\kappa(u)$ is positive the curve is said to be *convex*, if it is negative the curve is *concave*; in Fig. 3.14a the curve is convex at u_1 and concave at u_2. At the *inflection* point u_3, where the contour changes from convex to concave the local curvature

$\kappa(u_3)=0$ because a straight line passes through u_3 and two closely adjoining points, i.e. $\rho(u_3)=\infty$.

The total curvature of Y_1 (or the "integral curvature") is obtained as the integral of all values $\kappa(u)$, whereby the signs of $\kappa(u)$ must be observed:

$$C(Y_1)=\int_{Y_1} \kappa(u)\cdot dB \tag{3.145}$$

$$C(Y_1)=\int_{Y_1} [1/\rho(u)]\cdot dB \tag{3.146}$$

or, in terms of the subtended angle (eq. 3.143),

$$C(Y_1)=\int_{Y_1} [d\psi(u)/dB]\cdot dB$$

$$=\int_{Y_1} d\psi(u) \tag{3.147}$$

The last result (eq. 3.147) is interesting in so far as it tells us that the total or integral curvature of a planar curve is the total net angle subtended by the normal to the curve as we move it from the starting point to the end point, whereby positive and negative angles alternate as we pass through convex and concave regions respectively. For an open curve extending between the points u_0 and u_1, as shown in Fig. 3.14c, we find

$$C(Y_1;\, u_0,\, u_1)=\int_{u_0}^{u_1} d\psi(u)=\psi_{\text{net}} \tag{3.148}$$

For a simple closed curve, like the one shown in Fig. 3.14a, the tangent moves through a full 360°, so that its total curvature is

$$C(Y_1)=2\pi \tag{3.149}$$

If the domain contains some holes the total curvature of its inner boundaries is negative, i.e. -2π. The domain shown in Fig. 3.14d has one outer and two inner boundaries, and the total curvature of its boundary is accordingly $2\pi+2(-2\pi)=-2\pi$; that of the domain in Fig. 3.14e is $2\pi-2\pi+2\pi=2\pi$. The total curvature of a domain with one hole is 0.

An important parameter is the average curvature of a planar curve, which is obtained by averaging $\kappa(u)$ over the entire length of Y_1:

$$\bar{C}(Y_1)=\left[\int_{Y_1} \kappa(u)\cdot dB\right]\Big/\left[\int_{Y_1} dB\right]$$

$$=C(Y_1)/B \tag{3.150}$$

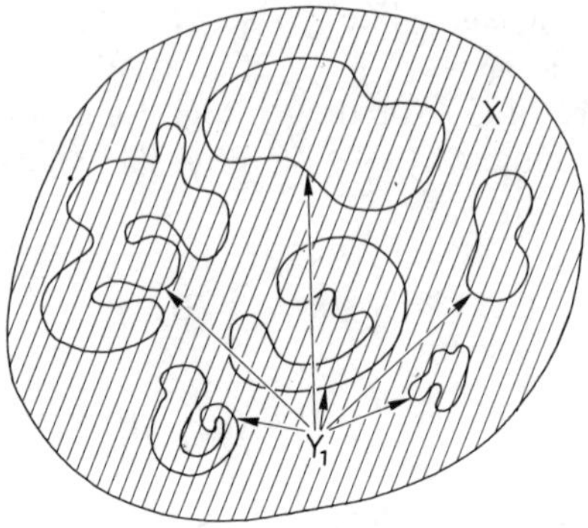

FIG. 3.15: Model for estimating curvature density on plane.

Finally, if Y_1 is contained in a planar specimen X of area $A(X)$ (Fig. 3.15), the density of integral curvature in X is given as

$$C_A = C(Y_1)/A(X) \tag{3.151}$$

If Y_1 consists of N simple closed loops surrounding simply connected domains it becomes evident from eq. (3.149) that

$$\begin{aligned} C_A &= 2\pi \cdot N(Y_1)/A(X) \\ &= 2\pi N_A \end{aligned} \tag{3.152}$$

because, for each loop, eq. (3.147) is 2π. But if N' loops are located inside some N^* domains, thus forming inner boundaries of the domains with total curvature -2π, $N(Y_1)$ is the difference between N^* and N'. This is related to the topological properties of the domains, dealt with in Section 3.5.

3.4.2. THE CURVATURE OF SURFACES

The local curvature of a surface Y_2 in three-dimensional space is defined in a similar way to that of a planar curve, essentially by the radius of circles closely fitting to three surface points; but since the surface has two dimensions the curvature at a surface point u on Y_2 is measured by two such radii. These are obtained as follows (Fig. 3.16):

Construct at u the normal z' to the surface, such that it points outwards from Y, and a plane T_z containing the normal z'. $T_{z'}$ intersects Y_2 at u and $T_{Z'} \cap Y_2$ is a planar curve on $T_{z'}$ for which we can find the local curvature $\kappa = 1/\rho$ at u, as before. As $T_{z'}$ is rotated around the normal κ will vary, and we obtain a curvature function $\kappa(\phi)$ over a range π. Choosing a suitable origin for ϕ, the theorem of Meusnier (Struik, 1961,

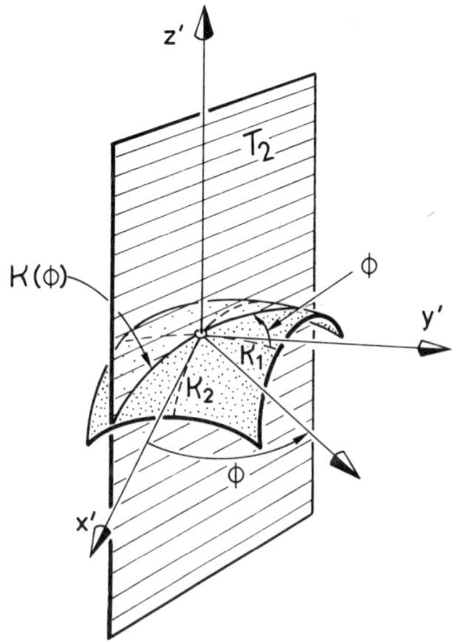

FIG. 3.16: Curvature of surface element on curved spatial surface ($T_2 = T_{Z'}$).

p. 81) gives us

$$\kappa(\phi) = \kappa_1 \cdot \cos^2 \phi + \kappa_2 \cdot \sin^2 \phi \tag{3.153}$$

where κ_1 and κ_2 are the largest and smallest values of κ as the plane is rotated; they are called the *principal curvatures* of the surface at u, and always occur at orientations perpendicular to each other.

Both κ_1 and κ_2 can be either positive or negative, and this defines three types of local surface curvature (Fig. 3.17):

convex surface: $\kappa_1 > 0$; $\kappa_2 > 0$
concave surface: $\kappa_1 < 0$; $\kappa_2 < 0$
saddle surface: $\kappa_1 > 0$; $\kappa_2 < 0$ (or $\kappa_1 < 0$; $\kappa_2 > 0$)

On any smooth surface of arbitrary shape all three types of local surface curvature may occur. A fourth type of local surface curvature can occur, called a "monkey saddle"; it has three principal curvatures, but these are actually zero, because the principal curves occurring on $T_{z'}$ have an inflection point at u. For this reason, monkey saddles are usually ignored (see De Hoff, 1968). They may, however, represent rather important structural features; in the lung, for example, they occur at the mouths of alveoli where the so-called entrance rings meet to form a hexagonal network bent around a cylinder. Indeed they occur in many networks of tubular structures (capillaries etc.) as well.

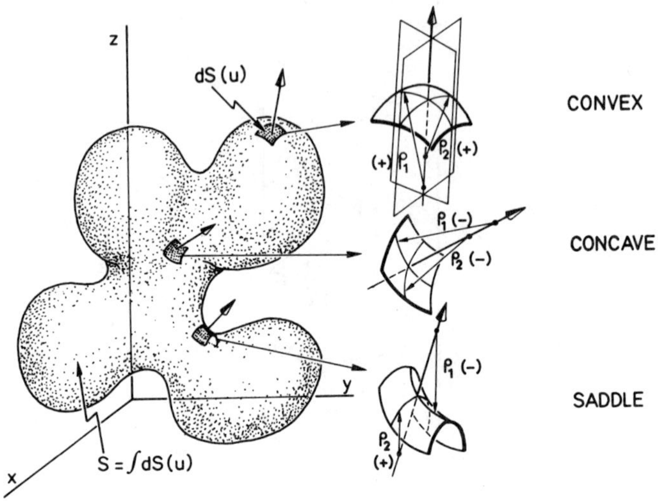

FIG. 3.17: Types of curvature on spatial surface defined by positive and/or negative radii of curvature.

The two principal local curvatures allow two types of local surface curvature to be defined: the *mean curvature* is the mean of the principal curvatures

$$\begin{aligned} K(u) &= \tfrac{1}{2}(\kappa_1 + \kappa_2) \\ &= \tfrac{1}{2}[(1/\rho_1) + (1/\rho_2)] \end{aligned} \tag{3.154}$$

The *Gaussian* or *total curvature* is their product

$$G(u) = \kappa_1 \cdot \kappa_2 = 1/(\rho_1 \cdot \rho_2) \tag{3.155}$$

With respect to the Gaussian curvature it is important to note that it can also be defined, in analogy to what we have seen for curves, in terms of the solid angle $d\Omega$ subtended by the infinitesimal surface element dS around u (Fig. 3.18)

$$G(u) = d\Omega/dS \tag{3.156}$$

With these definitions of local curvatures, i.e. the curvatures pertaining to a small infinitesimal surface element $dS(u)$ we can now obtain information on the curvature characteristics of the surface Y_2 by integration. Thus, the integral of mean curvature is

$$K(Y_2) = \int_{Y_2} \tfrac{1}{2}[\kappa_1 + \kappa_2] \cdot dS \tag{3.157}$$

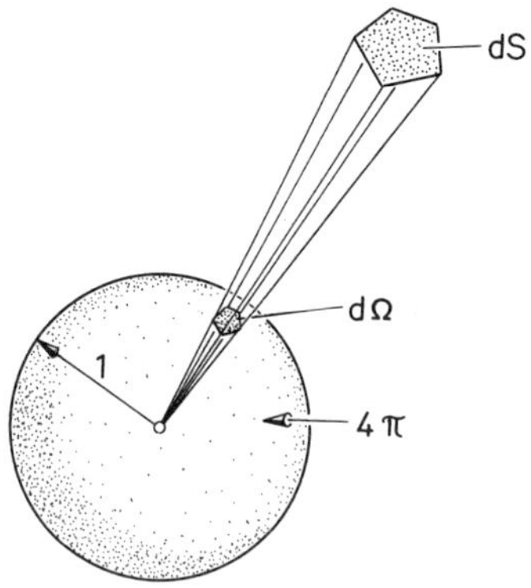

FIG. 3.18: Angle $d\Omega$ subtended by surface element dS of spatial surface defines Gaussian curvature.

and the integral of Gaussian curvature

$$G(Y_2) = \int_{Y_2} \kappa_1 \cdot \kappa_2 \cdot dS \tag{3.158}$$

where, again, one has to take note of the signs of κ_1 and κ_2. Note that the Gaussian curvature is positive for both convex and concave surface elements, but negative for saddle surfaces, whereas K is positive for convex and negative for concave surfaces, but can be either positive, negative or zero for saddle surfaces.

If the Gaussian curvature is integrated in terms of subtended solid angle (eq. 3.156) we obtain

$$\begin{aligned} G(Y_2) &= \int_{Y_2} [d\Omega/dS] \cdot dS \\ &= \int_{Y_2} d\Omega \end{aligned} \tag{3.159}$$

which, for simple closed surfaces of arbitrary shape, yields

$$G(Y_2) = 4\pi \tag{3.160}$$

A further interesting result relates to the special case of convex solids Y where we find, by the formula of Minkowski (1903)

$$K(Y_2 \mid \text{convex}) = 2\pi \cdot H(Y) \tag{3.161}$$

where $H(Y)$ is the mean caliper diameter of Y (see Section 5.5.2). Note that for a sphere of diameter $D = 2R$ we find by eq. (3.157):

$$\frac{1}{2}\int\left[\frac{1}{R}+\frac{1}{R}\right]dS = \frac{1}{R}\cdot S = 4\pi R = 2\pi D \tag{3.161a}$$

A physically important parameter describing a surface is its average mean curvature (or "mean mean" curvature) $\bar{K}$, which is simply obtained by averaging the local curvatures over the entire surface of Y_2:

$$\bar{K}(Y_2) = \frac{\int_{Y_2} \frac{1}{2}[\kappa_1 + \kappa_2]dS}{\int_{Y_2} dS} = \frac{K(Y_2)}{S(Y_2)} \tag{3.162}$$

The importance of this parameter derives from the fact that it determines the surface pressure p occurring at the interface Y_2 between Y and its surrounding; with γ as surface tension coefficient we have by the formula of Gibbs (1873)

$$p = 2\gamma \cdot \bar{K} \tag{3.163}$$

Finally, we consider a specimen X which contains some surfaces Y_2. The density of integral mean curvature is, evidently

$$K_V = K(Y_2)/V(X) \tag{3.164}$$

For the case of convex solids Y we derive from eq. (3.161) that

$$K_V = 2\pi \cdot N_V \cdot E\{H(Y)\} \tag{3.165}$$

where N_V is the number of particles Y per unit volume of X, and $E\{H(Y)\}$ is their average mean caliper diameter.

For simple particles, not necessarily convex, the density of Gaussian curvature is

$$G_V = 4\pi \cdot N(Y)/V(X) = 4\pi \cdot N_V \tag{3.166}$$

If these particles contain $N'(Y)$ holes, one finds (Miles and Davy, 1978)

$$G_V = 4\pi \cdot [N(Y) + N'(Y)]/V(X) \tag{3.167}$$

because the Gaussian curvature of the "concave" holes is also positive. This result can be exploited in deriving topological descriptors for complex solids (see Section 3.5).

3.4.3. MEAN CURVATURE DENSITY OF SURFACES AND CURVATURE DENSITY OF CURVES ON RANDOM SECTIONS

If the surface Y_2 of a solid Y is sectioned with a random plane T_2 the trace of Y_2 on T_2

$$T_2 \cap Y_2 = Y_1$$

is a planar curve (Fig. 3.19). Whereas the surface is described in quantitative terms by its surface area, $S(Y_2)$, and its integral of mean curvature, $K(Y_2)$, the equivalent descriptors of the curve are its length, $B(Y_1)$, and its integral curvature $C(Y_1)$. Assume Y to be contained in a specimen X; the structure is then defined by surface density, S_V, and density of mean curvature, K_V (eq. 3.164), whereas the corresponding parameters for the curve $Y_1 = Y_2 \cap T_2$ are the length density, B_A, and the curvature density, C_A (eq. 3.151).

The following fundamental stereological relationship exists between the density of integral mean curvature of a spatial surface and the integral curvature of its trace on a random section:

$$K(Y_2)/V(X) = C(Y_2 \cap T_2)/A(X \cap T_2) \tag{3.168}$$

or, in short

$$K_V = C_A \tag{3.169}$$

In terms of the average curvatures of the surface and the section trace one finds

$$\bar{K}(Y_2) = \frac{\pi}{4} \cdot \bar{C}(Y_2 \cap T_2) \tag{3.170}$$

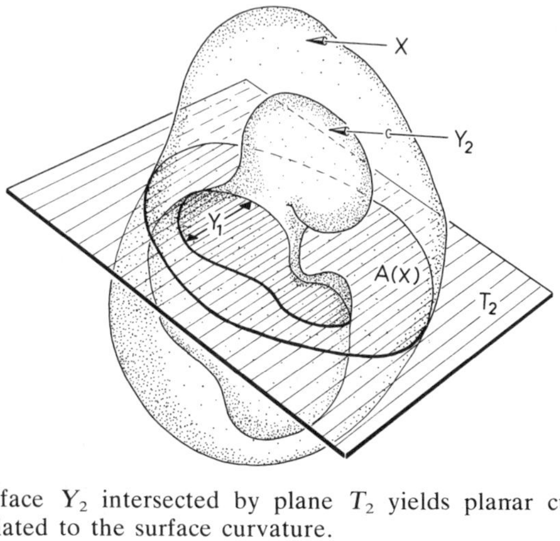

FIG. 3.19: Surface Y_2 intersected by plane T_2 yields planar curve Y_1 whose curvature is related to the surface curvature.

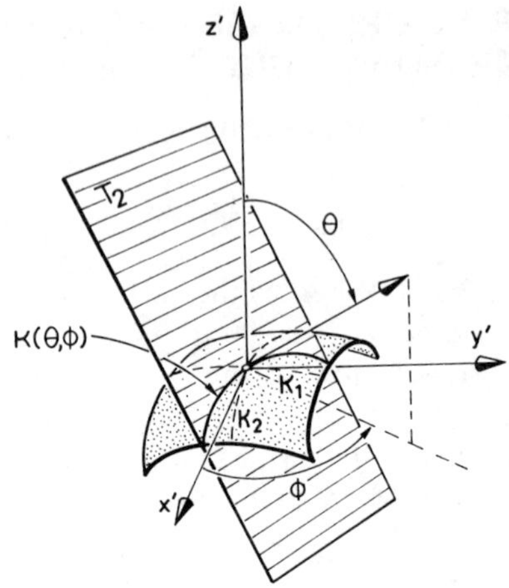

FIG. 3.20: Curvature of curve segment generated by random section (T_2) through point on surface element.

In order to derive these relationships, consider again a small (infinitesimal) surface element of area dS around a point u on Y_2 (Fig. 3.20). Construct a coordinate system x', y', z' at u with z' as the outward normal at u, as in Fig. 3.17, and with x' and y' lying in the two principal normal planes, i.e. those generating κ_1 and κ_2 respectively. The orientation of a random plane T_2 through u is evidently defined by the pair of angles (θ, ϕ) formed by the normal to T_2 with respect to the coordinate system x', y', z' (Fig. 3.20).

The theorems of Euler and Meusnier (see texts on differential geometry, such as Struik (1961, pp. 76, 81)) say that the curvature of the trace of dS on T_2 is

$$\kappa(dS \cap T_2 \mid \theta, \phi) = \frac{1}{\sin\theta}[\kappa_1 \cdot \cos^2\phi + \kappa_2 \sin^2\phi] \tag{3.171}$$

It is easily seen that eq. (3.171) becomes eq. (3.153) if T_2 contains z', because then $\theta = \pi/2$ and $\sin\theta = 1$.

In order to estimate the average curvature of $Y_2 \cap T_2$ in terms of $K(Y_2)$ we must first derive the average curvature of the intersection of the surface element dS with T_2; this will have to account for the probability of cutting dS as a function of the angles θ and ϕ. Thus

$$E\{\kappa(dS \cap T_2)\} = \frac{\int \kappa(dS \cap T_2 \mid \theta, \phi) \cdot \Pr\{dS \uparrow T_2 \mid \theta, \phi\} \cdot \Pr\{\theta, \phi\}}{\int \Pr\{dS \uparrow T_2 \mid \theta, \phi\} \cdot \Pr\{\theta, \phi\}} \tag{3.172}$$

The probability of hitting a small surface element dS has been derived in Section 3.2.1 (eq. 3.70):

$$\Pr\{dS \uparrow T_2 \mid \theta, \phi\} = H(dS \mid \theta, \phi)/H(X) \tag{3.173}$$

where $H(X)$ is the caliper diameter of the specimen and $H(dS \mid \theta, \phi)$ is the caliper diameter of dS at θ, ϕ; we have found it to be $h(dS) \cdot \sin\theta$ (Fig. 3.10) where $h(dS)$ is the diameter of dS, so that

$$\Pr\{dS \uparrow T_2 \mid \theta, \phi\} = [h(dS)/H(X)] \sin\theta \tag{3.174}$$

The probability of orientation was derived in Section 2.3 as (eq. 2.50)

$$\Pr\{\theta, \phi\} = \sin\theta \, d\phi \, d\theta/2\pi \tag{3.175}$$

Substituting eqs (3.171), (3.174) and (3.175) into eq. (3.172), and noting (1) that integration has to be over the angles θ and ϕ in the ranges $\{0 \leqslant \theta \leqslant \pi/2;\ 0 \leqslant \phi \leqslant 2\pi\}$ (see Section 2.3); (2) that the term $h(dS)/2\pi H(X)$ occurs in both the numerator and denominator and hence cancels, we find

$$E\{\kappa(dS \cap T_2)\} = \frac{\displaystyle\int_0^{\pi/2}\int_0^{2\pi} [\kappa_1 \cos^2\phi + \kappa_2 \sin^2\phi] \sin\theta \, d\phi \, d\theta}{\displaystyle\int_0^{\pi/2}\int_0^{2\pi} \sin^2\theta \, d\phi \, d\theta} \tag{3.176}$$

Performing the integrations leads to

$$E\{\kappa(dS \cap T_2)\} = \frac{\pi[\kappa_1 + \kappa_2]}{\pi^2/2} = \frac{2}{\pi}[\kappa_1 + \kappa_2] \tag{3.177}$$

and by eq. (3.154) this becomes

$$E\{\kappa(dS \cap T_2)\} = \frac{4}{\pi} \cdot K(dS) \tag{3.178}$$

The average curvature of the intersection between dS and T_2 is linearly related to the mean curvature of dS through a coefficient $4/\pi$.

The average curvature of a random intersection between Y_2 and T_2 is obtained by averaging of the length $B(Y_2 \cap T_2)$ of the trace; because the local curvature varies this requires integration:

$$E\{C(Y_2 \cap T_2)\} = \frac{\displaystyle\int_{Y_2} E\{\kappa(dS \cap T_2)\} \cdot dB}{\displaystyle\int_{Y_2} dB} \tag{3.179}$$

where $dB = dB(dS)$ is found by eq. (3.71) to be

$$dB(dS) = dS/h(dS)$$

so that, $h(dS)$ being a constant,

$$E\{C(Y_2 \cap T_2)\} = \frac{\int_{Y_2} E\{\kappa(dS \cap T_2)\} \cdot dS}{\int_{Y_2} dS} \tag{3.180}$$

Introducing eq. (3.178) we find

$$E\{C(Y_2 \cap T_2)\} = \frac{4}{\pi} \cdot \frac{\int_{Y_2} K(ds) \cdot dS}{\int_{Y_2} dS} \tag{3.181}$$

Here the ratio of the two integrals is nothing other than the definition of average mean curvature so that we find

$$E\{C(Y_2 \cap T_2)\} = \frac{4}{\pi} \bar{K}(Y_2) \tag{3.182}$$

which corresponds to eq. (3.170).

It is now a straightforward matter to derive eq. (3.169). By eq. (3.162) the integral of mean curvature, $K(Y_2)$ is

$$K(Y_2) = \bar{K}(Y_2) \cdot S(Y_2) \tag{3.183}$$

and its density in the unit volume of the specimen X is (eq. 3.164)

$$\begin{aligned} K_V &= K(Y_2)/V(X) \\ &= \bar{K}(Y_2) \cdot S_V \end{aligned} \tag{3.184}$$

Noting eq. (3.170), proven by eq. (3.182), and that

$$S_V = \frac{4}{\pi} B_A$$

we derive by substitution into eq. (3.184)

$$\begin{aligned} K_V &= \left\{\frac{\pi}{4} \bar{C}(Y_2 \cap T_2)\right\} \cdot \frac{4}{\pi} B_A \\ &= \bar{C}(Y_2 \cap T_2) \cdot B_A \end{aligned} \tag{3.185}$$

By eqs (3.150) and (3.151) the right hand term is the density of integral curvature of a plane curve, so that eq. (3.169) is demonstrated.

In view of a practical estimation of K_V on sections the sampling rules for sectioning outlined in Section 2.4 must be observed. Thus for an A-weighted section of X the ratio $C(Y_2 \cap T_2)/A(X \cap T_2)$ is an unbiased estimator of K_V (Miles and Davy, 1976b). With repeated measurements on several A-weighted random sections we find K_V by

$$K_V = E\{C(Y_2 \cap T_2)\}/E\{A(X \cap T_2)\} \tag{3.186}$$

in analogy to the results of the previous sections.

3.4.4. ESTIMATION OF MEAN MEAN CURVATURE BY TANGENT COUNTS ON A PLANE SECTION

In 1967 J. W. Cahn and R. T. De Hoff independently developed a method for estimating the mean mean curvature of a spatial surface by a tangent count on plane sections; the two papers were published simultaneously in the same journal (Cahn, 1967; De Hoff, 1967). Their formula is

$$\bar{K} = \frac{\pi}{4}\bar{C} = \frac{\pi}{2} \cdot T_A/I_L \tag{3.187}$$

where I_L is the number of intersections per unit test line length, and T_A is the net tangent count per unit section area, as outlined below.‡

Let us demonstrate the relation

$$\bar{C} = 2T_A/I_L \tag{3.188}$$

which is at the basis of eq. (3.187). Figure 3.21 shows a specimen section

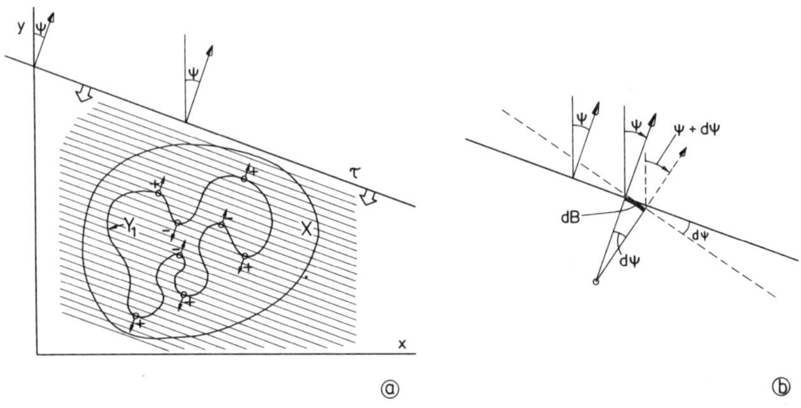

FIG. 3.21: Tangent count with sweeping test line at fixed orientation on planar curve Y_1 yields positive (+) and negative (−) tangent points.

‡ The formulas given here for mean curvature estimation apparently differ from those given by De Hoff (1968) by a factor of 2; however, De Hoff's $\bar{H}$ is defined as proportional to $[\kappa_1 + \kappa_2]$ instead of $\frac{1}{2}[\kappa_1 + \kappa_2]$ so that his $\bar{H} = 2\bar{K}$ in the present usage.

X of area $A(X)$ containing a curve Y_1 of length $B(Y_1)$, which we consider to be the result of a plane section intersecting a surface. If a test line τ is maintained at fixed orientation (its normal at angle ψ with respect to the x, y coordinate system) and swept across X it will at some points come to be tangent to Y_1. Tangents are always formed at points u^* of the curve where the normal to the curve, $Z(u^*)$, coincides with the normal to the sweeping test line, i.e. where

$$\psi\{Z(u^*)\} = \psi\{\tau\} \tag{3.189}$$

The probability of forming tangents is therefore clearly a matter of local curvature; one must therefore also distinguish between tangents to points of positive curvature (T_+) and tangents to points of negative curvature (T_-), as discussed in Section 3.4.1 (Fig. 3.14).

In eq. (3.143) we have defined local curvature as the ratio of the angle $d\psi$ subtended by the infinitesimal arc dB:

$$K(u) = d\psi(u)/dB \tag{3.190}$$

Referring to Fig. 3.21b we note that a short segment $dB(Y_1)$ will form tangents with all sweeping test lines of orientation between ψ and $(\psi + d\psi)$. It is therefore immediately evident that the probability for the sweeping test line τ to form a tangent with a segment of the curve, dB, is

$$\Pr\{\tau \bar{\cap} dB\} = d\psi/\pi \tag{3.191}$$

where π is simply the integral of $d\psi$ over the range of 180°, thus allowing dB to assume any orientation with equal likelihood. The total number of tangents formed between τ and Y_1 in one random sweep is

$$\begin{aligned} E\{T(\tau \bar{\cap} Y_1)\} &= \int_{Y_1} \Pr\{\tau \bar{\cap} dB\} \\ &= \tfrac{1}{2} \int_{Y_1} d\psi \end{aligned} \tag{3.192}$$

in which, according to eq. (3.147), the integral is the total curvature of Y_1, so that

$$E\{T(\tau \bar{\cap} Y_1)\} = C(Y_1)/\pi \tag{3.193}$$

With respect to the structure the reference system for the sweeping test line is the area over which it is swept, i.e. $A(X)$. Dividing both sides of eq. (3.193) by $A(X)$ leads to

$$\frac{E\{T(\tau \bar{\cap} Y_1)\}}{A(X)} = \frac{1}{\pi} \cdot \frac{C(Y_1)}{A(X)} \tag{3.194}$$

or

$$E\{T_A\} = C_A/\pi \tag{3.195}$$

i.e. the average tangent count per unit area is proportional to the density of integral curvature.

One point needs to be stressed. As mentioned above we will have to distinguish two types of tangents, T_+ at points of positive curvature, T_- at points of negative curvature, and the T entering eq. (3.194) and all the following are "*net* tangent counts", or specifically rewriting eq. (3.195) we have

$$C_A = \pi T_A = \pi[T_+ - T_-]/A(X) \tag{3.196}$$

This is evidently because in performing the integration over ψ in eq. (3.159) we had to observe the sign of the local curvature (see Section 3.4.1).

Using eq. (3.186) we can now see that a tangent count performed on A-weighted random sections of a three-dimensional structure X containing a surface Y_2 allows the density of the integral of mean curvature of Y_2 to be estimated:

$$K_V = \pi \cdot \frac{E\{T(Y_2 \cap T_2) \mid \tau\}}{E\{A(X \cap T_2)\}} = \pi \cdot T_A \tag{3.197}$$

The tangent count can also be used to estimate the mean curvature of Y_1. If we go back to eq. (3.193) we know from eq. (3.150) that the mean curvature is obtained by dividing the total curvature by the total length of Y_1:

$$\frac{E\{T(\tau \,\bar{\cap}\, Y_1)\}}{B(Y_1)} = \frac{1}{\pi}\,\bar{C}(Y_1) \tag{3.198}$$

or, considering again the area of X

$$\bar{C}(Y_1) = \pi \cdot \frac{E\{T(\tau \,\bar{\cap}\, Y_1)\}/A(X)}{B(Y_1)/A(X)} = \pi \cdot E\{T_A\}/B_A \tag{3.199}$$

A practical way of estimating B_A is by counting intersections per unit length of random test line, I_L; as discussed in Section 3.2.3

$$B_A = \tfrac{1}{2}\pi I_L$$

so that we find

$$\bar{C}(Y_1) = 2 \cdot T_A/I_L = 2 \cdot E\{T_A\}/E\{I_L\} \tag{3.200}$$

Considering finally that the mean mean curvature of a surface Y_2 is proportional to the mean curvature of its profile on random sections (eq. 3.170) we derive from eq. (3.200) an estimator for mean mean curvature:

$$\bar{K}(Y_2) = \tfrac{1}{2}\pi \cdot T_A/I_L = \frac{\pi}{2} \cdot \frac{E\{T(Y_2 \cap T_2) \mid \tau\} \cdot E\{L(X \cap T_1)\}}{E\{A(X \cap T_2)\} \cdot E\{I(Y_2 \cap T_1)\}} \tag{3.201}$$

The latter formulation signifies again that the estimate of $\bar{K}(Y_2)$ is to be obtained on A-weighted random samples of X.

The relations derived in this section, namely eqs (3.195) to (3.201) can be considered fundamental relations of stereology, in the sense that they have been derived with no restrictions with respect to shape or arrangement of the objects, curves on the plane or surfaces in space. They yield unbiased estimates if the sampling rules discussed in Section 2.4 are observed. In the short notation of stereology these fundamental relations read as follows:

for two-dimensional specimens,

$$C_A = \pi \cdot T_A \tag{3.195}$$

$$\bar{C} = \pi T_A / B_A = 2 \cdot T_A / I_L$$

for three-dimensional specimens,

$$K_V = C_A = \pi \cdot T_A \tag{3.197}$$

$$\bar{K} = \frac{\pi}{4} \cdot \bar{C} = \frac{\pi}{2} \cdot T_A / I_L \tag{3.201}$$

A number of most interesting results are derived from these fundamental relations for some special cases. In the case of planar specimens it was noted that for simple *closed loops of any shape*

$$T_{net} = T_+ - T_- = 2 \tag{3.202}$$

and accordingly, for N_A loops,

$$T_A(\text{closed loops}) = 2 \cdot N_A \tag{3.203}$$

$$C_A = 2\pi N_A \tag{3.204}$$

$$\bar{C} = 4 \cdot N_A / I_L \tag{3.205}$$

If a three-dimensional specimen is made of *discrete convex particles* their section profiles $Y_2 \cap T_2$ will always be convex closed loops, so that the following relations are derived:

$$K_V(\text{convex particles}) = 2\pi N_A \tag{3.206}$$

$$\bar{K} = \pi \cdot N_A/I_L \tag{3.207}$$

Now noting, as we shall show in Chapter 5, that

$$N_A = N_V \cdot E\{H(Y)\} \tag{3.208}$$

where $E\{H(Y)\}$ is the average mean caliper diameter of the particles, we derive, for convex particles, by substituting eq. (3.208) into eq. (3.206)

$$K_V = 2\pi \cdot N_V \cdot E\{H(Y)\} \tag{3.209}$$

From this we find

$$E\{H(Y)\} = \frac{1}{2\pi} \cdot K_V/N_V \tag{3.210}$$

This is the formula of Minkowsky (1903), which says that the integral mean curvature of a convex particle is directly proportional to its mean caliper diameter. This relation will be further discussed and exploited in Section 5.5.

3.5. Measurement of topological parameters of structure

One calls topological those properties of a structure which are unaffected by any deformations—stretching, swelling, distortions of any kind—which do not in any way disrupt the structure. Thus the topological properties of a sphere, a cube, a straight rod, a coiled sausage or a saucer are identical; a torus is topologically identical to a cup with a handle, etc. (Fig. 3.22). In

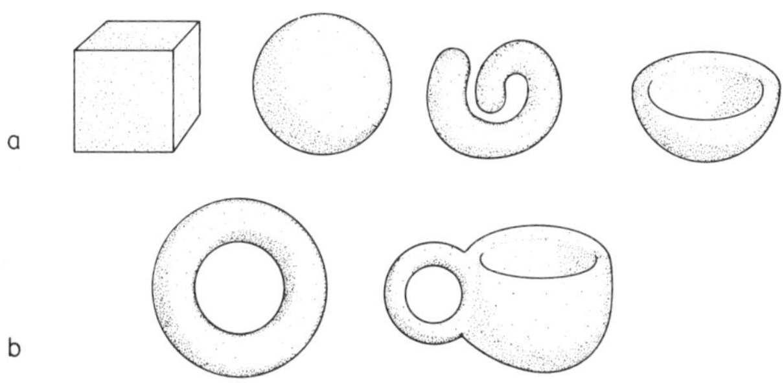

FIG. 3.22: Topological equivalence of (a) cube, sphere, sausage and saucer; (b) torus and cup with handle.

other words, shape and metric properties do not enter topological concepts. On the other hand, topological properties have to do with connectivity between the elements of a structure. Thus a pretzel is topologically different from a torus with only one hole, or a hollow sphere is topologically different from a solid sphere, although they may have the same size. Good descriptions of the topological properties of solids are found, for example, in Courant and Robbins (1941, 1962).

The topological properties are related to the total curvature of the structure. In eq. (3.167) we have shown that the density of Gaussian curvature G is related to the sum of the number of particles and the number of holes they contain. For a closed particle the Gaussian curvature is 4π (eq. 3.160), but if it contains n' holes, it is

$$G(Y_2 \mid n' \text{ holes}) = 4\pi(1+n') \tag{3.211}$$

This was because the integrated Gaussian curvatures for the convex outer surface and the concave inner surfaces are both positive (see Section 3.4.2) and equal to 4π, the net subtended solid angle, Ω_{net}, for closed surfaces.

In a torus we find convex and saddle surfaces, the Gaussian curvature of the latter being negative. It can be shown that the integrated Gaussian curvature of a simple torus (one loop) is zero (De Hoff, 1968). For a torus of higher order the net subtended solid angle depends on the genus g of the structure:

$$\Omega_{net} = 4\pi(1-g) \tag{3.212}$$

The topological genus can be described as the number of cuts by a closed loop that can be applied to a surface without cutting the surface into two disconnected parts (see, for example, Courant and Robbins, 1941, 1962). Thus the genus of a sphere is $g = 0$, that for a torus $g = 1$ (Fig. 3.23): a cut across the ring and sealing the two open ends transforms it into a sausage which is a "topological equivalent" to a sphere. A torus with two loops has

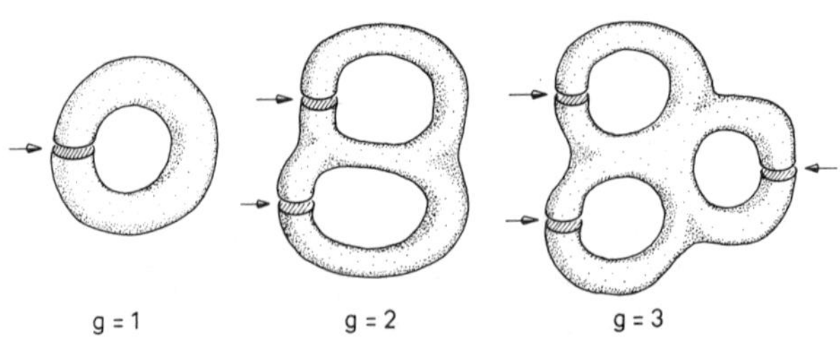

FIG. 3.23: Topological genus of various tori is given by number of cuts with closed curve that can be applied without splitting torus into two parts.

g = 2, one with three holes g = 3 etc. In a network of strings g measures the number of loops that form the network.

In eqs (3.211) and (3.212) the term in brackets is the characteristic of Euler–Poincaré, an important topological invariant (Hadwiger, 1957; Courant and Robbins, 1941, 1962; Santaló, 1976), which we shall not discuss any further. Further discussions of the topological properties of structures and their measures are found in De Hoff (1968), and De Hoff *et al.* (1972).

Since the net subtended solid angle is related to the topological properties of the structure and determines the Gaussian curvature (eq. 3.159) an estimate of G_V should yield some information on the topological nature of the structure. In Section 3.4.4 we have shown that the probability for a sweeping test line to form a tangent with a curve on a plane depends on the angle subtended by a small arc (eq. 3.191), and is hence a function of curvature. Since the local Gaussian curvature of a spatial surface is related to the solid angle subtended by a small surface element (eq. 3.156), it is intuitively plausible that the curvature of this element is related to the probability that a "sweeping plane" will form a tangent with this element. A "sweeping plane" is kept at fixed orientation and moved across the entire space containing the surfaces of interest (Fig. 3.24). Without going into the details of derivation, let us just state the result obtained by De Hoff (1968), that the number of tangents per

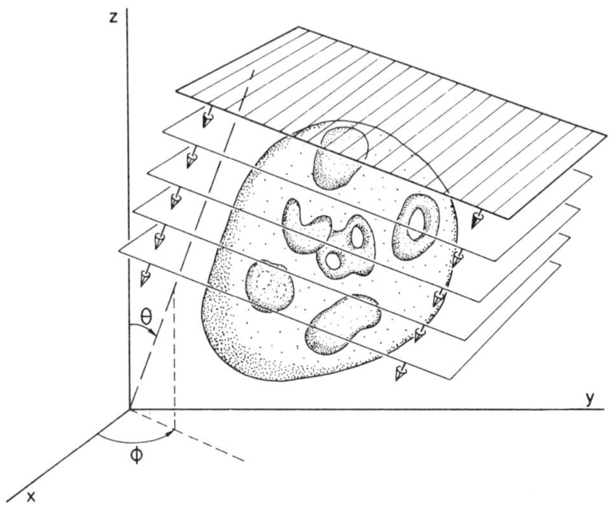

FIG. 3.24: Estimating topological properties of structure with sweeping test plane at fixed orientation in space.

unit volume, T_V, is related to the total subtended solid angle Ω_V contained in the unit volume by

$$\Omega_V = 2\pi \cdot T_V \tag{3.213}$$

Expressed in terms of net solid angle (integrated by observing the signs) and the net tangent count (concave and convex positive, saddle negative) one finds

$$\Omega_{V\text{net}} = 2\pi T_{V\text{net}} = G_V \tag{3.214}$$

Because of eqs (3.212) and (3.214) one now derives that T_V is related to the number of particles per unit volume, N_V, and the "genus per unit volume", g_V by

$$N_V - g_V = \tfrac{1}{2}\, T_{V\text{net}} \tag{3.215}$$

Thus if one is able to count the number of particles and the net number of tangents with a sweeping test plane one can determine the topological property of the structure "genus per unit volume" by

$$g_V = N_V - \tfrac{1}{2}\, T_{V\text{net}} \tag{3.216}$$

The crux of the matter is that one will have to measure these parameters on the spatial specimen and cannot rely on a random sectioning strategy; the best approximation would be to proceed to serial sectioning and tracking down the structural components section by section (De Hoff *et al.*, 1972). The reason is evident: integrated Gaussian curvature has dimension zero, and by eq. (2.11) only a "three-dimensional section" will be able to account for such a parameter. The same is true for all other topological parameters: since they are, by definition, non-metric properties, their dimension is evidently zero.

CHAPTER 4

Basic stereological principles and finite section thickness

One of the fundamental assumptions made in developing the basic stereological principles in Chapter 3 was that the planar sections of the specimen represented true planes, i.e. two-dimensional point sets of zero thickness; similarly, line sections should be one-dimensional point sets of no thickness, and test points should be true points. These conditions are in reality never fulfilled. The best approximation to a true planar section is probably a polished cut surface of an opaque block, as practised in metallurgy and mineralogy; but even here the image seen is rarely in a true plane, for the various components must be made to "stand out" by procedures such as etching etc. The "ultrathin" sections used in electron microscopy are in reality very thick slices when compared with the resolution afforded by the microscope—they are, in general, at least twice the thickness of a ribosome and about five times the thickness of a membrane! Clearly, finite section thickness presents the stereologist with very serious problems, problems which have been underrated in most practical stereological work—although the error introduced by finite section thickness has been recognized more than half a century ago by A. H. Holmes (1927), and is therefore generally dubbed as "Holmes effect".

The effect of finite section thickness on stereological measurements may be complex (Fig. 4.1). (1) The microscopic projection of the entire section content tends to make a greater sample of the structure content available for study than would be hit by a true planar section; accordingly, it can lead to overestimation of the structure content if the basic stereological methods are applied indiscriminately to the projected image of the slice. (2) Some components of structure may be projected on top of each other, so that their profiles can overlap; one may then miss some of

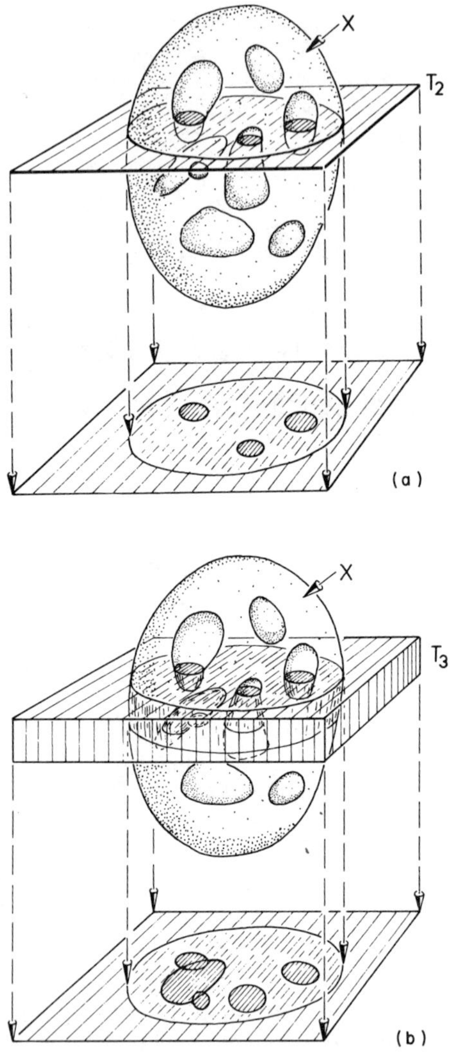

FIG. 4.1: Cutting specimen X with (a) plane section T_2 and (b) slice T_3.

the profiles on the micrographs, and this would lead to an underestimation of their density in the structure. (3) The contrast between components and matrix tends to become reduced with increasing section thickness particularly if the matrix is not "totally translucent"; this again has the tendency to obscure some component profiles, particularly if these are not "totally opaque", or if they represent grazing sections where the profiles do not extend all the way across the section. These three effects will occur simultaneously in any practical instance, so that one of the problems—largely unsolved as yet—is to determine the relative magnitude of each effect.

In this chapter we shall first present a set of equations which express the basic stereological parameters in terms of finite section thickness.

We shall begin with a special structural model made of loosely but randomly arranged discrete particles, and then proceed by presenting new general formulae developed by Miles (1976). Finally we shall discuss these relations with respect to some "useful" but restricted structural models in order to derive a set of correction coefficients which should allow the practitioner to at least estimate the order of magnitude of the error affecting his measurements, if the basic formulas pertaining to infinitely thin sections are used.

The reader is also directed to the detailed accounts on the stereology of projected images given by Underwood (1970, 1972).

4.1. Section thickness dependent estimates of V_V and S_V: discrete particle model

The probability of hitting an object Y with a random plane T_2 was found to be proportional to the caliper diameter, H, of the object, i.e. the measure of the object on the orthogonal complement to the section plane (Section 2.2). With Y enclosed in a space X we had

$$\Pr\{T_2 \uparrow Y\} = H(Y)/H(X) \tag{4.1}$$

In fact, $H(Y)$ measures the range over which T_2 can be moved in the perpendicular direction and still intersect Y (Fig. 4.2a). If we now cut the same object with a slice T_3 of thickness t the slice can be moved over a range $H(Y) + t(T_3)$ and still form an intersection with Y (Fig. 4.2b). Hence the probability of intersecting Y with T_3 is

$$\Pr\{T_3 \uparrow Y\} = [H(Y) + t(T_3)]/[H(X) + t(T_3)] \tag{4.2}$$

or, with $H(X) \gg t(T_3)$

$$\begin{aligned} \Pr\{T_3 \uparrow X\} &= [(H(Y) + t(T_3)]/H(X) \\ &= H(Y)/H(X) + t(T_3)/H(X) \\ &= \Pr\{T_2 \uparrow Y\} + t(T_3)/H(X) \end{aligned} \tag{4.3}$$

The basic stereological formulas presented in Chapter 3 were derived by integrations using probabilities of the type of eq. (4.1). If probabilities

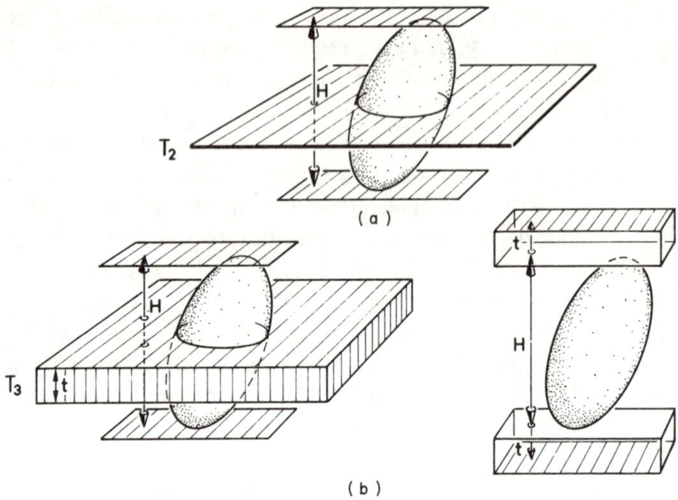

FIG. 4.2: Caliper or tangent diameter of convex particle when cut with plane (T_2) and with slice (T_3) of thickness t.

such as eq. (4.3) are used to derive section thickness dependent estimators of stereological parameters it is plausible that the resulting formulas will contain the basic estimator plus an additive term depending on section thickness and the parameters of the object.

In order to demonstrate this let us first consider a model structure made of discrete convex particles. We assemble the particles by a random process which allows all orientations and all positions except that the particles may not interpenetrate. The latter restriction implies that the particles must be very loosely arranged, i.e. their density in the unit volume must be very small. The structure is quantitatively characterized by

N_V = numerical density of particles
$\bar{v}$ = mean particle volume
$\bar{s}$ = mean particle surface
$\bar{H}$ = mean tangent diameter of the particles

so that

$$V_V = N_V \cdot \bar{v}, \qquad S_V = N_V \cdot \bar{s}$$

Cahn and Nutting (1959) found that the apparent areal density A'_A of the profiles on the projection of a thick slice of thickness t is

$$A'_A = V_V + \tfrac{1}{4} S_V \cdot t \tag{4.4}$$

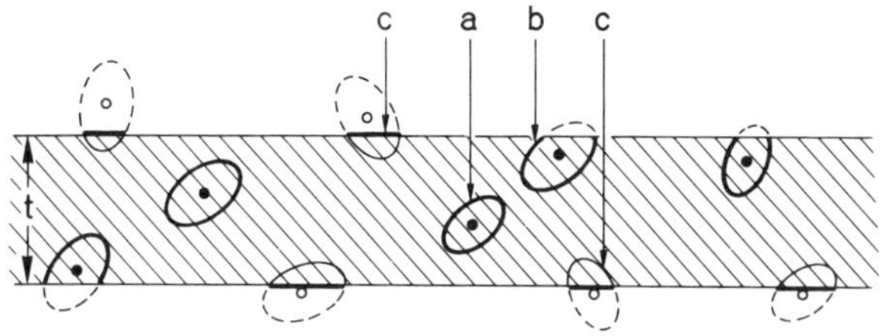

FIG. 4.3: Convex particles cut by slice of thickness t: types of profiles a, b, c observed on projected image.

This can be demonstrated as follows. If the model structure is cut by a slice of thickness t (Fig. 4.3) we can distinguish three types of particles: (a) those which are totally within the slice and are not cut by either cut surface; (b) those with their centre point inside but with some part cut by the upper or lower slice surface; (c) those which are cut by the slice but have their centre point outside. These three types of particles will contribute differently to the projected image.

Particles of type (a) will be fully projected, and by the formula of Cauchy (1908) we know that their average projected area is

$$\overline{a_p} = \tfrac{1}{4}s \tag{4.5}$$

There are $N_V(a)$ such particles per unit volume, and accordingly

$$N_A(a) = N_V(a) \cdot t \tag{4.6}$$

profiles are observed on the unit area of slice. It follows that the areal density of the projected areas is

$$A'_A(a) = \tfrac{1}{4}S_V(a) \cdot t \tag{4.7}$$

where we wrote

$$A'_A(a) = N_A(a) \cdot \overline{a_p} \tag{4.8}$$

$$S_V(a) = N_V(a) \cdot \bar{s} \tag{4.9}$$

If we now take particles (b) it can easily be seen that they also will project on to the observation plane with apparent areal density

$$A'_A(b) = \tfrac{1}{4}S_V(b) \cdot t \tag{4.10}$$

Note that they have lost the surface which protrudes above and below the slice surfaces; but this surface loss is compensated by the surface of particles (c) which is contained within the slice.

It remains to consider the particles (c). Since only a small cap of these particles is within the slice their contribution to the projected image is essentially determined by their cut surface. It is evident that the areal density of the cut surfaces (true planar profiles) of (*b*) and (*c*) particles taken together is equal to $2V_V$ if both slice surfaces are considered. But only half of these profiles are derived from particles (*c*) so that

$$A'_A(c) = V_V \tag{4.11}$$

The total apparent areal density is then

$$\begin{aligned} A'_A &= A'_A(c) + A'_A(b) + A'_A(a) \\ &= V_V + \tfrac{1}{4}t \cdot (S_V(b) + S_V(a)) \end{aligned} \tag{4.12}$$

Since the sum of the surface density of all particles with their centre within the slice (*a* and *b*) is S_V, the true surface density, we finally obtain the formula of Cahn and Nutting (1959)

$$A'_A = V_V + \tfrac{1}{4}S_V \cdot t \tag{4.13}$$

or

$$A'_A = N_V \cdot \left[\bar{v} + \frac{t}{4}\,\bar{s}\right] \tag{4.14}$$

The apparent areal density is hence larger than the volume density by the product of surface density with slice thickness. Equation (4.13) could be used to infer true volume density from

$$V_V = A'_A - \tfrac{1}{4}S_V \cdot t \tag{4.15}$$

if t can be measured and an estimate of S_V is available. However, in estimating S_V from the thick section available we have to cope with a similar effect of section thickness, as we will now show.

A formula similar to eq. (4.13) has also been derived for the surface density of convex particles by De Hoff (1968). It says that the apparent boundary length density B'_A observed on the projection plane is

$$\begin{aligned} B'_A &= \frac{\pi}{4} S_V + t \cdot N_V \cdot (\pi\bar{H}) \\ &= \pi \cdot N_V\left[\frac{\bar{s}}{4} + t \cdot \bar{H}\right] \end{aligned} \tag{4.16}$$

This equation is of the same form as eq. (4.13). The first term says that the contribution of type (*c*) particles to B'_A is their true boundary length density on section which, by eq. (3.66) is $\frac{1}{4}\pi S_V$. The second term indicates that all N_V particles having their centre within the slice of thickness t (particles *a* and *b*) contribute their mean projected circumference which, for convex solids, is $\pi\bar{H}$; this can easily be verified for spheres. As in the derivation of eq. (4.13) for V_V it must be considered that the "caps" of particles (c) included in the slice compensate for the loss of caps in particles (b).

Equation (4.16) can be solved for the required parameter, S_V:

$$S_V = \frac{4}{\pi} B'_A - 4 \cdot t \cdot \bar{H} \cdot N_V \tag{4.17}$$

Here again the first term is the measured quantity whereas the correction term $4 \cdot t \cdot \bar{H} \cdot N_V$ contains two unknowns, N_V and $\bar{H}$. This is only slightly improved if we consider that by eq. (5.85), demonstrated in Chapter 5,

$$N_V = N'_A/(t + \bar{H}) \tag{4.18}$$

so that we may write,

$$S_V = \frac{4}{\pi} B'_A - 4 \cdot N'_A \cdot \frac{t \cdot \bar{H}}{t + \bar{H}} \tag{4.19}$$

There is still one unknown left. If we substitute eq. (4.19) into eq. (4.15) we obtain

$$V_V = A'_A - t\left(\frac{1}{\pi} B'_A - N'_A \cdot \frac{t \cdot \bar{H}}{t + \bar{H}}\right) \tag{4.20}$$

Note that by eqs (3.55) and (3.103) we can substitute P'_P for A'_A, and $2I'_L$ for $(4/\pi)B'_A$ which leads to the following practical formulas in terms of point and intersection counts:

$$V_V = P'_P - t\left(\tfrac{1}{2} I'_L - N'_A \cdot \frac{t \cdot \bar{H}}{t + \bar{H}}\right) \tag{4.21}$$

$$S_V = 2I'_L - 4 \cdot N'_A \cdot \frac{t \cdot \bar{H}}{t + \bar{H}} \tag{4.22}$$

With three measured stereological quantities, $\{A'_A, B'_A, N'_A\}$ or $\{P'_P, I'_L, N'_A\}$, we could thus derive by eqs (4.19) to (4.22) the true values for volume and surface density. All that is required in addition is a measurement of section thickness (see Volume 1, Section 4.3.4.3) and an estimate of the average mean caliper diameter of the structural elements. Expressing t as a multiple of $\bar{H}$, such that

$$t = g \cdot \bar{H}; \qquad \bar{H} = t/g \tag{4.23}$$

the term containing t and $\bar{H}$ in eqs (4.19) to (4.22) becomes

$$\frac{t \cdot \bar{H}}{t + \bar{H}} = t \cdot \frac{1}{1 + g} \tag{4.24}$$

So, evidently, it suffices to have a relative estimate of the size of the elements.

If the areal density of profiles $A'_A = P'_P$ can be measured on two paired sections of different thickness t_1 and t_2 one can derive the following estimation formulas for V_V and S_V from eq. (4.13):

$$V_V = \frac{t_1 \cdot A'_A(t_1) - t_2 \cdot A'_A(t_2)}{t_1 - t_2} \tag{4.25}$$

$$S_V = \frac{A'_A(t_1) - A'_A(t_2)}{t_1 - t_2} \tag{4.26}$$

or, from eq. (4.17)

$$S_V = \frac{4}{\pi} \cdot \frac{t_2 \cdot B'_A(t_1) - t_1 \cdot B'_A(t_2)}{t_2 - t_1} \tag{4.27}$$

These formulas might prove to be quite useful in practice; the problems are mainly the doubling of the work required, and rather high requirements on the precision of the measurements of section thickness and of the stereological quantities, because, by being ratios of differences the formulas are quite sensitive to errors of measurement.

The limitation of the formulas presented in this section lies in the two model assumptions: (1) the structural elements have to be convex and discrete; (2) they need to be very sparse (V_V very small), so as to avoid their "collisions" in the random assembly of the structure and an overlap of profiles in the projected image.

4.2. The general formulas by Miles for thick sections

R. E. Miles has recently developed a new set of fundamental stereological equations which account for section thickness (Miles, 1976). He derived expressions for volume density, V_V, surface density, S_V, integrated mean curvature density, K_V, and integrated Gaussian curvature density, G_V, in terms of the stereological quantities areal density, A'_A, boundary length density, B'_A, and integrated curvature density, C'_A, measurable on the projection of a slice of thickness t onto a plane parallel to the slice. The approach is somewhat similar to that used in the preceding section, but the model conditions are much more general.

Miles considers a stochastic model made of an aggregate of random opaque convex particles, which we shall call "elements"; they are randomly distributed throughout a translucent matrix. The structure is characterized quantitatively by the following properties:

ν_V = numerical density of the elements
γ = mean volume of the elements
σ = mean surface of the elements
ζ = mean caliper diameter of the elements.

Note that this model structure is quantitatively defined by the same parameters as those introduced in Section 4.1; we are using Greek letters to show that these characteristics of the elements are not necessarily identical with the characteristics of the real objects. The reason is that the model of Miles now assumes unrestricted randomness for location and orientation of the particles: their centroids (or any other well defined particle point) are sited at the points of a Poisson process with density ν_V, and their orientations are independently isotropic about these points. Since neither position nor orientation are restrained in any way the elements may interpenetrate (Fig. 4.4), a seemingly unrealistic property of the model. However, this makes the model more general, because, in the end, structural "objects" are formed by the union of elementary particles which are no longer necessarily convex, as it had to be assumed in the preceding developments of formulae for section thickness effects (Section 4.1). The model resulting from this process is one of disjoint opaque objects each of which is the union of one or more overlapping elements (Fig. 4.5). The degree of overlap between particles will increase with their density.

It is evident that the stereological parameters characterizing the "real objects" must be smaller than those of the "elements" if they would not overlap. Thus the fictitious "volume density" of the elements being

$$\gamma_V = \nu_V \cdot \gamma \tag{4.28}$$

one finds the true volume density of the aggregate to be

$$V_V = 1 - \exp\{-\gamma_V\} = 1 - \exp\{-\nu_V \cdot \gamma\} \tag{4.29}$$

which is smaller than γ_V. For $\gamma_V = 0.5$, $V_V = 0.393$; for $\gamma_V = 0.2$, $V_V = 0.181$; for $\gamma_V = 0.05$, $V_V = 0.0488$, so that evidently the difference becomes smaller the smaller γ_V, because there will be less interpenetration. For the surface density of the real objects one finds

$$S_V = \nu_V \cdot \sigma \cdot \exp\{-\nu_V \cdot \gamma\} \tag{4.30}$$

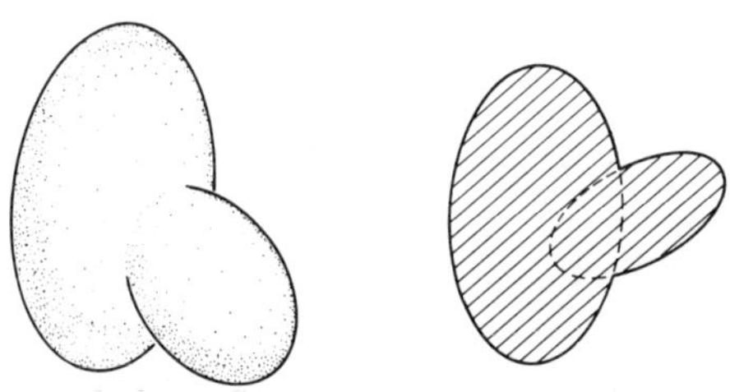

FIG. 4.4: Union of "interpenetrating" convex elements in 3D or "overlapping" profile projections in 2D.

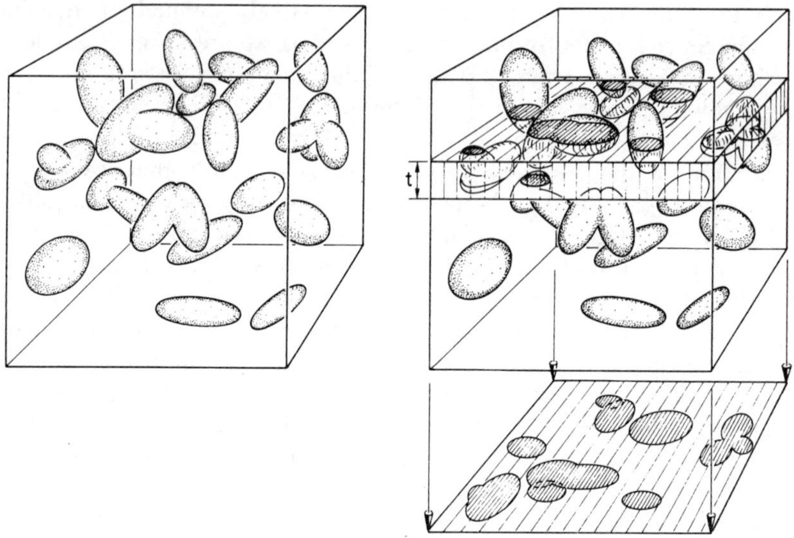

FIG. 4.5: A structure assembled from convex elements by the model of Miles (1976) is cut by a slice of thickness t whose content is projected onto the plane of observation.

for the total or integrated mean curvature per unit volume

$$K_V = \left[2\pi\nu_V \cdot \zeta - \frac{\pi^2}{32} \cdot (\nu_V \cdot \sigma)^2\right] \exp\{-\nu_V \cdot \gamma\} \tag{4.31}$$

and for the total Gaussian curvature per unit volume

$$G_V = \left[4\pi\nu_V - 2\pi\nu_V^2 \cdot \zeta \cdot \sigma + \frac{\pi^2}{96} \cdot (\nu_V \cdot \sigma)^3\right] \exp\{-\nu_V \cdot \gamma\} \tag{4.32}$$

Each of these formulae contains one common term, namely $\exp\{-\nu_V \cdot \gamma\}$ which, by eq. (4.29), is the volume density of the matrix in the real structure.

This structure is now cut by a slice of thickness t, and the content of this slice is projected perpendicularly onto the observation plane (Fig. 4.5), on which we will be able to discern areas covered with profiles from uncovered areas. The profiles will again be the union of one or more of the elements contained in the slice; some of them will be the result of overlap of two or more objects which lie one above the other (Fig. 4.5).

In the observation plane one can measure the fractional area covered by profiles, A'_A, the length density of their boundary, B'_A, and the total curvature of this boundary per unit area, C'_A. Miles has found that these quantities can be expressed in terms of the density and dimensions of the elements and of section thickness t:

$$A'_A = 1 - \exp\left\{ -\nu_V\left(\gamma + \frac{t}{4}\sigma\right)\right\} \tag{4.33}$$

$$B'_A = \pi \cdot \nu_V\left[\frac{\sigma}{4} + t \cdot \zeta\right]\exp\left\{ -\nu_V\left(\gamma + \frac{t}{4}\sigma\right)\right\} \tag{4.34}$$

$$C'_A = \left\{2\pi\nu_V(\zeta + t) - \frac{\pi^2}{2}\nu_V^2\left[\frac{\sigma}{4} + t\zeta\right]^2\right\}\exp\left\{ -\nu_V\left(\gamma + \frac{t}{4}\sigma\right)\right\} \tag{4.35}$$

If we compare the formulas for A'_A and B'_A with those given in eqs (4.13) and (4.16) with respect to the simple model of discrete particles, the similarity becomes evident; the difference lies in the term $\exp\{\cdot\}$ which is the areal density of the matrix projection. In fact, considering the relationship

$$e^{-x} = 1 - x + \frac{x^2}{2!} - \frac{x^3}{3!} \dots$$

one can write, for small x,

$$e^{-x} \simeq 1 - x$$

If this approximation is performed for eqs (4.33) and (4.34) one derives the corresponding eqs (4.13) and (4.16) pertaining to the sparse model made of discrete particles.

Miles (1976) proposes to use the three eqs (4.33) to (4.35) to estimate the quantities $\nu_V\left[\gamma + \frac{1}{4}\sigma\right]$, $\nu_V\left[\frac{\sigma}{4} + t \cdot \zeta\right]$ and $\nu_V[\zeta + t]$. Thus if t and any one of the other quantities describing the model structure are known one can solve for the others; introducing these quantities into eqs (4.30) to (4.32) one can then calculate the true stereological parameters describing the real structure.

Miles (1976) also proposes an alternative approach, namely obtaining the stereological measurements on projections of sections of different thickness, t_1 and t_2; the thickness of these sections need not be known! The estimates of the stereological parameters are then

$$\hat{V}_V = 1 - e^{-u} \tag{4.36}$$

$$\hat{S}_V = 4 \cdot v \cdot a_{21} \cdot e^{-u} \tag{4.37}$$

$$\hat{K}_V = [2\pi v b_{21} - \tfrac{1}{2}(\pi \cdot v \cdot a_{21})^2] \cdot e^{-u} \tag{4.38}$$

$$\hat{G}_V = [4\pi v c_{21} - 8\pi v^2 \cdot a_{21} \cdot b_{21} + \tfrac{2}{3}\pi^2(v \cdot a_{21})^3]e^{-u} \tag{4.39}$$

where with a_2 and a_1 (or b_2 and b_1 etc.) as obtained on sections t_2 and t_1 respectively, we have

$$a_{21}=a_2-a_1; \qquad b_{21}=b_2-b_1; \qquad c_{21}=c_2-c_1 \tag{4.40}$$

$$a=-\ln(1-A'_A) \tag{4.41}$$

$$b=\frac{1}{\pi}\frac{B'_A}{1-A'_A} \tag{4.42}$$

$$c=\frac{1}{2\pi}\frac{C'_A}{1-A'_A}+\frac{\pi b^2}{4} \tag{4.43}$$

$$u=\{a_{21}(c_1a_2-c_2a_1)+b_{21}(a_1b_2-a_2b_1)\}/(b_{21}^2-a_{21}\cdot c_{21}) \tag{4.44}$$

$$v=(c_1b_2-c_2b_1)/(b_{21}^2-a_{21}\cdot c_{21}) \tag{4.45}$$

Evidently, the beauty of this set of equations is that all quantities required can, in principle, be measured: A'_A by a point count, B'_A by an intersection count and C'_A by a tangent count using a sweeping test line (see Chapter 3). The problem is that these three quantities must be obtained twice on paired sections, and that the requirements on accuracy are rather high because the estimates essentially depend on differences. The computations appear involved, but this can easily be coped with; even a larger programmable desk-top calculator should suffice.

The impressive progress offered by this theory is its generality, no simplifying assumptions being required. The convex elements from which the model structure was constructed may have any shape distribution, and the resulting structure is an aggregate which will often be comparable to real structures; also, t need not necessarily be small. It is hoped that ways of applying this method in practice are soon found, because this will open a new epoch of stereology, particularly in the biological sciences.

4.3. Correction factors for section thickness effects

In Volume 1, Section 4.3.4 we have presented the notion that the effect of section thickness on the estimation of stereological quantities can be corrected by means of correction factors. With A'_A, B'_A, C'_A as the quantities measured on the projected image of a slice of thickness t we could thus find the true stereological quantities V_V, S_V and K_V by

$$V_V=K_t(V_V)\cdot A'_A \tag{4.46}$$

$$S_V=K_t(S_V)\cdot\frac{4}{\pi}B'_A \tag{4.47}$$

$$K_V=K_t(K_V)\cdot C'_A \tag{4.48}$$

or, in terms of points counts,

$$V_V = K_t(V_V) \cdot P'_P \tag{4.49}$$

$$S_V = K_t(S_V) \cdot 2 \cdot I'_L \tag{4.50}$$

$$K_V = K_t(K_V) \cdot \pi T'_A \tag{4.51}$$

The rationale is to derive the correction factors, K_t, on theoretical grounds based on the two models discussed in Sections 4.1 and 4.2, i.e. by solving the paired equations for V_V and A'_A etc. for some characteristics of the structural elements that can be estimated. For that purpose we need to specify the shape of the elements used to assemble the structure. We shall limit ourselves to three basic shapes (Fig. 4.6): spheres, thin discs and cylindrical tubules. We shall consider the size of these elements to be measurable by a dimension d which is the sphere diameter, the disc thickness and the tubule diameter respectively; the shape of the discs is defined by a dimensionless shape factor $\delta = D/d$ and that of the tubules by $\lambda = L/d$. With these definitions one can easily find the quantities which were used in deriving eqs (4.13) to (4.16) and (4.29) to (4.35). For the three models they are:

(a) *Spheres* of diameter d:

mean volume $$\bar{v} = \frac{\pi}{6} d^3 \tag{4.52}$$

mean surface $$\bar{s} = \pi d^2 \tag{4.53}$$

mean tangent diameter $$\bar{H} = d \tag{4.54}$$

(b) *Discs* of thickness d and diameter $D = \delta \cdot d$

mean volume $$\bar{v} = \frac{\pi}{4} d^3 \cdot \delta^2 \tag{4.55}$$

mean surface $$\bar{s} = \pi d^2 \cdot \delta\left(\frac{\delta}{2} + 1\right) \tag{4.56}$$

mean tangent diameter $$\bar{H} = \frac{\pi}{4} d\left(\delta + \frac{2}{\pi}\right) \tag{4.57}$$

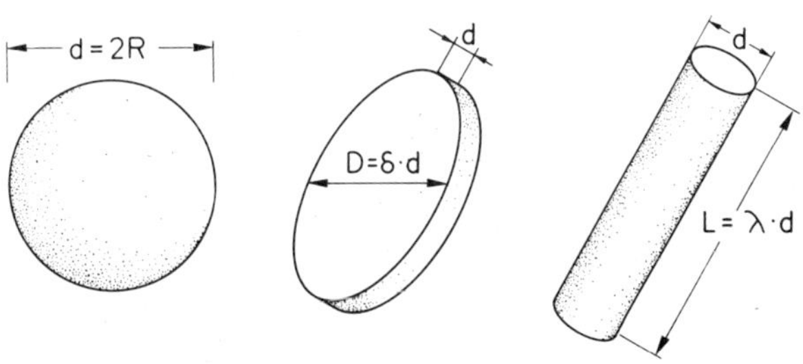

FIG. 4.6: Shape and dimensions of model elements.

(c) *Tubules* of diameter d and length $L = \lambda \cdot d$

mean volume $$\bar{v} = \frac{\pi}{4} d^3 \cdot \lambda \qquad (4.58)$$

mean surface closed $$\bar{s}_c = \pi d^2 \cdot (\lambda + \tfrac{1}{2}) \qquad (4.59)$$

mean surface open-end‡ $$\bar{s}_0 = \pi d^2 \cdot \lambda \qquad (4.60)$$

mean tangent diameter $$\bar{H} = \frac{\pi}{4} d\left(1 + \frac{2}{\pi} \cdot \lambda\right) \qquad (4.61)$$

In the case of the general model of Miles we shall set $\gamma = \bar{v}$, $\sigma = \bar{s}$ and $\zeta = \bar{H}$ (see Section 4.3.2).

To further simplify the relations we shall also express section thickness t in relation to the characteristic dimension d, such that

$$t = g \cdot d \qquad (4.62)$$

where g is relative section thickness, a dimensionless factor.

4.3.1. CORRECTION FACTORS FOR V_V AND S_V DERIVED FOR MODEL OF DISCRETE PARTICLES

Noting that, in the model made of discrete convex particles (Section 4.1), the true parameters of structure were

$$V_V = N_V \cdot \bar{v} \qquad (4.63)$$

$$S_V = N_V \cdot \bar{s} \qquad (4.64)$$

and that the corresponding observed values were

$$A'_A = N_V \cdot \left[\bar{v} + \frac{t}{4}\bar{s}\right] \qquad (4.65)$$

$$B'_A = N_V \cdot \pi\left[\frac{\bar{s}}{4} + t \cdot \bar{H}\right] \qquad (4.66)$$

we could now seek correction factors which should allow us to obtain the true from the observed value, by substituting eqs (4.63) to (4.66) into eqs (4.46) and (4.47). With respect to the models made of particles of well-defined shape we can now obtain the *correction factors for* V_V by introducing the appropriate particle dimensions into eqs (4.63) and (4.65)

‡ For corrections with respect to membrane tubules one would have to use the surface of an open-ended tubule.

and taking their ratio by eq. (4.46):

(a) *Spheres*:

$$V_V = N_V \cdot \frac{\pi}{6} d^3 \tag{4.67}$$

$$A'_A = N_V \cdot \left[\frac{\pi}{6} d^3 + \frac{t}{4} \pi d^2\right] = N_V \cdot \frac{\pi}{6} d^3 \left[1 + \frac{3}{2} \cdot \frac{t}{d}\right] \tag{4.68}$$

$$K_t(V_V) = 1/[1 + (3t)/(2d)] = 2/(2 + 3g) \tag{4.69}$$

This factor has been derived independently by several authors, notably Holmes (1927), Chayes (1956), Hennig (1957) and Cahn and Nutting (1959).

(b) *Discs*:

$$V_V = N_V \cdot \frac{\pi}{4} d^3 \cdot \delta^2 \tag{4.70}$$

$$A'_A = N_V \cdot \left[\frac{\pi}{4} d^3 \, \delta^2 + \frac{\pi}{4} t \cdot d^2 \cdot \delta\left(\frac{\delta}{2} + 1\right)\right]$$

$$= N_V \cdot \frac{\pi}{4} d^3 \left[\delta^2 + \frac{t}{d} \cdot \delta\left(\frac{\delta}{2} + 1\right)\right] \tag{4.71}$$

$$K_t(V_V) = \delta \Big/ \left[\delta + g \cdot \left(\frac{\delta}{2} + 1\right)\right] \tag{4.72}$$

(c) *Tubules*:

$$V_V = N_V \cdot \frac{\pi}{4} d^3 \lambda \tag{4.73}$$

$$A_A = N_V \frac{\pi}{4} d^3 \left[\lambda + \frac{t}{d} \cdot \left(\lambda + \frac{1}{2}\right)\right] \tag{4.74}$$

$$K_t(V_V) = \lambda \Big/ \left[\lambda + g \cdot \left(\lambda + \frac{1}{2}\right)\right] \tag{4.75}$$

Note that an interesting sequence of correction factors develops if we allow the tubules to be infinitely long ($\lambda \to \infty$) or the discs infinitely wide ($\delta \to \infty$); then

$$K_t(V_V) \text{ spheres} = 2/(2 + 3g) \tag{4.76}$$

$$K_t(V_V) \text{ discs} = 2/(2 + g) \tag{4.77}$$

$$K_t(V_V) \text{ tubules} = 1/(1 + g) \tag{4.78}$$

The correction factors for discrete particles tend to those derived by Hennig (1969) for infinitely large structures.

We should remark that these factors could also have been derived by slightly modifying eq. (4.65) to

$$A'_A = V_V\left\{1 + \frac{t}{4} \cdot \left(\frac{s}{v}\right)\right\} \tag{4.79}$$

The term in brackets { } is the reciprocal value of the correction factor. The surface-to-volume ratio of particles of many shapes is easily obtained from Table 5.2; if the value for spheres $(s/v) = (6/d)$ is introduced one immediately obtains the correction factor.

The *correction factors for* S_V are likewise obtained by introducing the particle dimensions into eqs (4.64) and (4.66) and taking the ratio by eq. (4.47):

(a) *Spheres*

$$S_V = N_V \cdot \pi d^2 \tag{4.80}$$

$$B'_A = \pi N_V \left\{\frac{\pi}{4} d^2 + t \cdot d\right\}$$

$$= \pi[N_V \pi d^2]\left\{\frac{1}{4} + \frac{t}{d}\right\} \tag{4.81}$$

$$K_t(S_V) = \pi d/(\pi d + 4t) = \pi/(\pi + 4g) \tag{4.82}$$

(b) *Discs*

$$S_V = N_V \pi d^2 \cdot \delta\left(\frac{\delta}{2} + 1\right) \tag{4.83}$$

$$B'_A = \frac{\pi}{4} \cdot N_V \pi d^2 \cdot \left\{\delta\left(\frac{\delta}{2} + 1\right) + g\left(\delta + \frac{2}{\pi}\right)\right\} \tag{4.84}$$

$$K_t(S_V) = \frac{\delta\left(\frac{\delta}{2} + 1\right)}{\delta\left(\frac{\delta}{2} + 1\right) + g\left(\delta + \frac{2}{\pi}\right)} \xrightarrow[\delta \to \infty]{} 1 \tag{4.85}$$

If the plates become infinitely wide $K \to 1$, i.e. there is no overestimation of S_V by section thickness.

(c) *Tubules*

$$S_V = N_V \pi d^2(\lambda + \tfrac{1}{2}) \tag{4.86}$$

$$B'_A = \frac{\pi}{4} \cdot N_V \pi d^2 \left\{ (\lambda + \tfrac{1}{2}) + g\left(1 + \frac{2\lambda}{\pi}\right) \right\} \tag{4.87}$$

$$K_t(S_V) = \frac{\lambda + \frac{1}{2}}{\lambda + \frac{1}{2} + g\left(1 + \frac{2\lambda}{\pi}\right)} \xrightarrow[\lambda \to \infty]{} \pi/(\pi + 2g) \tag{4.88}$$

This holds for cylinders with closed ends. If one would like to consider open-ended cylinders, one should note that $s = \pi d^2 \lambda$. As a consequence the term $\frac{1}{2}$, which accounted for the end surfaces, drops out and $K_T(S_V)$ for open-ended cylinders becomes

$$K_t(S_V) = \lambda \Big/ \left[\lambda + g\left(1 + \frac{2\lambda}{\pi}\right) \right] \tag{4.89}$$

Note that all these coefficients become 1 if section thickness vanishes.

The use to be made of these correction factors is further discussed in Volume 1, Section 4.3.4.

Before concluding this section we must reiterate one of the fundamental conditions of the model used to derive these coefficients: the random process of assembling the particles into the structure was restricted in the sense that the particles were not allowed to interpenetrate; nor did we account for any overlap of the projected images as they will almost certainly occur. The correction coefficients will therefore work reliably only if the particle aggregate is very loose, i.e. V_V is very low, perhaps at most a few per cent, and if there is no marked clustering of particles.

4.3.2. DERIVATION OF SECTION THICKNESS CORRECTION FACTORS FROM THE GENERAL MODEL OF MILES

The new formulae of Miles (1976), based on a more general structural model, can also be used to derive correction factors for section thickness effects on V_V, S_V and K_V (Weibel and Paumgartner, 1978). This will eliminate the restriction imposed on those derived in Section 4.3.1, namely that the structure must be made of discrete convex particles and their V_V be very small. Such improved correction factors could then be used for any real structure that fulfils the condition of being generated by a Poisson random process (Bolender *et al.*, 1978; Weibel and Paumgartner, 1978).

The approach is straight-forward; these correction factors are obtained as in Section 4.3.1, except that we introduce the formulae of Miles from eqs (4.29) to (4.35)

$$K_t(V_V)=\frac{V_V}{A'_A}=\frac{1-\exp\{-\nu_V\cdot\gamma\}}{1-\exp\left\{-\nu_V\left(\gamma+\frac{t}{4}\sigma\right)\right\}} \tag{4.90}$$

$$K_t(S_V)=\frac{S_V}{\frac{4}{\pi}B'_A}=\frac{\sigma\cdot\exp\{-\nu_V\cdot\gamma\}}{[\sigma+4t\zeta]\cdot\exp\left\{-\nu_V\left(\gamma+\frac{t}{4}\sigma\right)\right\}} \tag{4.91}$$

$$K_t(K_V)=\frac{K_V}{C'_A}$$

$$=\frac{\left[2\zeta-\frac{\pi}{32}\cdot\nu_V\cdot\sigma^2\right]\cdot\exp\{-\nu_V\cdot\gamma\}}{\left[2(\zeta+t)-\frac{\pi}{32}\cdot\nu_V(\sigma+4t\zeta)^2\right]\cdot\exp\left\{-\nu_V\left(\gamma+\frac{t}{4}\sigma\right)\right\}} \tag{4.92}$$

As an example of how these formulas can be exploited in practice we shall now derive specific correction factors applying to the three shapes: sphere, thin disc and long cylinder (Fig. 4.6), which are now assigned to the elements whose interpenetrating aggregate forms the real structure.

(a) *Spherical model elements*

The sphere element has diameter $d=2R$. The volume, surface and mean tangent diameter are, by eqs (4.52) to (4.54),

$$\gamma=\frac{\pi}{6}d^3$$

$$\sigma=\pi d^2$$

$$\zeta=d$$

As before we express the section thickness relative to d:

$$t=g\cdot d$$

Introducing these values into eqs (4.90) to (4.92) and setting

$$\nu_V\cdot\frac{\pi}{4}d^3=\xi \tag{4.93}$$

we obtain the following correction factors:

$$K_t(V_V) = \frac{1 - \exp\{-\xi \cdot \frac{2}{3}\}}{1 - \exp\{-\xi(\frac{2}{3} + g)\}} \tag{4.94}$$

$$K_t(S_V) = \frac{\pi \cdot \exp\{-\xi \cdot \frac{2}{3}\}}{\pi + 4g \cdot \exp\{-\xi \cdot (\frac{2}{3} + g)\}} \tag{4.95}$$

$$K_t(K_V) = \frac{\left[4 - \frac{\pi^2}{4}\xi\right] \cdot \exp\{-\xi \cdot \frac{2}{3}\}}{\left[4 + 4g - \xi\left(\frac{\pi}{2} + 2g\right)^2\right] \cdot \exp\{-\xi(\frac{2}{3} + g)\}} \tag{4.96}$$

Note that ξ has the character of a volume density, namely $\frac{3}{2}$ that of the spherical elements, not considering overlap.

(b) *Disc-shaped model elements*
The rough endoplasmic reticulum (RER) occurs in the form of broad cisternae; we have assumed them to be comparable to thin discs of thickness d and diameter D (Weibel and Paumgartner, 1978). The diameter does not matter very much, as these discs may partly interpenetrate to form cisternal complexes. By expressing both the section thickness t and the disc diameter D as a function of disc thickness d, such that

$$D = \delta \cdot d$$

$$t = g \cdot d$$

we have introduced two dimensionless parameters relative to cisternal width; the volume, surface and mean tangent diameter of these discs are hence, again by eqs (4.55) to (4.57),

$$\gamma = \frac{\pi}{4} d^3 \cdot \delta^2$$

$$\sigma = \pi d^2 \cdot \delta\left(\frac{\delta}{2} + 1\right)$$

$$\zeta = \frac{\pi}{4} \cdot d\left(\delta + \frac{2}{\pi}\right)$$

Substituting into eqs (4.90) to (4.92) and setting

$$\nu_V \cdot \frac{\pi}{4} d^3 = \xi \tag{4.97}$$

we obtain

$$K_t(V_V) = \frac{1 - \exp\{-\xi\delta^2\}}{1 - \exp\left\{-\xi\left[\delta^2 + g \cdot \delta\left(\frac{\delta}{2} + 1\right)\right]\right\}} \tag{4.98}$$

$$K_t(S_V) = \frac{\delta\left(\frac{\delta}{2} + 1\right)\exp\{-\xi\delta^2\}}{\left[\delta\left(\frac{\delta}{2} + 1\right) + g\left(\delta + \frac{2}{\pi}\right)\right]\exp\left\{-\xi\left[\delta^2 + g \cdot \delta\left(\frac{\delta}{2} + 1\right)\right]\right\}} \tag{4.99}$$

$$K_t(K_V) = \frac{\left[\pi\delta + 2 - \xi\left\{\frac{\pi}{2}\delta\left(\frac{\delta}{2} + 1\right)\right\}^2\right] \cdot \exp\{-\xi\delta^2\}}{\left[\pi\delta + 2 + 4g - \xi\left\{\frac{\pi}{2}\delta\left(\frac{\delta}{2} + 1 + 4g\right) + 2g\right\}^2\right] \cdot \exp\left\{-\xi\left[\delta^2 + g \cdot \delta\left(\frac{\delta}{2} + 1\right)\right]\right\}} \tag{4.100}$$

The coefficient ξ has the characteristics of the volume density of a small cylinder of diameter and length d circumscribed around the centroid of the elements; its value is about the volume density of the discs divided by δ^2.

(*c*) *Cylinder model*

The smooth endoplasmic reticulum (SER) of liver cells is predominantly a meshwork of tubules. We have chosen circular cylinders of diameter d and length $L = \lambda \cdot d$ as model elements. Using the values of γ, σ, ζ and ξ as given by eqs (4.58), (4.60), (4.61) and (4.97) respectively—with respect to their surface they are considered open-ended—we find the following factors:

$$K_t(V_V) = \frac{1 - \exp\{-\xi\lambda\}}{1 - \exp\{-\xi[\lambda + g(\lambda + \frac{1}{2})]\}} \tag{4.101}$$

$$K_t(S_V) = \frac{\lambda \cdot \exp\{-\xi\lambda\}}{\left[\lambda + g\left(1 + \frac{2}{\pi}\lambda\right)\right] \cdot \exp\{-\xi[\lambda + g(\lambda + \frac{1}{2})]\}} \tag{4.102}$$

$$K_t(KV) = \frac{\pi + 2\lambda - \xi\frac{\pi}{2}\lambda^2 \cdot \exp\{-\xi\lambda\}}{\left[\pi + 2\lambda + 4g - \xi\left(\frac{\pi}{2}\lambda + g\left[\frac{\pi}{2} + \lambda\right]\right)^2\right] \cdot \exp\{-\xi[\lambda + g(\lambda + \frac{1}{2})]\}} \tag{4.103}$$

The coefficient ξ has the same meaning as above; its value is about the volume density of the tubules divided by λ.

The practical use that can be made of such factors is discussed in Volume 1, Section 4.3.4; examples of their application are found in Bolender *et al.* (1978) and Weibel and Paumgartner (1978). The major problem of these correction factors lies in the fact that they include one quantity which is hard to assess: ξ. Since it occurs in both exponentials in the numerator and in the denominator it is, for the time being, adequate to estimate it approximately from a rough estimate of V_V and dividing it by δ^2 for discs or λ for cylinders. Computation of these factors can easily be realized on any small programmable computer, even on some pocket calculators. Graphs of some of these factors are presented in Volume 1, Section 4.3.4 (Figs 4.20–21).

4.4. Correction factors for spherical particles, allowing for loss of cap sections

So far we have only considered one effect of finite section thickness, namely "overprojection" of the slice content; this assumed that the particles were perfectly opaque and embedded in a perfectly translucent matrix. In practice this is rarely, if ever, the case: there is a certain difference in contrast between particles and matrix which allows the particles to be recognized; but if a particle is contained in the slice with a shallow cap section only (type (*c*) particle in Fig. 4.3) then the contrast difference between this profile and the matrix may become deficient and the profile lost. Since such caps do no longer contribute to the apparent areal or boundary densities, the application of the aforementioned correction factors would tend to over-correct the estimate.

The loss of cap sections evidently introduces a further complication: stereological measurements on projected images yield either overestimates or underestimates of the specimen parameters, depending on whether the "overprojection" of the slice content or the loss of cap sections predominate. For the time being this situation cannot be treated with the same generality as the correction for mere overprojection effects. We shall therefore deal with a special case, the volume and surface density of a population of discrete spherical particles, randomly dispersed in the specimen but without interpenetration; in the first instance we shall assume them of equal size, but then we shall consider the effect of size distribution on the derived correction factors. Since we consider this to be just a model case, we shall derive the correction factors from first principles, proceeding in detail through the entire derivation; at first, this will require a derivation of some of the correction factors encountered in previous sections, but through a different approach.

4.4.1. MODEL STRUCTURE AND PROJECTED IMAGE

We shall assume a structure containing a very large number of randomly dispersed spherical objects of diameter D. They are rather loosely spaced, so that there will be no overlap of projected profiles. A slice of thickness t intercepts part of this population of spheres. Figure 4.7 shows, in analogy to Fig. 4.3, that two different positions of these spheres with respect to the slice must be considered.

(a) Their centre point lies outside the section: in this case the radius of the projected circle, $r_1 \leq R$, corresponds to the radius of the circular cut surface which is located at either the upper or the lower surface of the slice. As the sphere gradually penetrates into the slice from $x=0$ to $x=R$, where the centre point reaches the slice surface, we obtain r_1 from

$$r_1(x)=[R^2-(R-x)^2]^{\frac{1}{2}} \tag{4.104}$$

(b) The centre point lies within the slice: in this case the radius of the projected image will be constant, namely the radius of the sphere's great circle, so that

$$r_2=R \tag{4.105}$$

These two functions are plotted in Fig. 4.8.

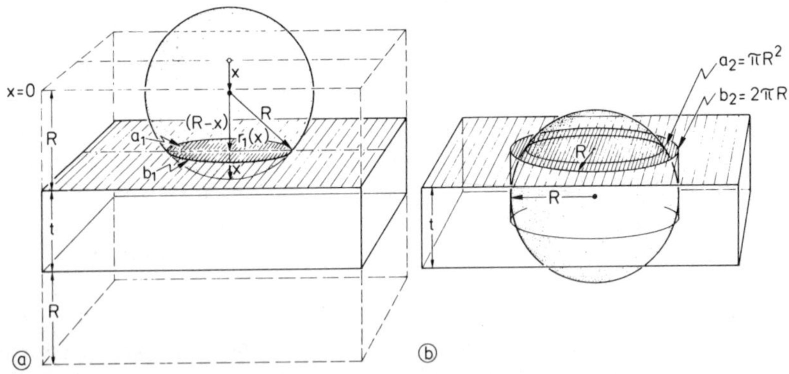

FIG. 4.7: Sphere of radius R cut by slice of thickness t. (a) If sphere centre is outside slice the profile area a_1 and boundary length b_1 depend on position x; (b) if sphere centre is within slice the profile area a_2 and boundary length b_2 are independent of position.

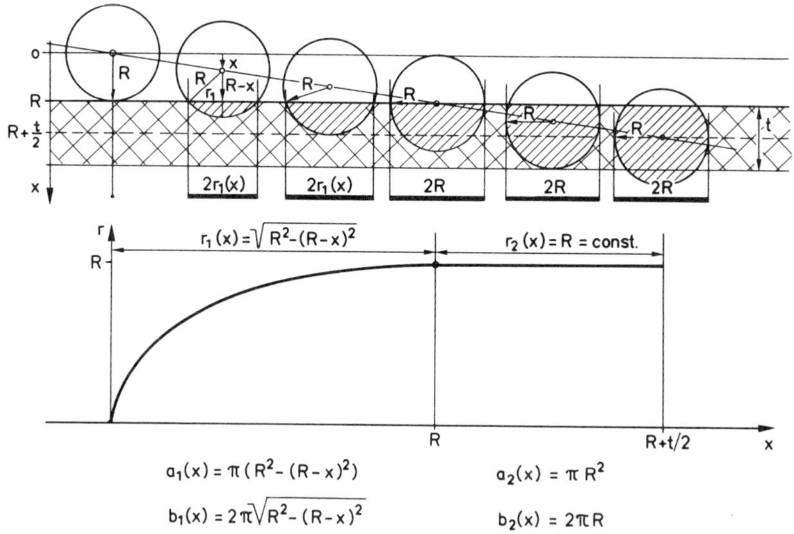

FIG. 4.8: Profile radius r as function of position x of sphere centre.

We must now introduce a number of relations between the true properties of the structure and the observations made on the projected image of the slice content.

Since the slice must be a representative sample of the structure, the volume density, V_V, the surface density, S_V, and the numerical density, N_V, of the objects contained in the slice are equal to the volume, surface and numerical densities in the entire structure. Since our model assumes the objects to be spheres of radius R the relation between these parameters can be expressed as follows:

$$V_V = N_V \cdot \tfrac{4}{3}\pi R^3 \tag{4.106}$$

$$S_V = N_V \cdot 4\pi R^2 \tag{4.107}$$

The areal density of projected profiles, A'_A, observed on the projected image is the sum of all the profile areas, a, found on the unit area of slice surface; this may be written as

$$A'_A = N'_A \cdot \bar{a} \tag{4.108}$$

where N'_A is the numerical density of profiles and $\bar{a}$ the expected or mean value of the profile area. Similarly, the observed boundary density of the projected profiles, B'_A, is defined as the sum of the profile boundaries, b, per unit area, or

$$B'_A = N'_A \cdot \bar{b} \tag{4.109}$$

where $\bar{b}$ is the expected or mean value of b for a single profile.

4.4.2. CORRECTION COEFFICIENTS FOR TOTALLY PROJECTED PROFILES (NO TRUNCATION)

In the following we shall first consider the case of completely opaque objects in a translucent matrix; all profiles are hence visible. We must find the parameters describing the projected image which are N'_A, A'_A and B'_A.

In Chapter 5 we shall find the apparent numerical density of profiles to be

$$N'_A = N_V(t + D) \tag{4.110}$$

This reflects the situation shown in Fig. 4.7 where profiles formed from particles with their centre outside the slice (a) are contributed in proportion to D, whereas those with their centres inside contribute profiles in proportion to t.

The area a and the boundary length b of the profiles depend on the profile radius r

$$a(x) = \pi \cdot r^2(x) \tag{4.111}$$

$$b(x) = 2\pi \cdot r(x) \tag{4.112}$$

which on the basis of eqs (4.104) and (4.105) may have different values whether the sphere centre is within or outside the slice. Placing $x = 0$ at the distance R above the slice (Figs 4.7a and 4.8), we find if $0 < x < R$

$$a_1(x) = \pi[R^2 - (R - x)^2] \tag{4.113}$$

$$b_1(x) = 2\pi \cdot [R^2 - (R - x)^2]^{\frac{1}{2}} \tag{4.114}$$

if $R < x < (R + t/2)$ (Fig. 4.7b)

$$a_2(x) = \pi R^2 \tag{4.115}$$

$$b_2(x) = 2\pi R \tag{4.116}$$

It is evident from Figs 4.7 and 4.8 that the function $r(x)$ must be symmetrical around the mid-plane of the section; it hence suffices to consider the range

$$0 < x < (R + t/2)$$

in order to derive the mean area and boundary length of the profiles by integration (Fig. 4.8). By integration of eqs (4.113) and (4.115) we thus find the mean profile area to be:

$$\bar{a} = \frac{\pi \cdot \int_0^R [R^2 - (R-x)^2]\, dx + \pi \int_R^{R+t/2} R^2\, dx}{R + t/2}$$

$$= \frac{R^2(\frac{4}{3}R + t)}{2R + t} \tag{4.117}$$

The mean boundary length is accordingly

$$\bar{b} = \frac{2\pi}{R + t/2} \left\{ \int_0^R [R^2 - (R-x)^2]^{\frac{1}{2}}\, dx + \int_R^{R+t/2} R\, dx \right\}$$

$$= \frac{2R\left(\frac{\pi}{2} R + t\right)}{2R + t} \tag{4.118}$$

Substituting eqs (4.110), (4.117) and (4.118) into eqs (4.108) and (4.109), we derive the following relationships for the apparent areal and boundary length densities observed on the projected image of the slice:

$$A'_A = N_V \cdot \pi R^2(\tfrac{4}{3}R + t) \tag{4.119}$$

$$B'_A = N_V \cdot 2\pi R\left(\frac{\pi}{2} R + t\right) \tag{4.120}$$

Note that A'_A is dimensionless and that B'_A has the dimension cm^{-1}.

It is now a straight-forward procedure to estimate the Holmes correction coefficients introduced in eqs (4.46) and (4.47), by dividing the true parameter by the apparent one. With $D = 2R$ and $g = t/D$, as above, we obtain for the volume density

$$K_t(V_V) = \frac{V_V}{A'_A}$$

$$= \frac{N_V \cdot \frac{4}{3}\pi R^3}{N_V \cdot \pi R^2 \cdot (\frac{4}{3}R + t)}$$

$$= \frac{4R}{4R + 3t} = \frac{2D}{2D + 3t} = \frac{2}{2 + 3g} \tag{4.121}$$

This value had been obtained by Holmes (1927) and by Hennig (1957).

For the surface density we find

$$K_t(S_V) = \frac{S_V}{2 \cdot I'_L} = \frac{S_V}{\frac{4}{\pi} B'_A}$$

$$= \frac{N_V \cdot 4\pi R^2}{4N_V \cdot 2R\left(\frac{\pi}{2} R + t\right)}$$

$$= \frac{2\pi R}{2\pi R + 4t} = \frac{\pi D}{\pi D + 4t} = \frac{\pi}{\pi + 4g} \tag{4.122}$$

This is the factor derived by Hennig (1957). It is evident that both these factors are identical with those derived from the general formulas and given in eqs (4.69) and (4.82).

4.4.3. CORRECTION COEFFICIENT FOR TRUNCATED PROFILES

We shall now use the same procedure to derive correction coefficients for Holmes effect on V_V and S_V by allowing for a loss of small profiles due to contrast deficiency.

The model is the same as before: a population of spheres of radius R is sampled with a slice of thickness t. Whereas before we assumed that the spheres began to supply profiles to the projection image as soon as they touched the slice surface ($x = 0$), we will now introduce the limitation that these will remain undetected until they have penetrated a distance h_0 into the slice (Fig. 4.9). If r_0 is the radius of this minimum detectable profile,

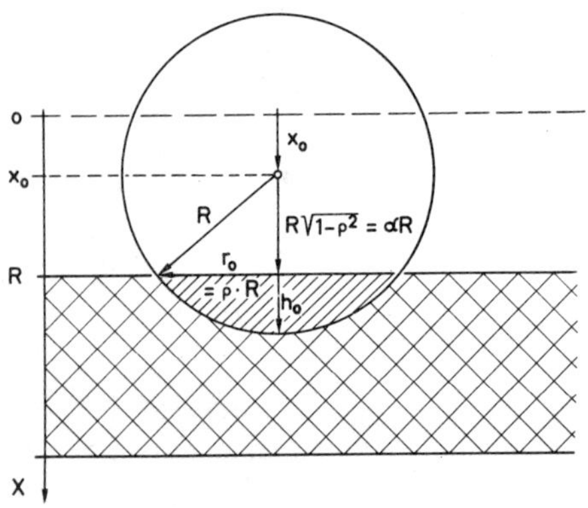

FIG. 4.9: Dimensions of minimal visible cap.

the coordinate of the centre point, which is equal to h_0, can be written as

$$x_0 = R - [R^2 - r_0^2]^{\frac{1}{2}} = h_0 \tag{4.123}$$

We now express both t and r_0 in terms of R:‡

$$t = g \cdot 2R \tag{4.124}$$

$$r_0 = \rho \cdot R, \qquad \rho \leq 1 \tag{4.125}$$

so that

$$h_0 = x_0 = R[1 - (1 - \rho^2)^{\frac{1}{2}}] = R(1 - \alpha) \tag{4.126}$$

where

$$\alpha = (1 - \rho^2)^{\frac{1}{2}} = \frac{1}{R}(R^2 - r_0^2)^{\frac{1}{2}} \tag{4.127}$$

a frequently occurring term, which is the relative distance from the centre point to the slice surface at which a minimum detectable cap is formed.

It is now evident that only those spheres which have their centre point in the range

$$x_0 < x < (t + 2R - 2x_0)$$

can contribute profiles which are detected on the projection image of the slice; the other profiles are lost (Fig. 4.10). In order to derive correction

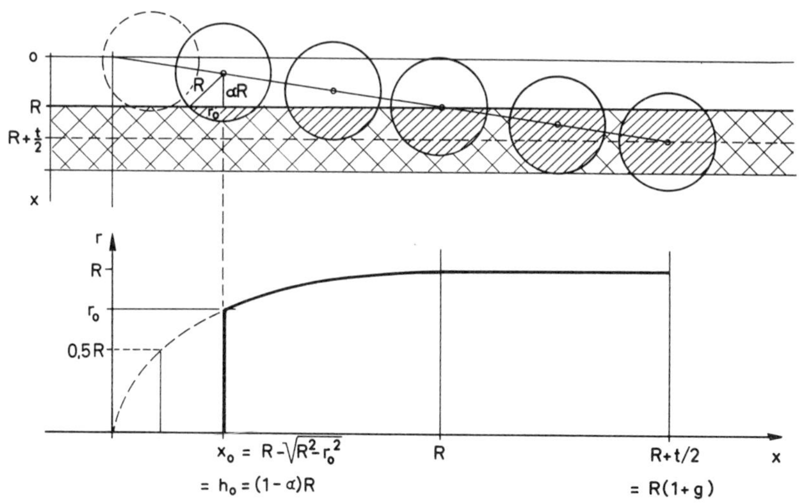

FIG. 4.10: Visible profile radius as function of position x of sphere centre accounting for loss of small cap sections.

‡ We have elected to measure the minimal visible cap by the radius of its cut surface, a parameter directly measurable, rather than by cap height as used by other authors (Floderus, 1944; Haug, 1967).

coefficients applying to this situation we must derive numerical densities N'_A and mean profile area and boundary length, $\bar{a}'$ and $\bar{b}'$, which consider this restricted range of sampling. We shall again show in Chapter 5 that the apparent numerical profile density is

$$N'_A = N_V(t + 2R - 2h_0) \tag{4.128}$$

the equation of Floderus (1944) and Haug (1967). Substituting from eqs (4.123) to (4.127) this can be written as

$$\begin{aligned} N'_A &= N_V \cdot 2R[g + (1 - \rho^2)^{\frac{1}{2}}] \\ &= N_V \cdot 2R(g + \alpha) \end{aligned} \tag{4.129}$$

The mean area and boundary length of detected profiles is now derived by integrating eqs (4.113) and (4.116) over the range of the restricted "half-superslice" (Fig. 4.10)

$$x_0 < x < (R + t/2)$$

or

$$(1 - \alpha)R < x < R(1 + g)$$

The expected profile area is then obtained from eqs (4.113) and (4.115):

$$\bar{a}' = \frac{\pi}{(\alpha + g)R} \left\{ \int_{(1-\alpha)R}^{R} [R^2 - (R - x)^2] \, dx + \int_{R}^{(1+g)R} R^2 \, dx \right\} \tag{4.130}$$

This is solved by straight-forward integration to yield

$$\bar{a}' = \pi R^2 \cdot \frac{\alpha - \frac{1}{3}\alpha^3 + g}{\alpha + g} \tag{4.131}$$

The apparent mean boundary length is obtained by integrating eqs (4.114) and (4.116):

$$\bar{b}' = \frac{2\pi}{(\alpha + g)R} \left\{ \int_{(1-\alpha)R}^{R} [R^2 - (R - x)^2]^{\frac{1}{2}} \, dx + \int_{R}^{(1+g)R} R \, dx \right\} \tag{4.132}$$

This yields

$$\bar{b}' = \pi R \cdot \frac{\alpha\rho + \sin^{-1} \alpha + 2g}{\alpha + g} \tag{4.133}$$

By substituting eqs (4.129) and (4.131) into eq. (4.108) the apparent areal density of profiles on the projected slice image is found to be

$$A'_A = N_V \cdot 2\pi R^2(\alpha - \tfrac{1}{3}\alpha^3 + g) \tag{4.134}$$

By analogy, substituting eqs (4.129) and (4.133) into eq. (4.109) yields the apparent boundary length density of profiles on the projected image of a truncating slice:

$$B'_A = N_V \cdot 2\pi R^2(\alpha\rho + \sin^{-1}\alpha + 2g) \tag{4.135}$$

It is now a straight-forward procedure to derive the correction coefficients for combined Holmes effect and truncation, by dividing the true values for V_V and S_V given in eqs (4.106) and (4.107) by the apparent areal and boundary densities respectively. It is thus found that the coefficient applying to volume density estimations is

$$\begin{aligned} K'_t(V_V) &= \frac{N_V \cdot \frac{4}{3}\pi R^3}{N_V \cdot 2\pi R^3(\alpha - \frac{1}{3}\alpha^3 + g)} \\ &= 2/(3 - \alpha^3 + 3g) \end{aligned} \tag{4.136}$$

Reintroducing for α the form given by eq. (4.127) we find

$$K'_t(V_V) = 2/[(2+\rho^2)(1-\rho^2)^{\frac{1}{2}} + 3g] \tag{4.137}$$

The correction coefficient applying to surface density estimations is

$$\begin{aligned} K'_t(S_V) &= \frac{\pi \cdot N_V \cdot 4\pi R^2}{4 \cdot N_V \cdot 2\pi R^2(\alpha\rho + \sin^{-1}\alpha + 2g)} \\ &= \frac{\pi}{2(\alpha\rho + \sin^{-1}\alpha + 2g)} \end{aligned} \tag{4.138}$$

or, in analogy with eq. (4.137):

$$K'_t(S_V) = \frac{\pi}{2[\rho(1-\rho^2)^{\frac{1}{2}} + \sin^{-1}(1-\rho^2)^{\frac{1}{2}} + 2g]} \tag{4.139}$$

These two coefficients are dimensionless; it is evident that they tend towards eqs (4.121) and (4.122) respectively if the truncation effect becomes vanishingly small, i.e. if $\rho \to 0$, and that they become 1 if also $g \to 0$.

The values of $K'_t(V_V)$ and $K'_t(S_V)$ are plotted against g in Fig. 4.11 for various values of ρ. It is first of all worth noting that the correction coefficients may be larger or smaller than 1, depending on whether overestimation or loss of cap sections predominates.

There are some limits on the values that ρ may assume. First,

$$\rho < 1$$

because it is reasonable to assume for the general case that all profiles which include the equatorial plane, i.e. all profiles derived from spheres having the centre point within the slice, will produce sufficient contrast to

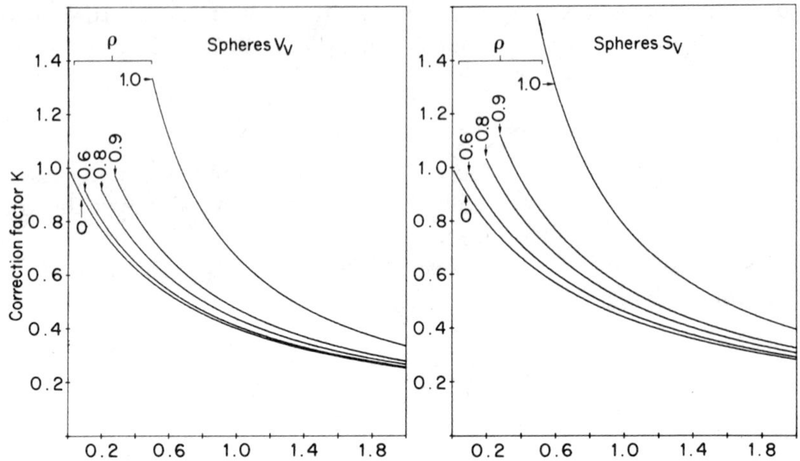

FIG. 4.11: Plot of correction coefficients for V_V and S_V for spherical particles as function of relative section thickness and relative radius ρ of smallest visible cap section.

be recognized, mainly since they have a sharp edge. If there are reasons to believe that more than half the sphere is needed to produce adequate contrast, then this correction cannot be applied. Secondly, it is also reasonable to assume that all profiles which extend through the entire section thickness will be recognized. One may therefore postulate that the largest height, h_0, of an unrecognized cap section should be

$$h_0 < t$$

Figure 4.9 shows that

$$h_0 = [1 - (1 - \rho^2)^{\frac{1}{2}}]R$$

and from this it follows that, with a certain minimal recognized profile diameter, the section thickness may only be such that

$$g < [1 - (1 - \rho^2)^{\frac{1}{2}}]/2$$

which defines the left-hand end of the ρ-isopleths in Fig. 4.11.

4.4.4. EFFECT OF SIZE DISTRIBUTION ON CORRECTION COEFFICIENTS

The correction coefficients derived so far are strictly valid only for a monodispersed population of spheres, i.e. if all spheres have the same diameter. If the sphere size is variable, various moments of the size distribution must be considered.

To derive correction coefficients for total projection (i.e. no truncation allowed) volume and surface density are defined as

$$V_V = \tfrac{4}{3}\pi \sum_j N_{Vj} \cdot R_j^3 \tag{4.140}$$

$$S_V = 4\pi \sum_j N_{Vj} \cdot R_j^2 \tag{4.141}$$

where N_{Vj} and R_j are numerical density and radius of spheres in size class j. Apparent numerical density is now (eq. 4.110) derived for each size class

$$N'_{Aj} = N_{Vj}(t + 2R_j) \tag{4.142}$$

and so is mean profile area and boundary length (eqs 4.117 and 4.118):

$$\bar{a}(j) = \frac{\pi R_j^2(\tfrac{4}{3}R_j + t)}{t + 2R_j} \tag{4.143}$$

$$\bar{b}(j) = \frac{2\pi R_j\left(\frac{\pi}{2} R_j + t\right)}{t + 2R_j} \tag{4.144}$$

Apparent areal and boundary densities now are

$$A'_A = \sum_j N'_{Aj} \cdot \bar{a}(j) \tag{4.145}$$

$$B'_A = \sum_j N'_{Aj} \cdot \bar{b}(j) \tag{4.146}$$

By appropriate substitutions and rearrangements we can derive, on the basis of eqs (4.46) and (4.47), the following correction coefficients:

$$K''_t(V_V) = \frac{4E(R^3)}{4E(R^3) + 3tE(R^2)} \tag{4.147}$$

$$K''_t(S_V) = \frac{2\pi E(R^2)}{2\pi E(R^2) + 4tE(R)} \tag{4.148}$$

$E(R^n)$ is the nth moment of the size distribution or the expected (or mean) value of R^n, such that

$$E(R^n) = \sum_j f_j \cdot R_j^n \tag{4.149}$$

where f_j is the relative frequency of size class j.

It is noted that eqs (4.147) and (4.148) become equal to eqs (4.121) and (4.122) respectively if there is only one size class of spheres, because then

$$E(R) = E^{\frac{1}{2}}(R^2) = E^{\frac{1}{3}}(R^3) = E^{1/n}(R^n)$$

The consideration of the truncation effect leads to more involved relationships which cannot be dealt with in detail here. The procedure is the same as above: we determine mean profile area and boundary length for each size class and derive apparent areal and boundary length densities by weighing the contribution of each size class with the factor

$$N'_A(j) = N_V(j) \cdot (2R_j + t - 2h_0) \tag{4.150}$$

By this procedure we derive as correction factor for volume density estimates

$$K'''_t(V_V) = \frac{4 \cdot E(R^3)}{4 \cdot E(R^3) + 3t \cdot E(R^2) - 6h_0^2 \cdot E(R) + 2h_0^3} \tag{4.151}$$

Instead of expressing the limit of visibility of cap sections by the minimum visible radius r_0 we have here retained minimum cap height h_0; the reason is that r_0 must be different for each size class of spheres whereas we can assume h_0 to be independent of size. It can be shown by appropriate substitutions that the factor given in eq. (4.151) reduces to eq. (4.136) or (4.137) if there is only one size of spheres.

Equation (4.151) can be simplified by introducing the dimensionless ratio of the nth moment to the nth power of the mean radius:

$$m_n = E(R^n)/[E(R)]^n; \qquad m_n \geq 1 \tag{4.152}$$

and by expressing section thickness t and minimal cap height h_0 as fractions of mean radius:

$$t = 2g \cdot E(R) = g \cdot \bar{D} \tag{4.153}$$

$$h_0 = \eta \cdot E(R) = \eta \cdot \bar{D}/2 \tag{4.154}$$

Substituting eqs (4.153) and (4.154) into eq. (4.151) and dividing by $[E(R)]^3$ we derive, by eq. (4.152),

$$K'''_t(V_V) = \frac{2 \cdot m_3}{2 \cdot m_3 + 3gm_2 - 3\eta^2 + \eta^3} \tag{4.155}$$

A size distribution dependent correction coefficient can hence be derived by considering the ratios of third and second moment to the first moment as weighting factors. Approximate values of m_2 and m_3 can be read from Figs 4.12 and 4.13, where they are given as a function of the coefficient of

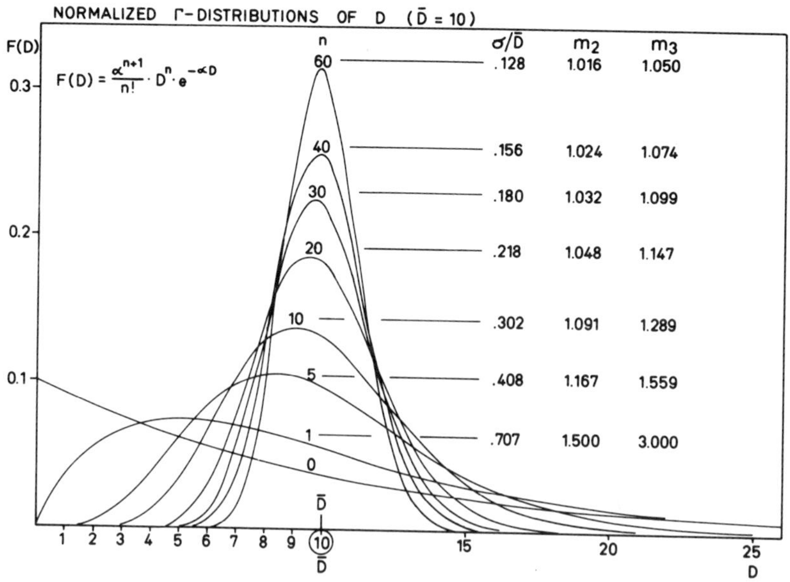

FIG. 4.12: Size distribution factors m_2 and m_3 (eq. 4.152) for normalized Γ-distributions of sphere diameter D.

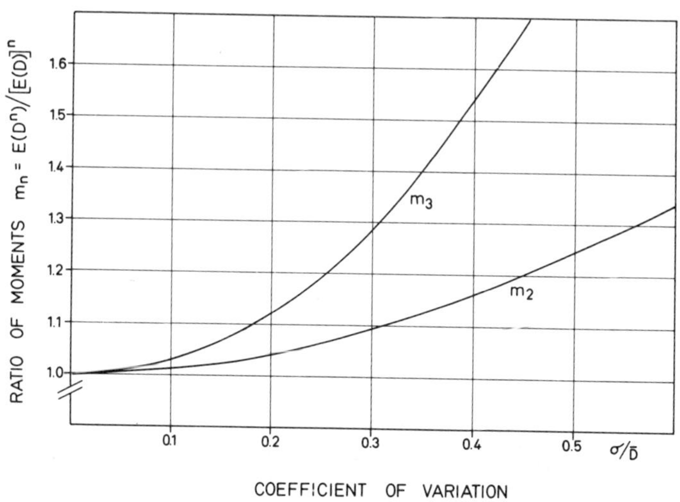

FIG. 4.13: Size distribution factors m_2 and m_3 as function of coefficient of variation.

variation CV = S.D./$\bar{D}$ calculated for normalized Γ-distributions of D (Fig. 4.12).

For the surface density correction factor no such simple relationship between moments of the size distribution and t and h_0 can be obtained, because integration of eq. (4.118) results in a mean boundary length per size class which contains the product

$$\alpha\rho = \frac{R_j - h_0}{R_j}(2R_j h_0 - h_0^2)^{\frac{1}{2}} \tag{4.156}$$

as well as

$$\sin^{-1}\alpha = \sin^{-1}\left(\frac{R_j - h_0}{R_j}\right) \tag{4.157}$$

The full equation for K_S looks like this:

$$K_t'''(S_V) = \frac{\pi E(R^2)}{2\left\{E\{R\sqrt{(2Rh_0 - h_0^2)}\} - h_0 \cdot E\{\sqrt{(2Rh_0 - h_0^2)}\} + E\left(R^2 \sin^{-1}\frac{R - h_0}{R}\right) + t \cdot E(R)\right\}} \tag{4.158}$$

Noting that

$$(2Rh_0 - h_0^2)^{\frac{1}{2}} = r_0 = \rho R \tag{4.159}$$

and

$$\frac{R - h_0}{R} = \alpha \tag{4.160}$$

this amounts to

$$K_t'''(S_V) = \frac{\pi \cdot E(R^2)}{2\{E(\rho R^2) - h_0 \cdot E(\rho R) + E(R^2 \cdot \sin^{-1}\alpha) + t \cdot E(R)\}} \tag{4.161}$$

Again, if the size of the spheres is uniform, eq. (4.161) reduces to eq. (4.138) or (4.139).

Equations (4.158) or (4.161) do not lend themselves easily to a practical evaluation of the effect of size distribution on the correction factor. Considering the uncertainties occurring in estimating all the parameters involved it may be permitted to proceed with some simplification. We assume that the radius of the minimal visible cap, r_0, can be expressed as a fraction of the mean sphere radius $E(R)$ such that

$$\rho \approx r_0/E(R) \tag{4.162}$$

On this basis we approximate the components of eq. (4.161) containing ρ by

$$E(\rho R^2) \approx \rho E(R^2) \tag{4.163}$$

$$E(\rho R) \approx \rho E(R) \tag{4.164}$$

$$E(R^2 \cdot \sin^{-1} \alpha) \approx [\sin^{-1}(1-\rho^2)^{\frac{1}{2}}] E(R^2) \tag{4.165}$$

We furthermore note that the cap height h_0 is related to ρ as

$$h_0 = [1-(1-\rho^2)^{\frac{1}{2}}] E(R) \tag{4.166}$$

and that we can write for section thickness

$$t = 2g \cdot E(R) \tag{4.167}$$

This then leads to the simplified correction factor

$$K_t'''(S_V) \approx \frac{\pi E(R^2)}{2\{[\rho + \sin^{-1}(1-\rho^2)^{\frac{1}{2}}] E(R^2) + [2g - \rho(1-(1-\rho^2)^{\frac{1}{2}})] \cdot [E(R)]^2\}} \tag{4.168}$$

Finally, dividing by $E(R)^2$ and introducing the dimensionless ratio of moments, as defined by eq. (4.152), this becomes

$$K_t'''(S_V) \approx \frac{\pi m_2}{2\{[\rho + \sin^{-1}(1-\rho^2)^{\frac{1}{2}}] m_2 + 2g - \rho[1-(1-\rho^2)^{\frac{1}{2}}]\}} \tag{4.169}$$

An appropriate coefficient can hence be derived if the ratio of second to first moment of the size distribution is introduced as a weighting factor; approximate values of m_2 can be read from Figs 4.12 and 4.13. This coefficient still reduces to eq. (4.139) if all spheres have the same size.

CHAPTER 5

Numerical density: shape and size of particles

5.1. Numerical density of particles: basic principle

In this chapter we will examine the relationship between the number of profiles seen on a random section and the number of objects—which in this case must be discrete particles—contained in the unit volume of the structure.

Assume a large number n_0 of particles (objects) to be enclosed in a very large space such that the number of particle centres contained in a sample of volume V_T is

$$N_V = n_0 / V_T \tag{5.1}$$

which we shall call the numerical density of the particles in the volume (Fig. 5.1).

We now section this sample with a test plane of area A_T; on this we observe n_p particle profiles which have their centre point within A; we define as numerical density of profiles on the test area

$$N_A = n_p / A_T \tag{5.2}$$

It is required to find the relationship between N_A and N_V.

We must first specify the following definitions and restrictions.

(1) The particles must be convex solids, so that a plane intersecting the particle at any orientation will produce one profile only.
(2) The particles must be randomly orientated within the space and their number must be so large that all orientations between particle and test plane are equally likely.

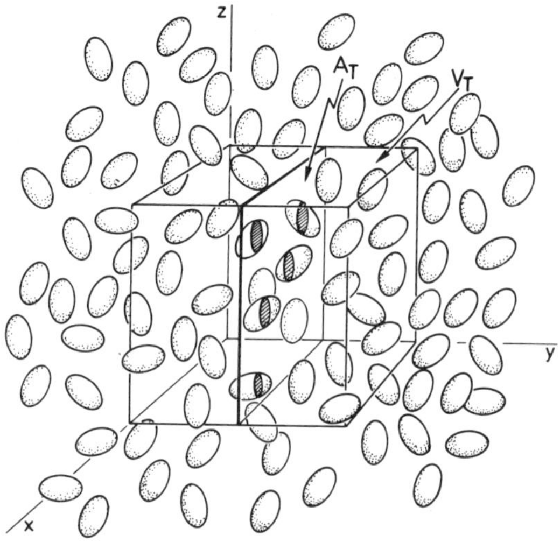

FIG. 5.1: Sectioning randomly oriented convex particles with a plane.

(3) The sample space is defined as a cube of volume

$$V_T = l^3 \tag{5.3}$$

placed into an x, y, z coordinate system. Its size is such that the number of particles it contains is very large. This sample space is randomly sectioned by a plane parallel to the x–z plane of the x, y, z coordinate system, so that the section area is

$$A_T = l^2 \tag{5.4}$$

If we single out one particle (Fig. 5.2) the probability that it will be cut by the random section plane is given by the ratio of the projections of both particle and containing space on to the orthogonal complement to the section plane (see Section 2.3), or to the ratio of tangent diameter $H(Y)$ to the extent of the sample space perpendicular to the section plane, l:

$$\Pr\{Y \uparrow T_2\} = H(Y)/l \tag{5.5}$$

The probability to cut any one of the n_0 particles is

$$\Pr\{Y \uparrow T_2\} = \bar{H}(Y)/l \tag{5.6}$$

where $\bar{H}(Y)$ is the mean of all the tangent diameters observed. In this case we kept the orientation of the section plane constant; $\bar{H}(Y)$ is hence the mean of all projections of the particles onto the orthogonal complement to the section plane. If the particles are present in great number,

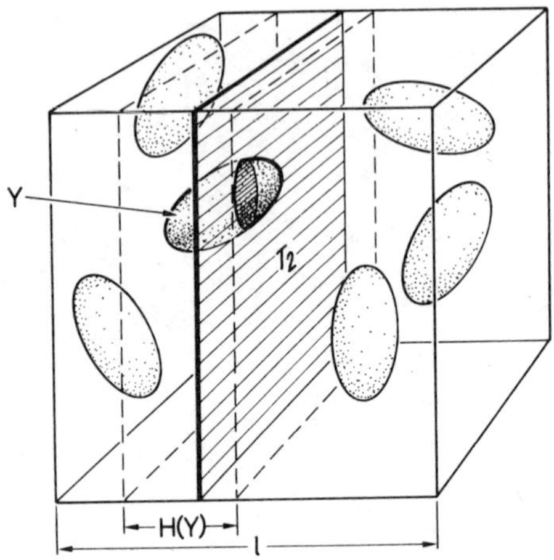

FIG. 5.2: Probability of cutting random convex particle.

and if they are allowed to assume any orientation in space with equal likelihood, then $\bar{H}$ is equal to the mean tangent diameter of the particles, as derived in eq. (2.54).

Having thus defined the average probability of cutting any one of the particles it now immediately follows that the expected number of profiles observed on the section is

$$E(n_p) = n_0 \cdot p' = n_0 \cdot \bar{H}/l \tag{5.7}$$

Dividing both sides of eq. (5.7) by l^2 we obtain

$$E(n_p)/l^2 = n_0 \cdot \bar{H}/l^3 \tag{5.8}$$

Since we had defined N_V and N_A as

$$N_V = n_0/l^3 \tag{5.9}$$

$$N_A = n_p/l^2 \tag{5.10}$$

it follows from eq. (5.8) that

$$E(N_A) = N_V \cdot \bar{H} \tag{5.11}$$

The number of profiles we expect to be found on the unit area of section is directly proportional to numerical particle density and to the mean tangent diameter of the particles, a fundamental relationship of very general validity for all possible convex bodies.

What we have described so far is the estimation of N_V for the "extended case" where one first obtains an IUR sample of tissue, the cubic sample of volume V_T, and then uses an arbitrary section, parallel to one cube face in our example, to estimate N_A. This is not a necessary restriction; the same result is obtained for the "restricted case" with a specimen X of volume $V(X)$ which contains $N(Y)$ convex particles making up the object phase Y and which is sectioned by a random plane T. The result one obtains in this case is

$$N_A = \frac{E\{N(Y \cap T)\}}{E\{A(X \cap T)\}} = \frac{1}{\bar{H}(Y)} \cdot \frac{N(Y)}{V(X)} = \frac{1}{\bar{H}} \cdot N_V \tag{5.12}$$

The particles need not be isotropic in X, but T must be an A-weighted random section for eq. (5.12) to be an unbiased estimator of N_V.

This formula is generally attributed to De Hoff and Rhines who gave, in 1961, a rigorous development of the relationship between N_A and N_V for the general cylinder and the ellipsoid. The formula is, however, already stated in the fundamental works of Wicksell, both for spheres (1925) and for ellipsoids (1926) where the notion of "mean tangent diameter" of a convex body is explicitly introduced. Wicksell does not, however, give a derivation of the formula.

In principle, this relationship could be used to estimate N_V from an experimental determination of N_A on random sections, if the mean tangent diameter of the particles were known. There are, however, considerable difficulties in the latter respect, because the determination of $\bar{H}$ depends, in general, on the knowledge of particle shape; this restricts its applicability.

For a spherical particle, the simplest case, the tangent diameter H corresponds to its diameter $D = 2R$, irrespective of orientation, since rotation of a sphere does not change the distance of parallel tangent planes. The average mean tangent diameter of a population of spheres of varying size is hence

$$\bar{H} = 2\bar{R} \tag{5.13}$$

the mean diameter of the spheres; methods for estimating $\bar{R}$ are given below.

For all non-spherical particles the formula can only be applied if the shape of the particles can be specified and if their average size or size distribution can be determined, thus permitting an estimation of $\bar{H}$.

It should be noted that the particle population may well represent a mixture of different sizes and shapes as long as $\bar{H}_j$ can be specified for each size and shape class and the relative frequency $f(j)$ of each class is known. Evidently, the total numerical density N_V is the sum of the numerical densities of the particles belonging to each class, N_{Vj}

$$N_V = \sum_j N_{Vj} \tag{5.14}$$

For each class the numerical profile density is

$$N_{Aj} = N_{Vj} \cdot \bar{H}_j \tag{5.15}$$

and the total profile density is

$$N_A = \sum_j N_{Aj}$$
$$= \sum_j N_{Vj} \cdot \bar{H}_j \tag{5.16}$$

Introducing the relative numerical frequency

$$f(j) = N_{Vj}/N_V \tag{5.17}$$

this yields

$$N_A = N_V \cdot \left[\sum_j f(j) \cdot \bar{H}_j\right] \tag{5.18}$$

where the term in brackets is a weighted average of the mean tangent diameters. It is evident that for many biological structures one may meet with great if not unsurmountable difficulties in estimating both $\bar{H}_j$ and $f(j)$ so that one may have to be satisfied with some gross approximations. (See also Chapters 6 and 8.)

5.2. Alternative methods for numerical density estimation based on simplifying assumptions

Because of the practical difficulties in obtaining a reliable estimate of the mean tangent diameter it is often desirable to have other methods available which work with directly obtainable stereological parameters such as volume or surface density in addition to numerical profile density. Such methods require, however, the introduction of some simplifying assumptions, particularly with respect to shape and size variation of the particles. A selection of such methods are presented in the following.

5.2.1. METHOD OF WEIBEL AND GOMEZ (1962)

The approach chosen in this method is to find a relationship between the numerical density of particles, N_V, the numerical profile density, N_A, and the volume density of the particles, V_V, as a measure of particle size.

Suppose first particles of well-defined shape and equal size, v being the volume of the individual particle. The volume density is evidently

$$V_V = N_V \cdot v \tag{5.19}$$

On a random section the areal density of the profiles is

$$A_A = N_A \cdot \bar{a} \tag{5.20}$$

where $\bar{a}$ is the average profile area produced by random sectioning of the particle. If the particle shape is known $\bar{a}$ must depend on v and on some measure of shape. A dimensionless relationship would be

$$v = \beta \cdot \bar{a}^{\frac{3}{2}} \tag{5.21}$$

The coefficient β solely depends on the shape of the particles and not on their size, as will be shown below. Solving eq. (5.21) for $\bar{a}$ and introducing this into eq. (5.20) we find for the particle volume

$$v = \beta \cdot \left(\frac{A_A}{N_A}\right)^{\frac{3}{2}} \tag{5.22}$$

From eq. (5.19) it hence follows that

$$N_V = \frac{V_V}{v} = \frac{1}{\beta} \cdot \frac{N_A^{\frac{3}{2}} \cdot V_V}{A_A^{\frac{3}{2}}} \tag{5.23}$$

Since the Delesse principle states that

$$V_V = A_A \tag{5.24}$$

eq. (5.23) reduces to

$$N_V = \frac{1}{\beta} \cdot \frac{N_A^{\frac{3}{2}}}{V_V^{\frac{1}{2}}} \tag{5.25}$$

which is valid if all particles have the same size and shape. The virtue of this formula is that N_A and V_V are two very easily determined measurements. It now remains (a) to define β as a function of shape, and (b) to estimate the effect of a variation in size of the particles.

(a) *The shape coefficient β*

Since the particle volume v is given as the measure of particle size we must find the mean profile area obtained by random sectioning. For a convex particle—and we must deal with this class of solids only—the profile area obtained depends on the orientation of the sectoning plane and on its position relative to the particle centre. At any fixed orientation (θ, ϕ) the mean profile area is

$$\hat{a}(\theta, \phi) = v/H(\theta, \phi) \tag{5.26}$$

where $H(\theta, \phi)$ is the tangent diameter of the particle measured at this orientation (Fig. 2.20a); this needs no further justification.

Allowing for all possible orientations we obtain, on the basis of eq. (2.51)

$$\bar{a} = \frac{1}{4\pi} \int_0^{\pi} \int_0^{2\pi} \hat{a}(\theta, \phi)[H(\theta, \phi) \cdot \sin\theta \, d\theta \, d\phi / E\{H(\theta, \phi)\}]$$

$$= \frac{v}{4\pi \cdot E\{H(\theta, \phi)\}} \int_0^{\pi} \int_0^{2\pi} \frac{1}{H(\theta, \phi)} [H(\theta, \phi) \cdot \sin\theta \, d\theta \, d\phi] \quad (5.27)$$

Because the integral of $\sin\theta \, d\theta \, d\phi$ is 4π this becomes

$$\bar{a} = v/E\{H(\theta, \phi)\} = v/\bar{H} \quad (5.28)$$

where $\bar{H}$ is the average mean tangent diameter of the particles.

A closely related general formula for the mean cross-sectional area of convex solids has been provided by integral geometry (Hadwiger, 1957), by using the integral of mean curvature K instead of the mean tangent diameter:

$$\bar{a} = 2\pi \cdot v/K \quad (5.29)$$

Equations (5.28) and (5.29) are consistent because of (eq. 3.161)

$$K = 2\pi\bar{H} \quad (5.30)$$

by Minkowski's theorem (Minkowski, 1903).

It is hence possible to estimate β for any convex particle, provided either $\bar{H}$ or K can be determined. Substituting either eq. (5.28) or eq. (5.29) into eq. (5.21) we find:

$$\beta = \sqrt{(\bar{H}^3/v)} = \sqrt{\{K^3/(8\pi^3 \cdot v)\}} \quad (5.31)$$

For the simple case of a sphere of radius R

$$v = \tfrac{4}{3}\pi R^3 \quad (5.32)$$

$$\bar{H} = 2R \quad (5.33)$$

$$K = 4\pi R \quad (5.34)$$

Introducing these parameters into eq. (5.31) we find

$$\beta(\text{sphere}) = \sqrt{(6/\pi)} \approx 1.382 \quad (5.35)$$

The calculation of β for other simple convex solids can make use of the fact that the integral of mean curvature K has been determined by integral geometry (Hadwiger, 1955); the reader is referred to the fundamental treatise of Hadwiger (1957) for their derivation. This is done here for ellipsoids of revolution and right circular cylinders (see also Table 5.1).‡

‡ The author is grateful to Dr H. R. Gnägi for working out these relations.

Table 5.1

Shape coefficients for particle counting methods by eqs (5.73), (5.75) and (5.22)

Particle shape	γ_1	γ_2	γ_3	β
Sphere	2	12.57	4.19	1.38
Spheroid prolate (2:1:1)	2.76	21.4	8.38	1.58
oblate (2:2:1)	1.71	8.67	2.09	1.55
Cube	1.5	6	1	1.84
Octahedron	1.17	3.46	0.47	1.86
Icosahedron	1.74	8.66	2.18	1.55
Dodecahedron	2.64	18.05	7.66	1.55
Tetrakaidecahedron	3	26.8	11.3	1.55

Right circular cylinders

Right circular cylinders are generated by revolving a rectangle around one of its axes (Fig. 5.3). The radius of the cylinder being R, its length is

$$L = \lambda R \tag{5.36}$$

where λ is a dimensionless shape factor measuring the relative elongation of the cylinder; it may assume any values in the range

$$0 < \lambda < \infty \tag{5.37}$$

yielding any cylinder from a very thin circular disc to a very long rod. The volume and integral of mean curvature are

$$V_c = 2\pi\lambda R^3 \tag{5.38}$$

$$K_c = 2\pi[\tfrac{1}{2}\pi + \lambda]R \tag{5.39}$$

so that the mean profile area becomes

$$\bar{a}_c = \frac{4\pi\lambda}{\pi + 2\lambda} R^2 \tag{5.40}$$

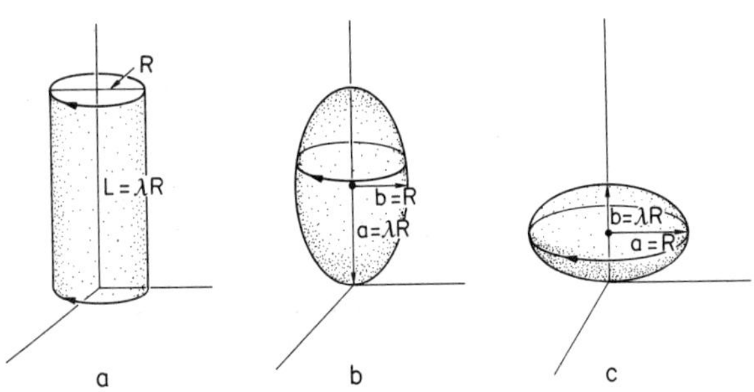

FIG. 5.3: Solids generated by revolution: (a) cylinder, (b) prolate ellipsoid, (c) oblate ellipsoid.

From this the value of β_c for right circular cylinders is found as

$$\beta_c = \left[\frac{\left(\lambda + \frac{\pi}{2}\right)^3}{2\pi\lambda}\right]^{\frac{1}{2}} \tag{5.41}$$

It is hence solely dependent of the shape factor λ. Values of β_c can be read from the graph in Fig. 5.4 for values of λ between 0.1 and 4.

Ellipsoids of revolution

Ellipsoids of revolution are generated by revolving an ellipse about one of its axes. Depending on whether the long or the short axis is taken as the axis of revolution, prolate or oblate ellipsoids result (Fig. 5.3). Let the hemi-axis which is revolved be of length R, and the hemi-axis coinciding with the axis of revolution have length λR, whereby λ is again a shape factor measuring "elongation" of the ellipsoid. If $0<\lambda<1$ an oblate ellipsoid is formed, whereas a prolate ellipsoid is characterized by $1<\lambda<\infty$. It is evident that $\lambda = 1$ refers to a sphere.

The volume of both prolate and oblate ellipsoids is given by

$$v_e = \frac{4\pi}{3}\lambda R^3 \tag{5.42}$$

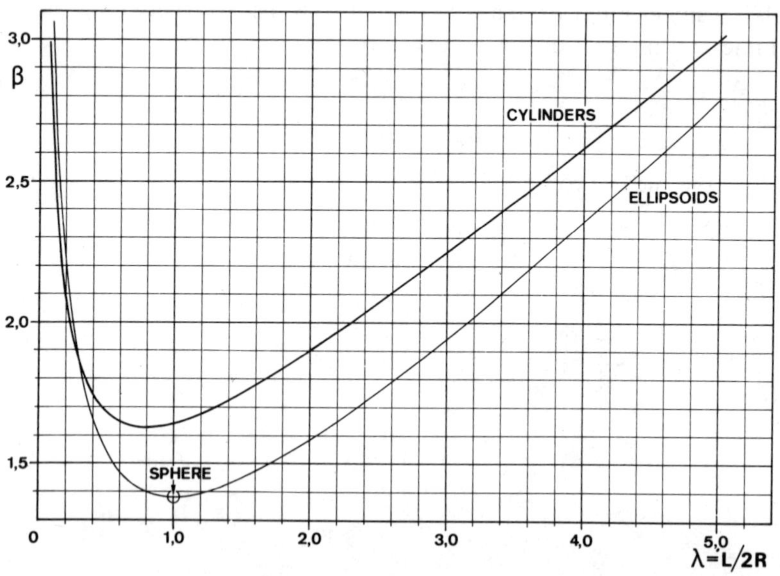

FIG. 5.4: Graph of shape coefficient β for ellipsoids and cylinders as function of relative length λ. (Reproduced with permission from Weibel, 1969.)

but the integrals of mean curvature will be different. For oblate ellipsoids

$$K_{oe} = 2\pi\left[\lambda + \frac{\arccos\lambda}{(1-\lambda^2)^{\frac{1}{2}}}\right]R, \qquad (0<\lambda<1) \tag{5.43}$$

and for prolate ellipsoids

$$K_{pe} = 2\pi\left[\lambda + \frac{\ln[\lambda + (\lambda^2-1)^{\frac{1}{2}}]}{(\lambda^2-1)^{\frac{1}{2}}}\right]R, \qquad (1<\lambda<\infty) \tag{5.44}$$

The terms in brackets depend solely on λ and are hence shape dependent. To simplify the further procedures we may introduce a conditional coefficient

$$\alpha = \begin{cases} \lambda + \dfrac{\arccos\lambda}{(1-\lambda^2)^{\frac{1}{2}}}, & 0<\lambda<1 \\ \lambda + \dfrac{\ln[\lambda + (\lambda^2-1)^{\frac{1}{2}}]}{(\lambda^2-1)^{\frac{1}{2}}}, & 1<\lambda<\infty \end{cases} \tag{5.45}$$

whereby $\alpha = 2$ when $\lambda = 1$. With the general formula

$$K_e = 2\pi\alpha R \tag{5.46}$$

the mean profile area is from eq. (5.29)

$$\bar{a} = \frac{4\pi}{3} \cdot \frac{\lambda}{\alpha} R^2 \tag{5.47}$$

It is seen that when $\lambda = 1$ this becomes equal to eq. (5.35) for the sphere. Introducing eqs (5.42) and (5.47) into eq. (5.21) we derive β for ellipsoids of revolution as

$$\beta_e = [(3\alpha^3)/(4\pi\lambda)]^{\frac{1}{2}} \tag{5.48}$$

where α must be introduced from eq. (5.45) for either the oblate or the prolate case, i.e. depending on whether $\lambda < 1$ or $\lambda > 1$. Values for β_e, which is solely dependent on shape, can be read from the graph in Fig. 5.4.

(b) *Effect of particle size distribution*

Equation (5.26) was derived under the assumption that all particles had equal size. We must now examine what effect a size distribution of particles may have on the applicability of this formula. The basic premises are again that the volume density is related to particle volume and number:

$$V_V = N_V \cdot \bar{v} \tag{5.49}$$

but this time the mean particle volume must be considered, whereas the areal density is related to profile number and mean area

$$A_A = N_A \cdot \bar{a} \tag{5.50}$$

Particle volume and profile area are connected by

$$\bar{v} = \beta' \cdot \bar{a}^{\frac{3}{2}} \tag{5.51}$$

and we must find in what way the particle size distribution may cause β' to differ from β.

We introduce as a measure for particle size the mean tangent diameter H_j and express the frequency of each size class by its numerical density

$$N_{Vj} = f_j \cdot N_V \tag{5.52}$$

where f_j is the relative frequency, and

$$\sum_j f_j = 1$$

For each size class we find a set of shape coefficients, k_2 and k_3, which relate the particle volume and its mean profile area to H_j, such that

$$v_j = k_3 \cdot H_j^3 \tag{5.53}$$

$$\bar{a}_j = k_2 \cdot H_j^2 \tag{5.54}$$

The mean particle volume then is

$$\bar{v} = \frac{\sum N_{Vj} \cdot v_j}{N_V} = k_3 \cdot \sum f_j \cdot H_j^3 \tag{5.55}$$

and the mean profile area

$$\bar{a} = \frac{\sum N_{Aj} \cdot \bar{a}_j}{N_A} = \frac{k_2 \cdot \sum N_{Aj} \cdot H_j^2}{N_A} \tag{5.56}$$

According to eqs (5.11) and (5.18), N_{Aj} and N_A depend on the numerical particle density and on the mean tangent diameter:

$$N_{Aj} = N_{Vj} \cdot H_j \tag{5.57}$$

$$N_A = N_V \cdot \bar{H} = N_V \cdot \sum f_j \cdot H_j \tag{5.58}$$

Introducing this into eq. (5.56) we find

$$\bar{a} = \frac{k_2 \sum N_{Vj} \cdot H_j^3}{N_V \sum f_j H_j} = \frac{k_2 \sum f_j H_j^3}{\sum f_j H_j} \tag{5.59}$$

We now obtain β' by

$$\beta' = \frac{\bar{v}}{\bar{a}^{\frac{3}{2}}} = \frac{k_3[\sum f_j H_j^3] \cdot [\sum f_j \cdot H_j]^{\frac{3}{2}}}{k_2^{\frac{3}{2}}[\sum f_j \cdot H_j^3]^{\frac{3}{2}}}$$

$$= \frac{k_3}{k_2^{\frac{3}{2}}} \cdot \left[\frac{\sum f_j H_j}{(\sum f_j H_j^3)^{\frac{1}{3}}}\right]^{\frac{3}{2}} \tag{5.60}$$

It is easy to show by substituting eqs (5.55) and (5.56) into eq. (5.21) that

$$\beta = \frac{k_3}{k_2^{\frac{3}{2}}} \tag{5.61}$$

The term in brackets [] is hence a coefficient defining the effect of size distribution on this counting principle. It evidently depends on the ratio of two moments of the size distribution:

$$\sum f_j H_j = \bar{H} = H_1 \tag{5.62}$$

is the arithmetic mean of the mean tangent diameter or its first moment, whereas

$$(\sum f_j H_j^3)^{\frac{1}{3}} = H_3 \tag{5.63}$$

is the cube root mean cube of the mean tangent diameter, or its third moment. The final formula applying to a population of particles of equal shape but varying size is

$$N_V = \frac{1}{\beta} \cdot \left[\frac{H_3}{H_1}\right]^{\frac{3}{2}} \cdot \frac{N_a^{\frac{3}{2}}}{V_V^{\frac{1}{2}}} \tag{5.64}$$

or

$$N_V = \frac{k(H)}{\beta} \cdot \frac{N_A^{\frac{3}{2}}}{V_V^{\frac{1}{2}}} \tag{5.65}$$

where

$$k(H) = \left[\frac{H_3}{H_1}\right]^{\frac{3}{2}} = \left[\frac{(\sum f_j H_j^3)^{\frac{1}{3}}}{\sum f_j H_j}\right]^{\frac{3}{2}} \tag{5.66}$$

Neglect of this size distribution coefficient, as was the case in our original communication (Weibel and Gomez, 1962), would lead to an underestimation of N_V, because $H_3 > H_1$. It now remains to estimate the magnitude of $k(H)$ for some typical cases.

Let us first deal with a general type of distribution function for which the moments can be easily derived, such as the Γ-distribution

$$F(H) = \frac{\alpha^{n+1}}{n!} \cdot H^n \cdot e^{-\alpha H} \tag{5.67}$$

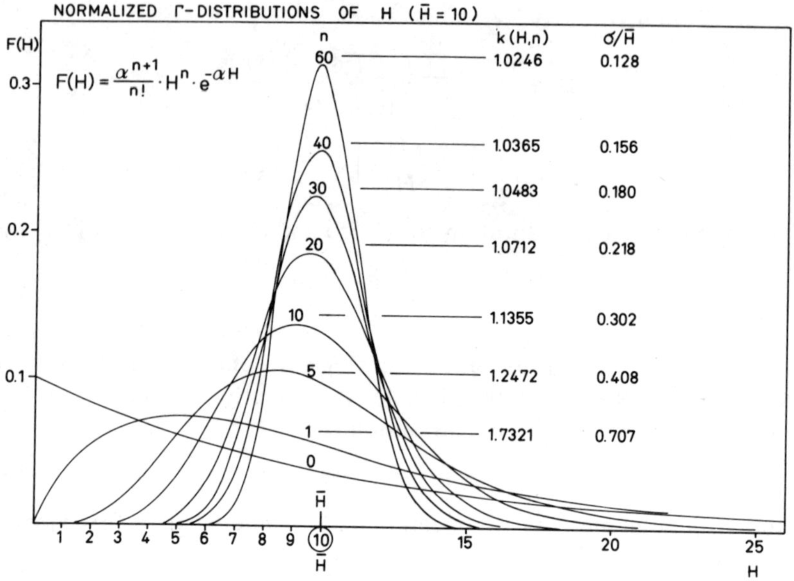

FIG. 5.5: Estimation of size distribution coefficient K for Γ-distributions.

As shown in Fig. 5.5 the shape of the size distribution can be varied by choosing different values of n; for $n = 0$, $F(H)$ is an exponential function, a quite unlikely distribution, but as n becomes larger $F(H)$ approaches the normal distribution. For this set of distributions one finds $k(H)$ by

$$k(H, n) = \left[\left(1 + \frac{1}{n+1}\right) \cdot \left(1 + \frac{2}{n+1}\right)\right]^{\frac{1}{2}} \tag{5.68}$$

It is seen from Fig. 5.5 that $K(H, n)$ decreases rather rapidly as n increases and tends to 1 as $n \to \infty$; for $n = 20$, $k(H) = 1.07$, for example.

Figure 5.6 relates this value of $k(H)$ to the coefficient of variation $\sigma/\bar{H}$, as calculated for the Γ-distributions of Fig. 5.5 It is seen that $k(H)$ reaches the value 1.1 for a coefficient of variation of 26%. This graph can be used to estimate the size distribution effect on counting particles with this method.

5.2.2. METHOD OF DE HOFF (1968) FOR CONSTANT SIZE PARTICLES

If reliable assumptions about shape can be made and if the size distribution of particles is negligible, then N_V and the size of the particles can be estimated by a combination of stereological principles through simple

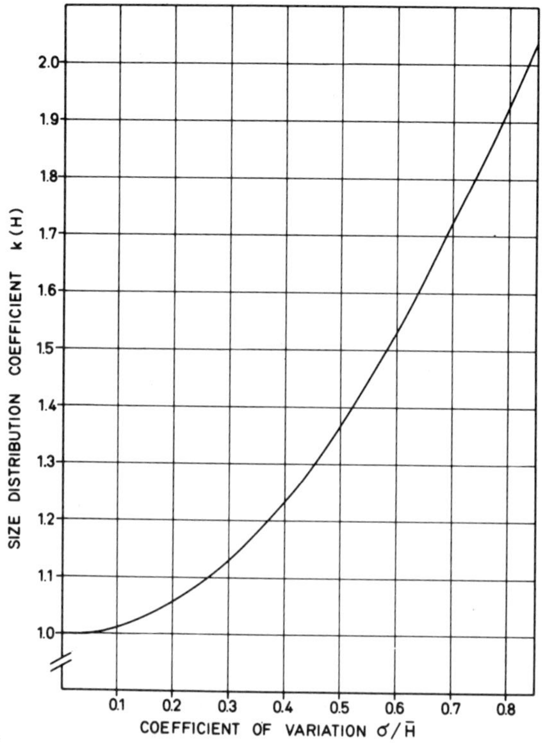

FIG. 5.6: Size distribution coefficient K as function of coefficient of variation (for Γ-distributions).

equations. We shall follow De Hoff's (1968) demonstration. Under these conditions we can assume that the geometric properties of the particles can be described by a linear dimension x and some shape dependent but dimensionless coefficient γ, such that

$$\begin{aligned} \bar{H} &= \gamma_1 \cdot x \\ s &= \gamma_2 \cdot x^2 \\ v &= \gamma_3 \cdot x^3 \end{aligned} \tag{5.69}$$

Table 5.1 gives a list of these coefficients for various shapes; they can be easily verified on the example of spheres, where x corresponds to the radius.

Since the aggregate of particles is, by definition, composed of N_V particles of equal size we can redefine the basic stereological equations as

Table 5.2

Mean tangent diameter ($\bar{H}$), volume (v) and surface area (s) for convex particles of various shapes. After Hilliard (1967) and De Hoff and Rhines (1961)

Particle shape	Characteristic dimension	$\bar{H}$	v	s	$\frac{v}{s}$
Sphere	radius R	$2R$	$\frac{4}{3}\pi R^3$	$4\pi R^2$	$0.333 \cdot R$
Ellipsoid, prolate	axes $2a \geq 2b$ ratio $q = b/a$ excentricity ε	$b\left[\frac{1}{q}+\frac{q}{\sqrt{(1-q^2)}}\ln\left(\frac{1+\sqrt{(1-q^2)}}{q}\right)\right]$	$\frac{4}{3}\pi ab^2$	$2\pi\left(b^2+\frac{a^2}{\varepsilon}\operatorname{arcsinh}\frac{b\varepsilon}{a}\right)$	—
Ellipsoid, oblate	axes $2a \geq 2b$ ratio $q = b/a$ excentricity ε	$a\left[q+\frac{\arcsin\sqrt{(1-q^2)}}{\sqrt{(1-q^2)}}\right]$	$\frac{4}{3}\pi a^2 b$	$\pi\left(2a^2+\frac{b^2}{\varepsilon}\ln\frac{1+\varepsilon}{1-\varepsilon}\right)$	—
Cylinder	l, r	$\frac{1}{2}(l+\pi r)$	$\pi l r^2$	$2\pi r(l+r)$	—
Hemisphere	R	$\left(1+\frac{\pi}{4}\right)R$	$\frac{2}{3}\pi R^3$	$3\pi R^2$	$0.222 \cdot R$

Polyhedra	edge lengths:				
Paralleliped	a, b, c	$\frac{1}{2}(a+b+c)$	abc	$2(ab+bc+ca)$	—
Cube	a	$\frac{3}{2}a$	a^3	$6a^2$	$0.166 \cdot a$
Tetrahedron	a	$a\left(\frac{3}{\pi} \cdot \arccos \sqrt{\frac{1}{3}}\right)$ $= 0.91226 \cdot a$	$a^3 \cdot \frac{\sqrt{2}}{12}$	$a^2 \cdot \sqrt{3}$	$0.068 \cdot a$
Octahedron	a	$a\left(\frac{6}{\pi} \arccos \sqrt{\frac{2}{3}}\right)$ $1.17548 \cdot a$	$a^3 \cdot \frac{\sqrt{2}}{3}$	$a^2 \cdot 2\sqrt{3}$	$0.136 \cdot a$
Dodecahedron	a	$a\left[\frac{15}{\pi} \arccos \sqrt{\left(\frac{1}{2}+\frac{\sqrt{5}}{10}\right)}\right]$ $2.64312 \cdot a$	$a^3 \cdot \frac{15+7\sqrt{5}}{4}$	$a^2 \cdot 3\sqrt{\{5(5+2\sqrt{5})\}}$	$0.371 \cdot a$
Icosahedron	a	$a\left[\frac{15}{\pi} \arccos \sqrt{\left(\frac{2+4/\sqrt{5}}{3+3/\sqrt{5}}\right)}\right]$ $1.7421 \cdot a$	$a^3 \cdot \frac{5(3+\sqrt{5})}{12}$	$a^2 \cdot 5\sqrt{3}$	$0.252 \cdot a$
Tetrakaideca-hedron	a	$3a$	$a^3 \cdot 8\sqrt{2}$	$a^2 \cdot 6(1+2\sqrt{3})$	$0.422 \cdot a$

follows:

$$V_V = N_V \cdot v = N_V \cdot \gamma_3 \cdot x^3 = P_P \tag{5.70}$$

$$S_V = N_V \cdot s = N_V \cdot \gamma_2 \cdot x^2 = 2 \cdot I_L \tag{5.71}$$

$$N_V = N_A/\bar{H} = N_A/(\gamma_1 \cdot x) \tag{5.72}$$

By combining these equations we may obtain for example (De Hoff, 1968):

$$N_V = \frac{N_A^2}{2 \cdot I_L} \cdot \frac{\gamma_2}{\gamma_1^2} \tag{5.73}$$

and

$$x = \frac{2I_L}{N_A} \cdot \frac{\gamma_1}{\gamma_2} \tag{5.74}$$

An alternative formula is obtained by dividing eq. (5.71) by eq. (5.70) and appropriate substitution:

$$N_V = N_A \cdot \frac{2I_L}{P_P} \cdot \frac{\gamma_1 \gamma_2}{\gamma_3} \tag{5.75}$$

Implicit in this relation is the volume-to-surface ratio (v/s) of the particles:

$$(v/s) = P_P/(2I_L) \tag{5.76}$$

It can easily be shown that (v/s) is directly related to the mean diameter by

$$\bar{H} = \frac{\gamma_3}{\gamma_1 \cdot \gamma_2} \cdot (v/s) \tag{5.77}$$

so that eq. (5.75) is equivalent to the basic formula (5.11). Values of (v/s) for some shapes are given in Table 5.2.

Aherne (1967) proposed using the volume-to-surface ratio (v/s) in a somewhat different way to infer N_V indirectly without counting profiles. The formula is

$$N_V = V_V \cdot \left[\frac{k}{(v/s)}\right]^3 \tag{5.78}$$

where $k = v^{\frac{2}{3}}/s$ is a dimensionless coefficient of shape, which is similar in nature to β used above. If the counts necessary to obtain V_V and (v/s) are performed with a short-line test system of the type M21 (Volume 1, Appendix 3) the formula would become

$$N_V = \left[\frac{64 \cdot k^3}{d^3 P_T}\right] \cdot \frac{I_i^3}{P_i^2} \tag{5.79}$$

It is seen that the point and intersection counts have to be raised to the square and third power respectively which would tend to exaggerate errors of measurement.

5.3. Counting in thick slices

5.3.1. BASIC PRINCIPLES

The number of particles is a dimensionless parameter; in very general terms it is related to the topological characteristic of Euler–Poincaré (see Section 3.5) which is related to the connectivity of the solids: for discrete convex solids the Eulerian characteristic is 1, so that the numerical density N_V is equal to the density of Eulerian characteristic of the structure (Giger, 1970).

These introductory remarks were to show that when we count objects we consider a dimensionless parameter related merely to their topological properties. In eq. (2.12) we had found that a trace is only formed upon sectioning if the sum of the dimension of object parameter and the dimension of the section is ≥ 3. The Eulerian characteristic having dimension 0 a section of dimension 3 is required to obtain a trace. Strictly speaking the density of Eulerian characteristic or of number can therefore only be estimated in thick slices.

This would be possible if each particle were marked by one characteristic point which is detectable in the slice, say its centre of gravity or one and only one point on the surface (Fig. 5.7). Then it would follow that a count of these points, N, on a test area A_T at the surface of the slice of

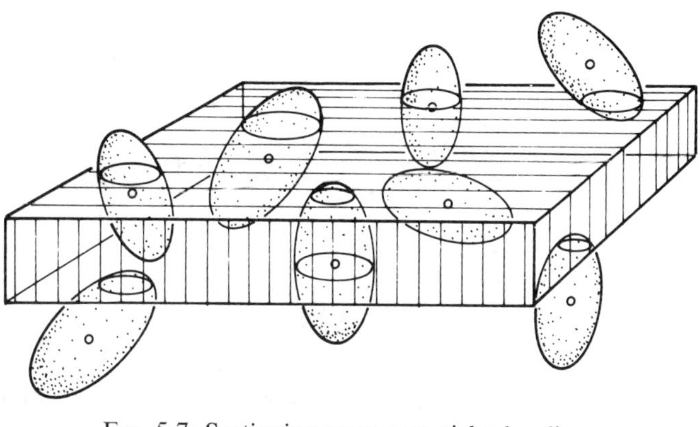

FIG. 5.7: Sectioning convex particles by slice.

thickness t is related to the numerical density by

$$N_V = \frac{N}{A_T \cdot t} = \frac{N_A}{t} \tag{5.80}$$

This is, however, not a situation which will occur in practice: particles are solids (Y_3) and the slice (T_3) will contain profiles which are themselves "slices" of the particles because

$$Y_3 \cap T_3 = Z_3 \tag{5.81}$$

Looking at the slice surface we observe a projection of these profiles and are usually not able to determine whether they contain the centre of gravity or not. The chance of hitting a particle by the slice depends not only on N_V but also on the mean tangent diameter $\bar{H}$ of the particle.

In Fig. 5.8 three particles which produce a profile are shown: one of them has its centre within the slice, but the centres of the others are above or below the slice surfaces respectively. Profiles are evidently formed by particles whose centre points are contained in a "super-slice" of thickness $(t+2x)$ and volume

$$V' = (t+2x)A_T \tag{5.82}$$

and we have to determine x which must depend on the size of the particle. It is easy to see that x is determined by the probability that the surface plane of the slice intersects the particle, and this probability is related to the mean tangent diameter $\bar{H}$ of the particle (Fig. 5.9), as discussed in Section 2.3. If the upper surface plane intersects the upper half of the particle the centre point is within the slice, if the lower half is

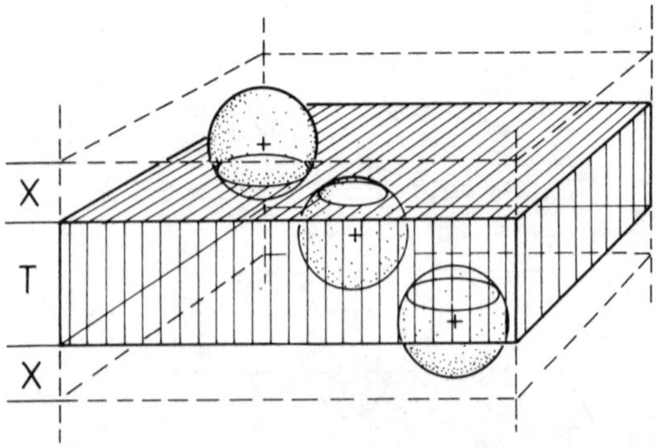

FIG. 5.8: Spheres hit by slice have their centre in "super-slice" of thickness $(T+2X)$.

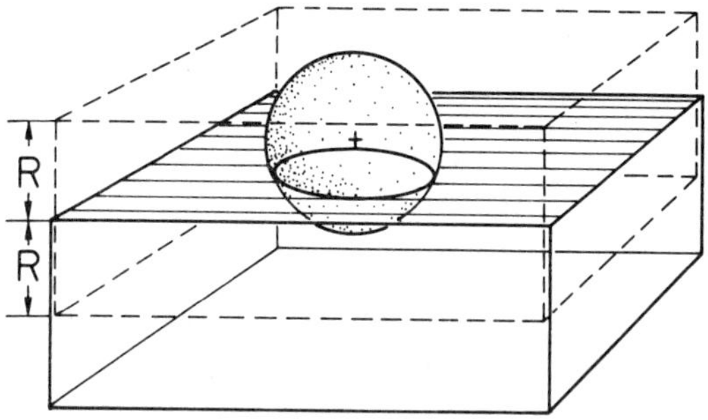

FIG. 5.9: Range in which centre may lie for sphere to be cut.

intersected it will lie outside; therefore

$$x = \tfrac{1}{2}\bar{H} \tag{5.83}$$

and the volume of the "super-slice" is

$$V' = (t + \bar{H})A_T \tag{5.84}$$

From this it follows that the number N of profiles in the slice, as counted on a slice area A_T, is related to N_V by

$$N_V = \frac{N}{A_T(t + \bar{H})} = \frac{N_A}{t + \bar{H}} \tag{5.85}$$

This relationship was first derived by Abercrombie (1946). It is evident that eq. (5.85) reduces to eq. (5.11) if $t \to 0$.

This principle assumes that a profile becomes recognizable as soon as the particle begins to penetrate into the slice. In practice this will rarely be the case: very small cap sections are very thin as compared to the section thickness (Fig. 5.10). Unless the particle is absolutely opaque and the matrix absolutely transparent such cap sections will not be detectable until they have reached a certain height h_0. This means that the detectable profiles have originated from particles whose centre point lies in a reduced "super-slice" of volume

$$V'' = (t + \bar{H} - 2h_0)A_T \tag{5.86}$$

so that, accounting for a loss of caps whose height is $< h_0$, eq. (5.85) becomes

$$N_V = N_A/(t + \bar{H} - 2h_0) \tag{5.87}$$

as first recognized by Floderus (1944).

The minimal detectable cap height h_0 is not easy to determine; it is however related to the radius r_0 of the smallest recognizable profile. For

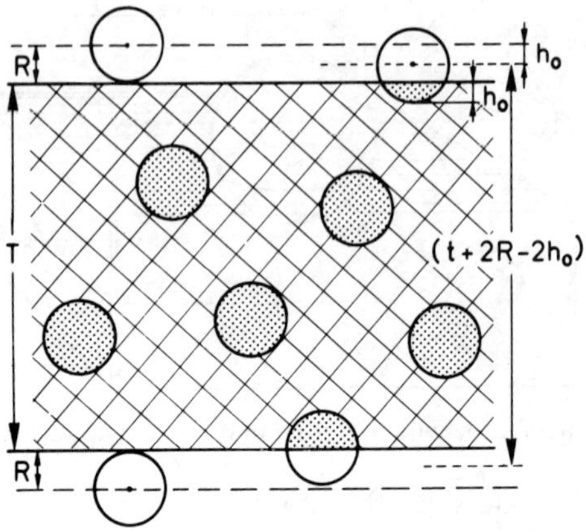

FIG. 5.10: Loss of caps reduces size of "super-slice".

all practical purposes we can use a spherical model to establish the relationship between h_0 and r_0, as long as r_0 is considerably smaller than half the mean tangent diameter of the particle. On the basis of a spherical model we find that (Fig. 5.11)

$$h_0 = R - \sqrt{(R^2 - r_0^2)} \tag{5.88}$$

where $R \sim \bar{H}/2$. This rough estimation of h_0 is valid only if the particle size does not vary too much. In the case of an appreciable size distribution h_0 must be determined for each size class and weighted. Haug (1967) has discussed the errors introduced by neglecting to consider size distribution.

5.3.2. METHOD OF AHERNE (1967) FOR COUNTING IN THICK SLICES

Aherne (1967) proposed a slight modification to eq. (5.11), replacing the mean particle diameter by an estimate of particle perimeter. The final

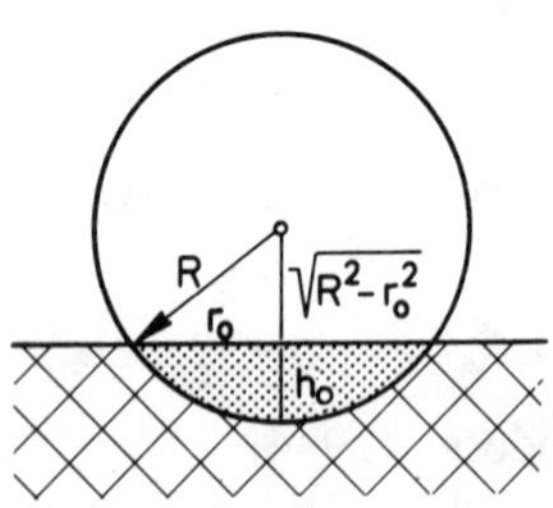

FIG. 5.11: Relation of cap height h_0 to profile radius r_0.

formula is

$$N_V = N_A \cdot \frac{2}{J \cdot d + 2t} \tag{5.89}$$

where the term $(J \cdot d)$ needs to be explained; it is related to an experimental estimate of the maximum equatorial perimeter of the particles, b_e. Since b_e is a linear trace its length could be estimated by applying a grid of parallel lines, spaced at distance d, to the equatorial plane of the particle by counting intersections between the perimeter and the lines; these intersections are called J which is hence the mean number of intersections per maximum equatorial perimeter. The following relation holds

$$b_e = \frac{\pi}{2} \cdot J \cdot d \tag{5.90}$$

We therefore have

$$J \cdot d = \frac{2}{\pi}\, b_e \tag{5.91}$$

which is proportional to the mean diameter of the particle in the maximal equatorial plane. This is a somewhat unusual measurement, but it is convenient because of the following condition. Assume a particle is intersected by a section of thickness t; according to Plummer (1940) the chance that a point within the object (Y) shall also be within the section (T_3) is

$$\Pr\{Y \cap T_3\} = \frac{\pi t}{b_e + \pi t} = \bar{\phi} \tag{5.92}$$

This also corresponds to the mean fraction $\bar{\phi}$ of the particle volume contained in the section. From here eq. (5.89) is readily developed. The volume density within the slice is

$$V_V = \frac{N_p \cdot \bar{\phi} \cdot v}{A \cdot t} \tag{5.93}$$

where N_p is profile number, v particle volume and A section area. Also,

$$N_V = V_V/v = N_A \cdot \bar{\phi}/t \tag{5.94}$$

Substituting using eqs (5.91) to (5.93) we obtain eq. (5.89).

The main problem with this method is the estimation of J which is, in fact, no easier than the estimation of mean particle diameter. Aherne

(1967) argues that in the case of spheres, b_e is proportional to $\bar{b}$, the mean perimeter of profiles, such that

$$b_e = \frac{4}{\pi} \cdot \bar{b} \tag{5.95}$$

It will be shown below that the same proportionality—evidently—exists between the sphere radius R and mean profile radius $\bar{r}$

$$R = \frac{4}{\pi} \cdot \bar{r} \tag{5.96}$$

one of the fundamental relationships. However, this is only true for infinitely thin sections. The proposition of Aherne is to apply a test line grid of spacing d to the section, to count the intersections I of this grid with the profile boundaries and to determine

$$\bar{I}_p = I/N_p \tag{5.97}$$

the mean intersection number per profile; this can be used either directly as an approximation of J, or by multiplying it by a factor between 1 and $1.27 = 4/\pi$. The reason why $\bar{I}_p$ may be a reasonable approximation of J is that in a thick slice many of the sphere profiles will indeed display the maximum perimeter.

Freedman (1974) has critically reviewed the conditions under which this formula produces acceptable estimations of N_V. He arrives at the conclusion that contrast deficiency with the ensuing loss of cap sections, as well as size variations, may influence the result in such a way as to yield biased results. However, within a well-defined range the method yields good results.

Mayhew (1972) has subjected several of the counting methods presented here to an experimental test and has found them to yield results in good agreement.

5.3.3. THE TWO SLICES METHOD OF EBBESON AND TANG (1965)

An ingenious method for counting in thick slices has been elaborated by Ebbeson and Tang (1965). The essence of it is that particle profiles are counted on two sections of different thickness, $t_1 > t_2$, yielding numerical profile densities N_{A1} and N_{A2} respectively. The numerical particle density is then obtained by

$$N_V = (N_{A1} - N_{A2})/(t_1 - t_2) \tag{5.98}$$

The derivation is straight-forward. From eq. (5.85) we write

$$N_{A1} = N_V(t_1 + \bar{H}) \tag{5.99}$$

$$N_{A2} = N_V(t_2 + \bar{H}) \tag{5.100}$$

Since N_V and $\bar{H}$ are properties of the material they must be equal for both sections, and hence

$$N_{A1} - N_{A2} = N_V(t_1 - t_2) \tag{5.101}$$

so that the mean particle diameter need not be known.

The same result is actually obtained if we allow for a loss of cap sections. By eq. (5.87) we write

$$N_{A1} = N_V(t_1 + \bar{H} - 2h_0) \tag{5.102}$$

$$N_{A2} = N_V(t_2 + \bar{H} - 2h_0) \tag{5.103}$$

If one can assume that the height of the smallest visible cap, h_0, is unaffected by the section thickness one again obtains by subtraction eq. (5.101), and consequently eq. (5.98). This is perhaps a critical point in the method, for it is to be expected that the contrast conditions deteriorate as t increases so that certain caps that are recognized in the thin section are missed in the thicker section. The consequence is that the difference between t_1 and t_2 should be kept as small as possible. Ebbeson and Tang (1965) did some counting experiments in the light microscope using $t_1 = 15\ \mu$m and $t_2 = 10\ \mu$m and obtained results that compared very well with the method of Floderus (1944) and with some other methods.

5.4. Counting along test lines

The preceding methods were based on an estimation of the number of profiles contained within a test area. Hilliard (1967) has pointed out that N_V can also be obtained by an analysis of intercepts formed by test lines with the profiles. The details of this principle will be discussed below in connection with the estimation of the moments of size distribution (Chapter 7); evidently the number of particles is related to the 0-moment of their distribution.

The method works as follows (Fig. 5.12). A certain number of parallel test lines are deposited on the section; the lines will form intercepts of varying length with the profiles. To determine the number, it suffices to fix a very small unit, ΔL, and to count the intercepts which fall into that unit; call this number ΔN. For a general convex body Hilliard (1967) finds

$$N_V = 2 \cdot \Delta N / \{\alpha_0 \cdot [df(\lambda)/d\lambda]_{\lambda=0} \cdot L_T(\Delta L)^2\} \tag{5.104}$$

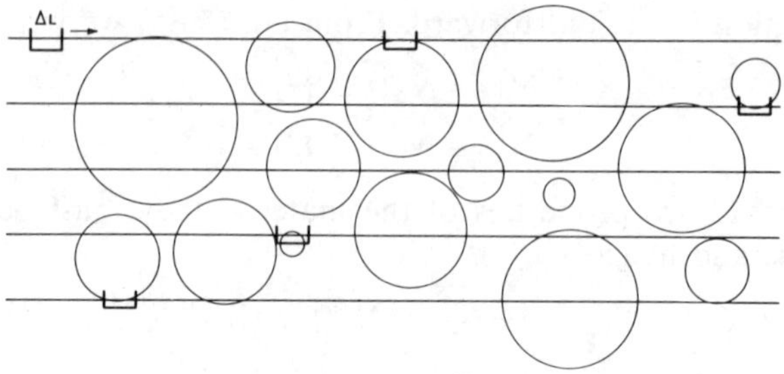

FIG. 5.12: Counting minimal chords (<ΔL) along test lines.

The coefficient α_0 is a shape factor related to the average projected area of the particles; with λ as relative intercept length ($\lambda = L/D$ where D is the largest particle diameter) and $f(\lambda)$ as the frequency distribution of λ, the term $[df(\lambda)/d\lambda]_{\lambda=0}$ is the slope of the frequency distribution near the origin which is again dependent on shape. These shape factors are difficult to determine (Hilliard, 1968) except for very few specific shapes. For spherical particles $\alpha_0 = \pi/4$ and $[df(\lambda)/d\lambda]_{\lambda=0} = 2$, so that eq. (5.104) becomes

$$N_V(\text{spheres}) = \frac{4}{\pi} \cdot \Delta N/(L_T \cdot (\Delta L)^2) \tag{5.105}$$

This relation is implicit in the method of Spektor (1950) for estimating the moments of particle size distribution from an analysis of chord or intercept lengths.

The method of Hilliard (1968) is the more accurate the smaller ΔL is chosen. In the limit this amounts indeed to a counting of tangent points (see Section 3.4.4); in the case of convex particles each profile has two tangents, and it is evident that an estimate of the number of tangents per unit area of section, or per unit length of test line, is proportional to particle number.

This method is particularly useful with automatic scanning instruments. In manual analysis, using only few test lines, a great number of particles will not be accounted for, mainly if ΔL is chosen to be very small. The major problem is, however, that it is only applicable to spherical particles, because the factor $[df(\lambda)/d\lambda]_{\lambda=0}$ is, at least so far, indeterminate for arbitrary or variable shapes.

5.5. Shape and mean tangent diameter of convex solids

In geometric probability considerations, as well as in estimating the number of particles, the notion of mean tangent diameter is pre-eminent.

It is hence appropriate to discuss this geometric measure for a number of typical solids.

It has already been said that the mean tangent diameter of spheres is equal to their diameter. With all solids which can be described by several axes, the relationship will not be as simple. Here, we will treat the cases of solids of revolution, cylinders and ellipsoids, and of polyhedra.

5.5.1. SOLIDS OF REVOLUTION: ELLIPSOIDS AND CYLINDERS

If a solid is generated by revolving a curve about an axis of symmetry the calculation of $\bar{H}$ is greatly simplified for the following reason.

Assume a prolate ellipsoid to be generated by revolving an ellipse about the z-axis of the x, y, z coordinate system (Fig. 5.13a). In this position the tangent diameter, $H(0, 0)$ corresponds to the diameter of the major axis, $2a$, of the ellipse or of the ellipsoid. The orientation of the ellipsoid can be changed in two ways, according to our model.

(1) The axis of rotational symmetry can be inclined to the z-axis while remaining in the y–z plane thus forming an angle θ to the z-axis; this will tend to decrease $H(\theta)$ until it becomes equal to the diameter of the minor axis, $2b$, when $\theta = \pi/2$, i.e. when the axis of rotation is parallel to the y-axis.

(2) The inclined axis can be revolved about the z-axis moving away from the $y - z$ plane by the angle ϕ (Fig. 5.13b). This does not, however, change the tangent diameter $H(\theta, 0)$ imposed by inclining the ellipsoid by the angle θ. This is easily verified: varying the angle θ moves

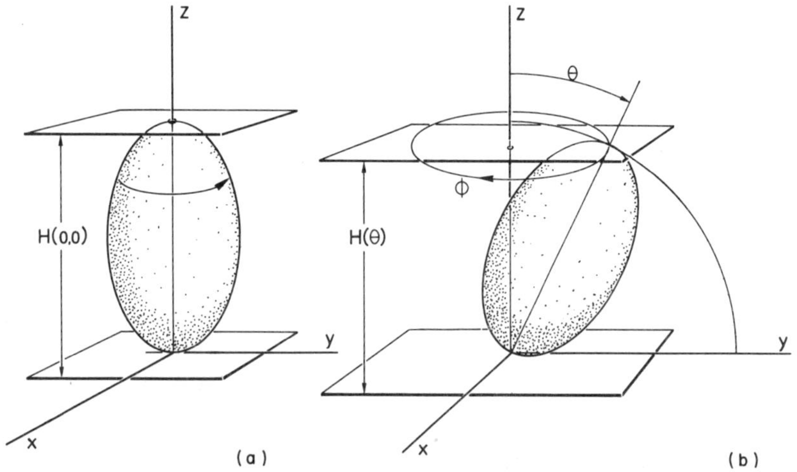

FIG. 5.13: Mean caliper diameter of ellipsoids.

the tangent points on horizontal circles within the two tangent planes and thus does not displace them. $H(\theta, \phi)$ is hence independent of ϕ and is exclusively defined by θ.

The original general definition of $\bar{H}$ may hence be simplified by integrating out ϕ, as proposed by De Hoff and Rhines (1961):

$$\bar{H} = \int_0^{\pi} \int_0^{2\pi} H(\theta) \frac{\sin \theta \, d\phi \, d\theta}{4\pi} = \frac{1}{2} \int_0^{\pi} H(\theta) \sin \theta \, d\theta \qquad (5.106)$$

or

$$\bar{H} = \int_0^{\frac{1}{2}\pi} H(\theta) \sin \theta \, d\theta \qquad (5.107)$$

since it suffices to integrate over one octant of the sphere because of the symmetry of the body. This formula can be used for calculating $\bar{H}$ for prolate or oblate ellipsoids, as well as for cylinders, generated by revolving a rectangle. This is particularly easy because the formulation of $H(\theta)$ can be obtained from the distance between parallel lines tangent to the two-dimensional curve from which the solid was generated by revolution.

5.5.1.1. *Right circular cylinders*

Let us perform this for a cylinder of length l and radius r, generated by revolving a rectangle of area $l \cdot 2r$ about the z-axis (Fig. 5.14). Now allow the cylinder to be gradually inclined, measuring the angle of inclination θ between z-axis and cylinder axis. At any angle θ the distance between the tangent planes $H(\theta)$ can be broken into three parts: a length z_1 which is

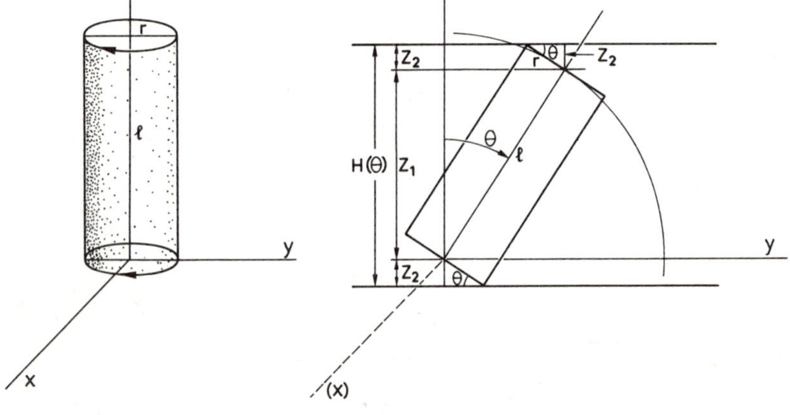

FIG. 5.14: Mean caliper diameter of cylinders.

the projection of the cylinder axis onto the z-axis, and two identical lengths z_2 which correspond to the projection of the cylinder radius onto the z-axis. It is evident that

$$\begin{aligned} z_1 &= l \cdot \cos\theta \\ z_2 &= r \cdot \sin\theta \end{aligned} \tag{5.108}$$

so that

$$H(\theta) = l\cos\theta + 2r\sin\theta \tag{5.109}$$

This can be substituted into eq. (5.107) in order to derive the mean tangent diameter by integration over θ:

$$\begin{aligned} \bar{H} &= \int_0^{\frac{1}{2}\pi} (l\cos\theta)\cdot\sin\theta\, d\theta + \int_0^{\frac{1}{2}\pi} (2r\sin\theta)\sin\theta\, d\theta \\ &= l\int_0^{\frac{1}{2}\pi} \cos\theta\cdot\sin\theta\, d\theta + 2r\int_0^{\frac{1}{2}\pi} \sin^2\theta\, d\theta \\ &= l[\tfrac{1}{2}\sin^2\theta]_0^{\frac{1}{2}\pi} + 2r[\tfrac{1}{2}\theta - \tfrac{1}{4}\sin 2\theta]_0^{\frac{1}{2}\pi} \end{aligned}$$

$$\bar{H} = \tfrac{1}{2}l + \tfrac{1}{2}\pi r \tag{5.110}$$

The mean tangent diameter of cylinders is hence easily calculated from length and radius. It should be noted that the two measures are additive; one can therefore easily show that the average mean tangent diameter, $\bar{H}$, can be calculated from the average length, l, and the average radius, $\bar{r}$, of the cylinder population:

$$\bar{H} = \tfrac{1}{2}\bar{l} + \tfrac{1}{2}\pi\bar{r} \tag{5.111}$$

The two measures may thus vary independently.

5.5.1.2. *Ellipsoids of revolution*

The procedure for calculating $\bar{H}$ for ellipsoids is the same as for cylinders. An ellipse, whose major hemi-axis is a and minor hemi-axis b, is placed in the y–z plane of the coordinate system such that the centre of the ellipse is at the origin and one of the axes coincides with the z-axis; this ellipse is then revolved about z (Fig. 5.3). If the major axis a coincides with z and forms the axis of revolution a *prolate ellipsoid* is produced; if the minor axis coincides with z an *oblate ellipsoid* results. The formulas for $\bar{H}$ are different for the two types of ellipsoids of revolution.

Prolate ellipsoids

It has been shown (Salmon, 1954) that for prolate ellipsoids the distance between parallel tangent planes is

$$H_p(\theta) = 2(a^2 \cos^2 \theta + b^2 \sin^2 \theta)^{\frac{1}{2}} \tag{5.112}$$

If this is introduced into eq. (5.107) the mean tangent diameter of prolate ellipsoids is defined as

$$\bar{H}_p = \int_0^{\frac{1}{2}\pi} 2(a^2 \cdot \cos^2 \theta + b^2 \cdot \sin^2 \theta)^{\frac{1}{2}} \cdot \sin \theta \, d\theta \tag{5.113}$$

If this integration is performed—we shall spare the reader the lengthy procedure (it is easily verified)—one obtains (De Hoff and Rhines, 1961):‡

$$\bar{H}_p = b\left[\frac{a}{b} + \frac{b/a}{[1-(b/a)^2]^{\frac{1}{2}}} \cdot \ln\left(\frac{1+[1-(b/a)^2]^{\frac{1}{2}}}{b/a}\right)\right] \tag{5.114}$$

It is noted that $\bar{H}_p$ depends on the "thickness" of the prolate ellipsoid, measured by the minor hemi-axis b, and on the axial ratio (b/a), the only measure defining the term in the brackets. If we call the axial ratio of the generating ellipse

$$b/a = q \tag{5.115}$$

we may rewrite eq. (5.114), according to De Hoff and Rhines (1961), as

$$\bar{H}_{pj} = 2b_j \cdot k_p(q) \tag{5.116}$$

where the index j is introduced to mean ellipsoids of a size class j. The coefficient $k_p(q)$ corresponds to the term in brackets reduced by $\frac{1}{2}$. De Hoff and Rhines (1961) have calculated these coefficients and have produced the graph shown in Fig. 5.15 which allows $k_p(q)$ to be read off without going through the calculations.

It now remains to be shown how one can estimate the axial ratio q; we follow again the study by De Hoff and Rhines (1961), where further details about justifications and development can be found. First, one can arrive at a reasonable estimate of q by searching on the sections for those profiles whose axial ratios q' show the greatest deviation from unity; such profiles are sectioned parallel to the long axis, the axis of revolution, and their axial ratio will be equal to q. For many purposes this will suffice.

‡ Note that DeHoff and Rhines (1961) designated the *diameters* of the ellipse with b and a, whereas we use these symbols for the hemi-axis; hence the differences of a factor of 2 in some equations.

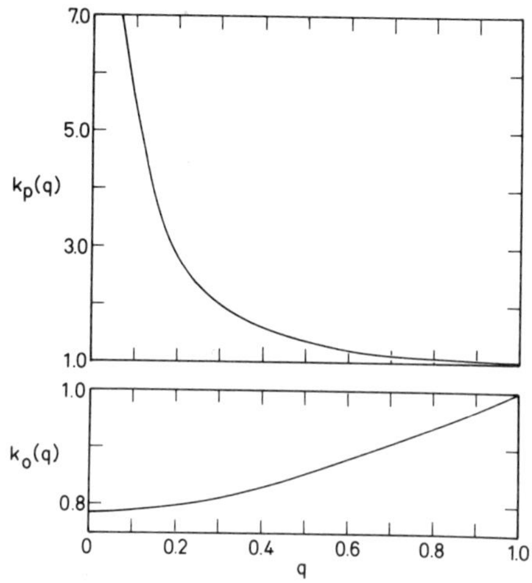

FIG. 5.15: Shape dependent coefficients k_p and k_o for prolate and oblate ellipsoids respectively. (Reproduced with permission from De Hoff and Rhines, 1961.)

For more precision one may use the following procedure. Let F_p be defined as the ratio of the minor axis of profiles, b', to their major axis, a':

$$F_p = b'/a' \tag{5.117}$$

It is then found that the average of these ratios (formed for each profile!) is

$$\bar{F}_p = q \cdot k_p(q) \tag{5.118}$$

and depends exclusively on the shape but not on the size of the ellipsoid. The relation between $\bar{F}_p$ and q is shown in Fig. 5.16 which can be used to read off q.

The next question is: how can b be estimated from the profiles? De Hoff and Rhines introduced the measurable quantity

$$Z_p = 1/(2b') \tag{5.119}$$

as the reciprocal value of the minor axes of elliptical profiles, $2b'$, and showed that its mean is

$$\bar{Z}_p = \pi/(4b) \tag{5.120}$$

The size of the ellipsoids is hence estimated from the harmonic mean of the size of the profiles. In practice this is problematic in as much as the harmonic mean is greatly influenced by the very small profiles which often remain undetected. It is therefore advisable to keep in mind that the values of b' measured should cover the entire range from b to 0; if the distribution of b' is truncated in the lower part due to inadequate recognition of small profiles it should be completed, as discussed in Volume 1, Section 5.2.3.

In principle one could now estimate b from eq. (5.120) and $k_p(q)$ from the graphs of Fig. 5.15 to calculate $\bar{H}_p$ from eq. (5.114). This would then allow one to estimate the number of ellipsoids from

$$N_{Vp} = N_A/\bar{H}_p \tag{5.121}$$

De Hoff and Rhines (1961) also give the following relationships which apply for a population of particles of constant shape (q) but variable size (b_j):

$$\bar{Z}_p = \frac{\sum_j N_{Vj} b_j \cdot k_p(q) \cdot \pi/(2b_j)}{\sum_j N_{Aj}} = \frac{\pi \cdot k_p(q)}{2} \cdot \frac{N_V}{N_A} \tag{5.122}$$

$$N_V = (2N_A \cdot \bar{Z}_p)/(\pi \cdot k_p(q)) \tag{5.123}$$

They also show that the mean particle size can be estimated from

$$\bar{b} = \pi/(4 \cdot \bar{Z}_p) \tag{5.124}$$

$$\bar{a} = \bar{b}/q = \pi/(4q \cdot \bar{Z}_p) \tag{5.125}$$

whereby q is determined from $\bar{F}_p$ (Fig. 5.16).

Oblate ellipsoids

These are produced by revolving an ellipse whose *minor* hemi-axis, b, coincides with the z-axis (Fig. 5.3). It again suffices to consider the variation in the distance of parallel tangents to the generating ellipse while its minor axis is inclined in the y–z plane. The mean tangent diameter is then found to be (De Hoff and Rhines, 1961):

$$\bar{H}_o = a\left[\frac{b}{a} + \frac{\sin^{-1}[1-(b/a)^2]^{\frac{1}{2}}}{[1-(b/a)^2]^{\frac{1}{2}}}\right] \tag{5.126}$$

or

$$\bar{H}_{oj} = 2a_j \cdot k_o(q) \tag{5.127}$$

The shape factor $q = b/a$ is again taken as the ratio of minor to major axis of the generating ellipse. Values of $k_o(q)$ are given in the lower graph of Fig. 5.15.

One may again estimate q from the profiles whose q' deviates most from unity, or use the factor

$$\bar{F}_o = b'/a' = q(k_o(q) - q) \tag{5.128}$$

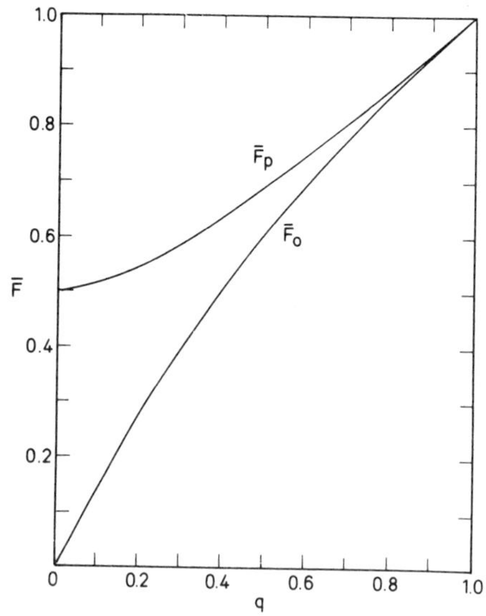

FIG. 5.16: Shape dependent coefficients $\bar{F}_p$ and $\bar{F}_o$ for prolate and oblate ellipsoids respectively. (Reproduced by permission from De Hoff and Rhines, 1961.)

The size measure $\bar{Z}_o$ is

$$\bar{Z}_o = 1/b' = \pi q \cdot k_o(q)/(4a) \tag{5.129}$$

which hence allows estimation of the *major* diameter of the ellipsoids—corresponding to the diameter of the circular equator generated by revolution—from the harmonic mean of the *minor* axes of the elliptic profiles; the same caution applies as above. From these measures one finds

$$N_{Vo} = \frac{2N_A \cdot \bar{Z}_o}{q \cdot k_o^2(q)} \tag{5.130}$$

$$\bar{a}_o = \frac{q^2 \cdot k_o(q)}{4\bar{Z}_o} \tag{5.131}$$

$$\bar{b}_o = \frac{q^2 k_o(q)}{4\bar{Z}_o} \tag{5.132}$$

This development has shown that different formulae apply to prolate and to oblate ellipsoids, but both particles produce ellipses on a section. How can one decide with which type of particles one is dealing? This can be resolved if the size of the *largest circular profiles* are compared with the *largest elliptical profiles* having the smallest axial ratio q'. With prolate ellipsoids we would expect the diameter of the largest circular profile to correspond to the minor axis of the largest ellipse; with oblate ellipsoids, on the other hand, it should be of the size of the major axis of the ellipse.

For the sake of completeness we should the mention the relationships between ellipsoidal size and shape and particle number proposed by Wicksell in 1926, since this was probably the first treatment of this problem. On page 166 of his mémoir he introduces the following relations (whereby we changed his notation to that used here):

$$N_V = N_A/\bar{H} \tag{5.133}$$

For prolate ellipsoids he finds

$$\bar{H}_p = 2b \cdot k'_p \tag{5.134}$$

with

$$k'_p = \frac{1}{2}\left[\frac{1}{(1-\varepsilon^2)^{\frac{1}{2}}} + \frac{(1-\varepsilon^2)^{\frac{1}{2}}}{2\varepsilon}\ln\frac{1+\varepsilon}{1-\varepsilon}\right] \tag{5.135}$$

where ε is the excentricity of the ellipse. For oblate ellipsoids:

$$\bar{H}_o = a \cdot k'_o \tag{5.136}$$

$$k'_o = \frac{1}{2}\left[(1-\varepsilon^2)^{\frac{1}{2}} + \frac{1}{\varepsilon}\sin^{-1}\varepsilon\right] \tag{5.137}$$

The formal similarity between these relations and those presented above is evident at first glance. In fact, if the excentricity

$$\varepsilon = (1-(b/a)^2)^{\frac{1}{2}} = (1-q^2)^{\frac{1}{2}} \tag{5.138}$$

is introduced it can be shown that Wicksell's coefficients are identical to those independently derived by De Hoff and Rhines thirty-five years later (eqs 5.114 and 5.126).

5.5.2. MEAN TANGENT DIAMETER OF THE GENERAL CONVEX SOLID AND OF CONVEX POLYHEDRA

In his fundamental work on convex solids Minkowski (1903) has demonstrated that the mean tangent diameter of any convex body is given by

$$\bar{H} = \frac{1}{4\pi}\int\left(\frac{1}{\rho_1} + \frac{1}{\rho_2}\right) ds \tag{5.139}$$

where ρ_1 and ρ_2 are the principal radii of curvature at every point of the surface, and integration has to be performed over the entire surface area of the body. A proof of this relationship is given by Kendall and Moran (1963). This formula would allow $\bar{H}$ to be calculated for any convex particle provided the integral of mean curvature can be estimated or measured. This is discussed in another chapter (Section 3.4) in connection with De Hoff's method of measuring curvature by a stereological method.

Minkowski's formula has one limitation, namely it excludes the existence of singularities at the surface. Singularities are defined as surface points through which more than one tangent plane can be fitted to the body; this is the case with edges or corners of polyhedra, so that Minkowski's formula is strictly not applicable to this wide class of solids. Hilliard (1967) has, however, proposed a small modification in defining the nature of edges and corners. Take an edge of a polyhedron to be l_i long and to have a dihedral angle $\alpha_i < \pi$ between the faces meeting at this edge (Fig. 5.17). We will now replace this sharp edge by a smoothly curved surface by fitting a cylinder of very small radius r into the angle of the edge. The cylindrical surface is parallel to the edge and hence has a length of l_i; it meets the flat faces of the polyhedron tangentially and it is seen from Fig. 5.17 that the angle subtended by the cylindrical surface at the edge is $(\pi - \alpha_i)$. The area of the cylindrical edge surface is thus $(\pi - \alpha_i)r \cdot l_i$. Now, the radii of curvature of the cylindrical surface are constant:

$$\rho_1 = r, \qquad \rho_2 = \infty \tag{5.140}$$

It follows that the contribution of this cylindrical edge to the integral for $\bar{H}$ is

$$\frac{1}{4\pi} \cdot \frac{(\pi - \alpha_i)r \cdot l_i}{r} = \frac{1}{4\pi} \cdot (\pi - \alpha_i)l_i \tag{5.141}$$

which is independent of the value of r chosen. Allowing for all edges of a

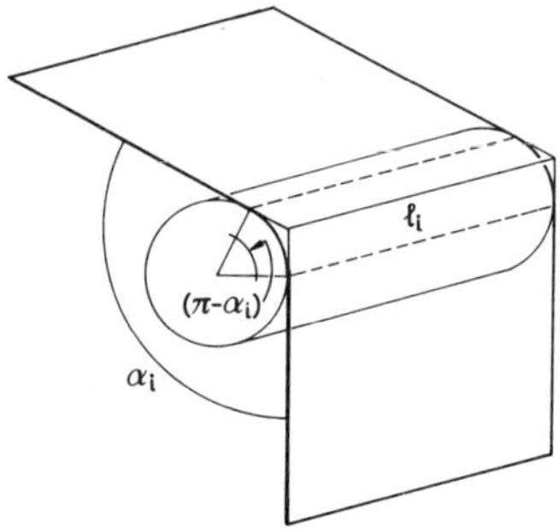

FIG. 5.17: Approximation of polyhedral edge by small cylinder.

polyhedron, eq. (5.139) becomes

$$\bar{H}=\frac{1}{4\pi}\left\{\int\left(\frac{1}{\rho_1}+\frac{1}{\rho_2}\right)ds+\sum_i(\pi-\alpha_i)l_i\right\} \tag{5.142}$$

in which the integration is over the smooth surfaces and summation over the edges. But it is evident that smooth surfaces of a polyhedron are, by definition, flat, so that

$$\rho_1=\rho_2=\infty \tag{5.143}$$

$$\frac{1}{\rho_1}+\frac{1}{\rho_2}=0 \tag{5.144}$$

Hence the integral in eq. (5.142) becomes zero and we are left with

$$\bar{H}=\frac{1}{4\pi}\sum_i(\pi-\alpha_i)l_i \tag{5.145}$$

as the mean tangent diameter of convex polyhedra; this result was first obtained by Santaló (1936).

For regular polyhedra, which are characterized by being bounded by identical faces and by all edges having the same length, the mean tangent diameter is (Hilliard, 1967)

$$\bar{H}_r=a\cdot\frac{E}{2\pi}\cos^{-1}\left[\frac{\cos(\pi/f)}{\sin(\pi/n)}\right] \tag{5.146}$$

where a is the edge length, E the total number of edges, f the number of faces meeting at each vertex, and n the number of sides to each face.

By using these equations Hilliard (1967) has worked out the formulas for computing the mean tangent diameter of some typical convex bodies. They are compiled in Table 5.2. For many practical purposes it may often be adequate to approximate the actual particles investigated by one of these shapes—e.g. prism, cube or icosahedron for epithelial cells—and to estimate $\bar{H}$ from the corresponding formula and some independent information on their shape; the latter could, for instance, be the volume to surface ratio from which the edge length can be derived.

CHAPTER 6

Size distribution of section profiles when spheres are randomly sectioned

In this and the following chapters we shall discuss the relationship between the size distribution of profiles on sections and the size or the size distribution of particles in the specimen. To introduce the topic we shall first examine how the profile sizes are distributed when a single sphere of radius R is randomly sectioned many times.

6.1. Sectioning a single sphere

Consider the following model situation (Fig. 6.1). A sphere of radius R is placed in an x, y, z coordinate system with its centre at the origin. For the sake of clarity we can consider it embedded in a cube of matrix material. We shall now place a large number of equidistant horizontal section planes (parallel to the x–y plane) into this block of material. It is evident that only sections which lie between the upper and lower parallel tangent planes will cut the sphere; their number will hence be proportional to the mean tangent diameter, which in the case of a sphere, is equal to $2R$. Only these sections need now be considered.

Each section plane will produce a circular profile whose radius r will depend on the position of the section plane with respect to the z-axis:

$$r(z) = \sqrt{(R^2 - z^2)} \tag{6.1}$$

because the fundamental equation of the sphere's main circle in the y–z

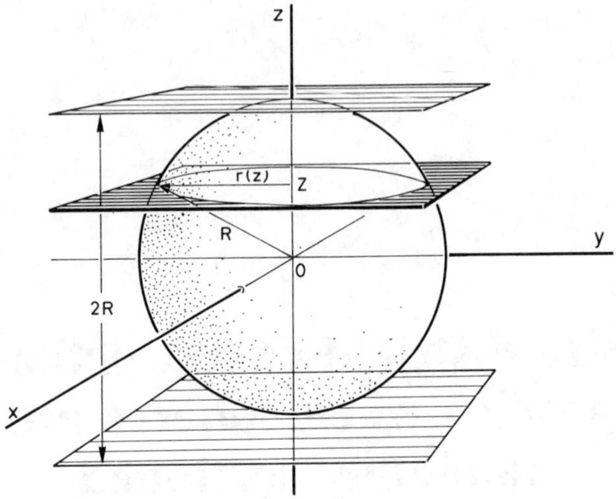

FIG. 6.1: Sectioning sphere of radius R yields profile of radius $r(z)$ depending on distance z of section from sphere centre.

plane is

$$y^2 + z^2 = R^2 \tag{6.2}$$

and evidently

$$r = y \tag{6.3}$$

Note that we can ignore the x-axis, and restrict this analysis to that of the sphere's main circle in the y–z plane: we can assume the sphere to have been generated by revolving this "master" circle about the z-axis; a section plane parallel to the x–y plane intersects the master circle yielding a chord of length y which is equal to the radius r of the circular section profile of the sphere. In fact, it will suffice to consider one quadrant of this main circle only (Fig. 6.2).

The size distribution of profiles is characterized by determining the probability density with which a profile radius r_i falls into the range

$$r > r_i > (r - dr) \tag{6.4}$$

Since we are sectioning the sphere by horizontal planes the probability measure must be related to the vertical distance between the plane which will generate a profile of radius r and that which generates a profile of radius $(r - dr)$. If we designate this distance as dz it follows immediately that the probability of having r_i in the class $\{r, (r - dr)\}$ is

$$\Pr\{r > r_i > (r - dr)\} = \begin{cases} dz/R & \text{for} \quad r \leq R \\ 0 & \text{for} \quad r > R \end{cases} \tag{6.5}$$

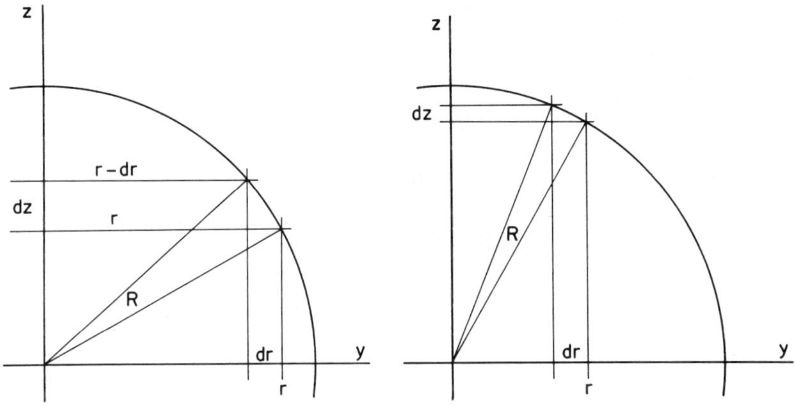

FIG. 6.2: Range dz varies with z if profile radius class dr is kept constant.

This is a straight-forward application of the rules of geometric probability outlined in Section 2.2: the probability of sectioning any object with a plane is the ratio of the tangent diameter of the object to the tangent diameter of the containing space. In the present case the "containing space" (or reference space) is the hemisphere from centre to pole, whence its orthogonal diameter is R; the "object" is the slice of the sphere which will allow profile radii to fall into the specified range, and its thickness dz is the measure orthogonal to the sampling plane which is equivalent to the tangent or caliper diameter.

Figure 6.2 shows that the slice thickness dz is not constant but rather depends on r. It can be determined by either of two approaches:

(a) The slice thickness dz is the difference between the values of z where the two limiting sections intersect the z-axis:

$$dz = z(r - dr) - z(r) \tag{6.6}$$

From the equation of the circle we derive

$$z(r) = \sqrt{(R^2 - r^2)} \tag{6.7}$$

$$z(r - dr) = \sqrt{(R^2 - (r - dr)^2)} \tag{6.8}$$

and hence

$$dz = \sqrt{(R^2 - (r - dr)^2)} - \sqrt{(R^2 - r^2)} \tag{6.9}$$

In this case the probability density is

$$\Pr\{r > r_i > (r - dr) \mid R\} = \frac{1}{R}[\sqrt{(R^2 - (r - dr)^2)} - \sqrt{(R^2 - r^2)}] \tag{6.10}$$

(b) The ratio dz/dr is evidently the gradient of the circle at $x = r$. It can be obtained by differentiating eq. (6.7) with respect to r:

$$\frac{dz}{dr} = \frac{-r}{\sqrt{(R^2 - r^2)}} \tag{6.11}$$

The probability density then becomes

$$\phi(r \mid R)dr = \frac{r}{R\sqrt{(R^2 - r^2)}}\, dr \tag{6.12}$$

whereby we have changed the sign to give positive probabilities and have extended the argument of ϕ to note that it depends on both the sphere size and the profile size. It should be noted that eq. (6.12) becomes eq. (6.10) if $\phi(r \mid R)$ is integrated between the class limits $\{r, r - dr\}$ of the histogram:

$$\Pr\{r, r - dr\} = \int_{r-dr}^{r} \phi(r \mid R)dr = \Pr\{r \mid R\} \tag{6.13}$$

Figure 6.3a shows the form of the curve $\phi(r \mid R)$. It starts at the origin and increases to end asymptotically at the $r = R$ line. However, the area under the curve is finite; if equation (6.12) is integrated over the range 0 to R the result is unity, as it should be because $\phi(r)$ is a probability density. A more realistic representation of this function is perhaps a histogram as shown in Fig. 6.3b whereby dr may be chosen as fine as desired ($dr \to 0$ at the limit). The values $\Pr\{r \mid R\}$ of the histogram columns is obtained by eq. (6.10). Note that r is the upper class limit.

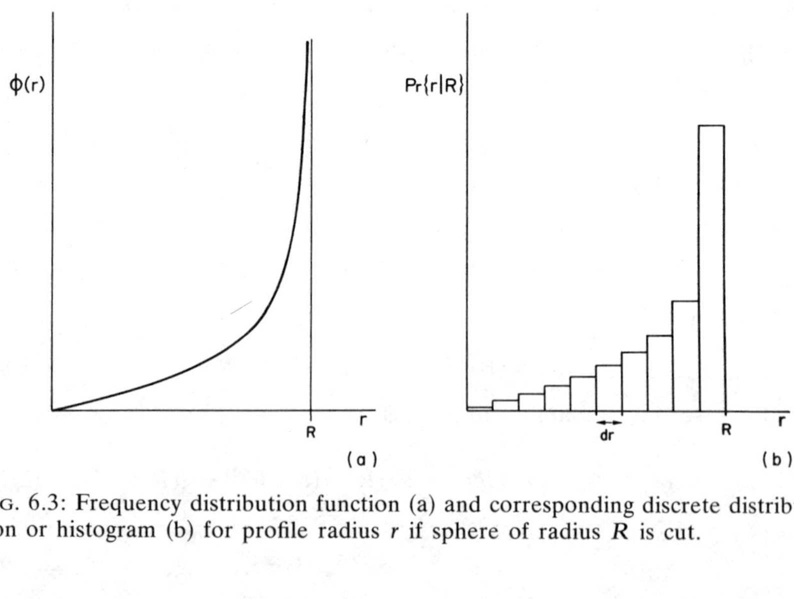

FIG. 6.3: Frequency distribution function (a) and corresponding discrete distribution or histogram (b) for profile radius r if sphere of radius R is cut.

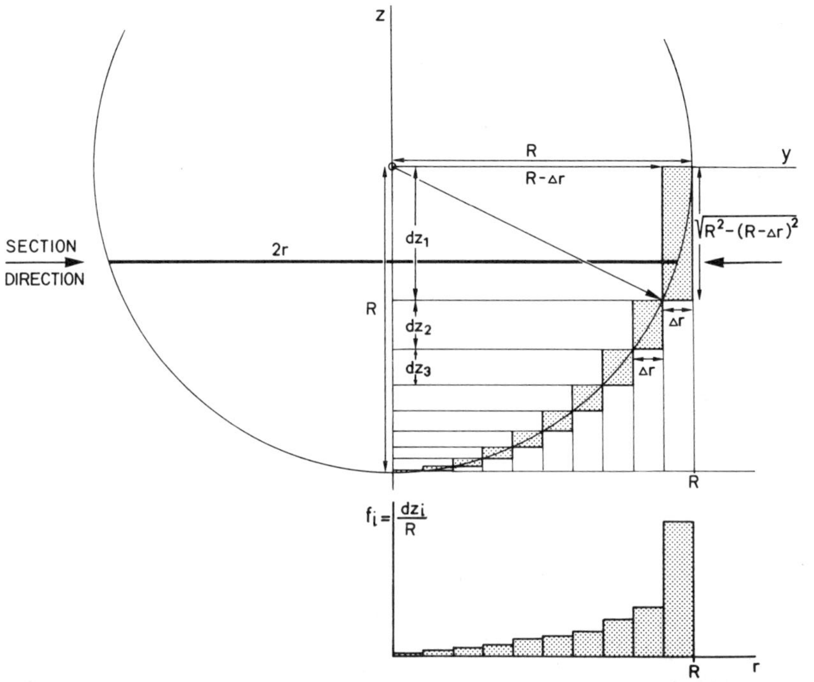

FIG. 6.4: Graphical derivation of profile size frequency histogram.

Figure 6.4 shows how this histogram can be derived graphically by a simple procedure which also makes the principle plausible. To this end, the horizontal sphere radius R is subdivided into a certain number of equal units of size Δr. The boundaries between the units are drawn vertically, i.e. parallel to the z-axis, to cross the circle. The intersection points are then carried horizontally to the z-axis. The distances between these horizontal lines are proportional to the frequencies of profiles falling into the corresponding size classes. By measuring dz_i one can directly construct the histogram. It can furthermore be noted that the histogram formed between the bottom tangent to the circle and the intersection points of upper class limits and circle represents the relative number of profile radii which are smaller or equal to the radius of that class; it hence represents the summation histogram which reaches 1 in the largest class. It is seen from this that the integral distribution derived from eq. (6.12)

$$\psi(r \mid R)\, dr = \int_0^r \frac{r}{R\sqrt{(R^2 - r^2)}}\, dr \tag{6.14}$$

must have the shape of a concave circle of radius R tangent to the origin.

Two interesting results of this distribution should be mentioned.

(a) If one integrates eq. (6.12) between $R/2$ and R one finds that 86.6 per cent of all profiles must be larger than half the maximal diameter, a result first mentioned by Elias and Pauli (1966).

(b) The mean radius of profiles can be derived from

$$\bar{r} = \int_0^R \phi(r) \cdot r \cdot dr$$

$$= \int_0^R \frac{r^2}{R\sqrt{(R^2 - r^2)}} dr \tag{6.15}$$

Performing the integration one obtains

$$\bar{r} = \tfrac{1}{4}\pi R \tag{6.16}$$

This result was obtained by a simple graphical procedure in Volume 1, Section 2.8.1 (Fig. 2.39).

This analysis has shown that, although all profiles are equally likely to occur, the probability density of profile *radii* varies with profile size: larger profiles are more likely to be found on sections than smaller ones. This does not contain any contradiction. In defining the probability of obtaining a certain profile we proceed at sectioning the sphere at equal intervals along the z-axis normal to the sectioning plane: the increments $dr(z)$ in profile radius will then become smaller as the section plane moves down (Fig. 6.5a). In defining the probability density of measuring a certain profile radius we assume equal increments of r by dr: this produces larger intervals $dz(r)$ between the slice boundaries as r becomes larger. In the first case we use the *sampling space* as reference, in the

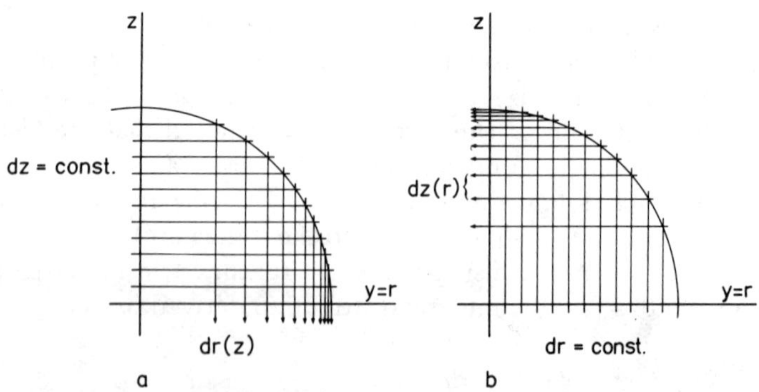

FIG. 6.5: Variation of $dr(z)$ or $dz(r)$ if dz or dr are kept constant, respectively.

second case, however, the *space of measurement.* These two conditions must be clearly distinguished.

One further comment. We have assumed here that one sphere was sectioned very many times. This amounts to the same thing as a very large number of spheres of identical size being randomly embedded in a large block and this block being sectioned once, or a few times at great distances.

6.2. Size distribution of profiles when spheres of varying size are randomly sectioned

We shall now consider the more realistic situation where our structure is composed of spheres of varying size, and ask what type of size distribution of the profiles will result if this population is randomly sectioned. The final question will be: how can we extract or reconstruct the size distribution of the particles from the measured size distribution of profiles? We shall proceed by first deriving the profile size distribution resulting from some particular particle size distributions in order to gain insight into the procedures necessary to resolve the inverse practical case.

From what we have learned so far we can presume *a priori* that three factors may influence in some way or another the number of profiles contributed to each profile class:

(a) The numerical density of spheres occurring in each class of the sphere size distribution;
(b) the probability of hitting a sphere of a certain size with a random sectioning plane, which is related to its (mean tangent) diameter;
(c) the frequency distribution of profiles derived from each class of spheres, as developed in the preceding section.

Let us look at this systematically.

Consider first a large cube of tissue which contains spheres of only one size class, $D_1 = 2R_1$ (Fig. 6.6a). Let their numerical density, i.e. the number of sphere centres per unit volume of the cube, be N_{V1}. If this cube is sectioned we know from eq. (5.11) that the number of profiles with their centres within the unit area of section, the numerical profile density N_{A1}, is

$$N_{A1} = N_{V1} \cdot D_1 = N_{V1} \cdot 2R_1 \tag{6.17}$$

If the profile radii (or diameters) are measured, the number of profiles per unit section area which fall into the size class r to $(r - \Delta r)$ is determined

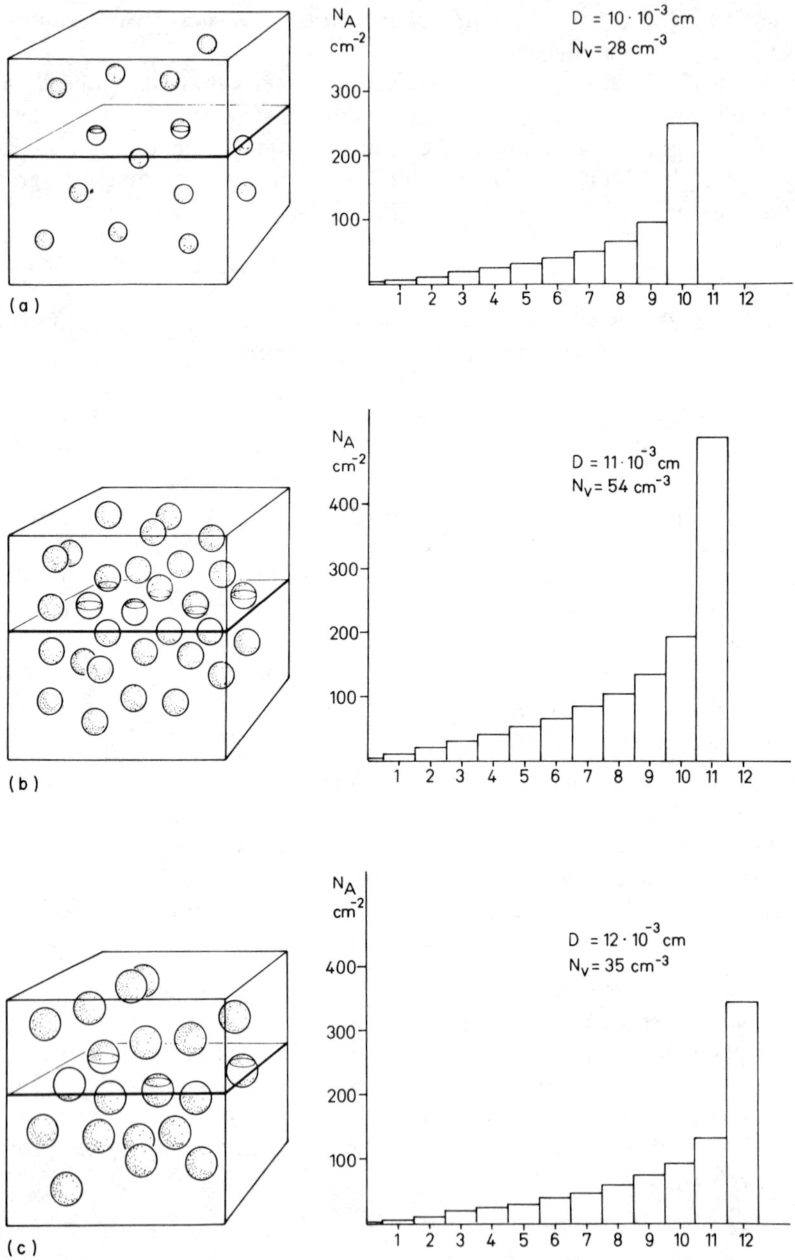

FIG. 6.6: Profile size histograms for three classes of spheres of different size and numerical density.

by the profile size frequency $\Pr\{r \mid R\}$ given in eq. (6.10):

$$\begin{aligned} N_{A1}(r) &= N_{A1} \cdot \Pr\{r \mid R\} \\ &= N_{V1} \cdot 2R_1 \cdot \left\{ \frac{1}{R_1} [\sqrt{(R_1^2 - (r-\Delta r)^2)} - \sqrt{(R_1^2 - r^2)}] \right\} \\ &= N_{V1} \cdot 2 \cdot [\sqrt{(R_1^2 - (r-\Delta r)^2)} - \sqrt{(R_1^2 - r^2)}] \end{aligned} \quad (6.18)$$

From this we can construct a histogram (Fig. 6.6a), plotting on the ordinate this time not frequencies, but numbers of profiles of a certain size class occurring per unit area of section; the area under the histogram is hence the total number of profiles N_{A1}, and the height of each column is given by the product of numerical frequency of spheres with the profile size frequency.

Consider a second cube of same size but containing larger spheres of diameter $D_2 = 2R_2$, whereby we assume R_2 to be an integer multiple of Δr, at a numerical density N_{V2} (Fig. 6.6b). The resulting profile density is evidently

$$N_{A2} = N_{V2} \cdot 2R_2 \quad (6.19)$$

and the numerical density of profiles per size class (the profile radius scale being the same as above) is

$$N_{A2}(r) = 2 \cdot N_{V2}[\sqrt{(R_2^2 - (r-\Delta r)^2)} - \sqrt{(R_2^2 - r^2)}] \quad (6.20)$$

This yields again a histogram, extending now from 0 to R_2 and having an area equal to N_{A2}.

A third cube contains N_{V3} spheres of $D_3 = 2R_3$ so that

$$N_{A3} = N_{V3} \cdot 2R_3 \quad (6.21)$$

$$N_{A3}(r) = 2 \cdot N_{V3} \cdot [\sqrt{(R_3^2 - (r-\Delta r)^2)} - \sqrt{(R_3^2 - r^2)}] \quad (6.22)$$

which yields a third histogram extending from 0 to R_3 (Fig. 6.6c).

This has produced three different histograms for each of the particle populations; they were determined by the numerical density of the particles, by the size of the particles and by the frequency distribution of profile sizes.

We now mix these populations by placing all three classes of spheres into the same cube (Fig. 6.7). The total numerical density of particles is evidently

$$N_V = N_{V1} + N_{V2} + N_{V3} = \sum N_{Vi} \quad (6.23)$$

and the total number of profiles on the unit area of a section must be

$$N_A = N_{A1} + N_{A2} + N_{A3} = \sum N_{Ai} \quad (6.24)$$

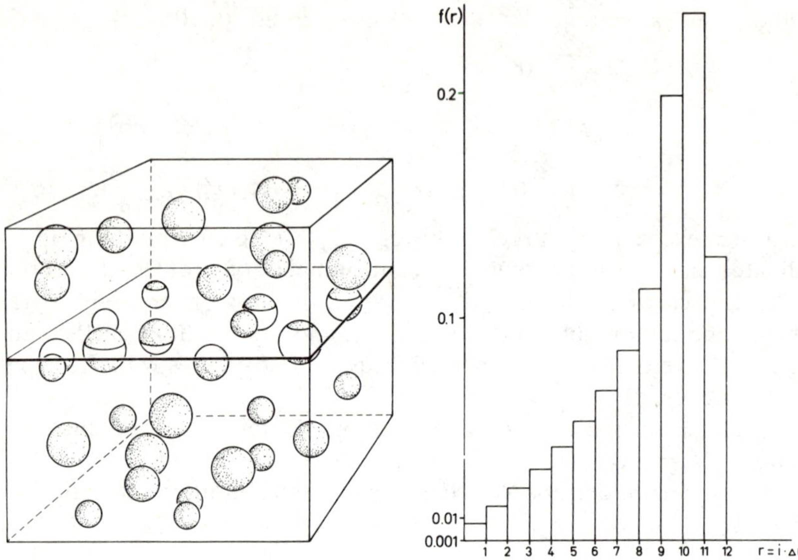

FIG. 6.7: Profile size histogram for mixture of three sphere size classes, obtained by addition of the three histograms in Fig. 6.6.

It is also evident that the numerical density of profiles falling into the size class r to $(r-\Delta r)$ must be

$$N_A(r) = N_{A1}(r) + N_{A2}(r) + N_{A3}(r) \tag{6.25}$$

which means that the corresponding columns of the histograms must be added to produce a compound histogram whose area must be N_A.

Before we can go on we must introduce a convenient definition of the relationship between sphere and profile radius. We assume that the sphere and profile size classes have the same interval, hence $\Delta R = \Delta r$; the sphere radius can then be expressed as an integer multiple of ΔR or Δr

$$R_j = j \cdot \Delta R = j \cdot \Delta r \tag{6.26}$$

Likewise we will define the radius of a profile size class as an integer multiple of Δr:

$$r_i = i \cdot \Delta r \tag{6.27}$$

We must further note that r_i is the radius corresponding to the *upper* class limit of profiles, and that R_j must hence be equal to the largest profile obtainable from these spheres; it hence corresponds to the upper class limit of the largest profile class. This had already been discussed above. Note that we explicitly consider a discrete distribution of sphere sizes.

The contribution of spheres of class j to profiles of radius $(r_i, r_i - \Delta r)$ hence falling into class i is, according to eq. (6.18):

$$N_{Aj}(r_i) = 2N_{Vj} \cdot [\sqrt{(R_i^2 - (r_i - \Delta r)^2)} - \sqrt{(R_j^2 - r_i^2)}] \tag{6.28}$$

This can be entered in eq. (6.25) in order to define the number of profiles in class i but we must first observe that not all sphere classes may contribute to $N_A(r)$: it is evident that spheres whose radius $R_j < r_i$ cannot give rise to any profiles of size class $(r_i, r_i - \Delta r)$. It is hence evident that in the following summation of the contributions of sphere size classes j to the profile number in profile class i we must consider only spheres of size

$$R_j \geq r_i \tag{6.29}$$

Accordingly, the number of profiles in class i is obtained by summing the contributions of all spheres in this size range

$$N_A(r_i) = \sum_{j=1}^{m} N_{Aj}(r_i) \tag{6.30}$$

which by eq. (6.28) becomes

$$N_A(r_i) = \sum_{j=1}^{m} 2N_{Vj} \cdot [\sqrt{(R_j^2 - (r_i - \Delta r)^2)} - \sqrt{(R_j^2 - r_i^2)}] \tag{6.31}$$

where m is the number of the largest sphere size class.

We now introduce the following definitions of frequencies: the frequency of spheres of a size class R_j is

$$F(R_j) = N_{Vj}/N_V \tag{6.32}$$

and the frequency of observed profiles in size class $(r, r - \Delta r)$ is

$$f(r_i) = N_A(r_i)/N_A \tag{6.33}$$

Considering also that, according to eq. (5.18),

$$N_A = N_V \cdot 2\bar{R} \tag{6.34}$$

where $\bar{R}$ is the mean radius of all spheres, we can proceed to the following development

$$\begin{aligned} N_A(r_i)/N_A &= 2\sum_{j=i}^{m} N_{Vj} \cdot [\sqrt{(R_j^2 - (r - \Delta r)^2)} - \sqrt{(R_j^2 - r^2)}]/(2\bar{R} \cdot N_V) \\ &= \frac{1}{\bar{R}} \sum_{j=i}^{m} \frac{N_{Vj}}{N_V} \cdot [\sqrt{(R_j^2 - (r_i - \Delta r)^2)} - \sqrt{(R_j^2 - r^2)}] \end{aligned} \tag{6.35}$$

or

$$f(r_i) = \frac{1}{\bar{R}} \sum_{j=i}^{m} F(R_j) [\sqrt{(R_j^2 - (r_i - \Delta r)^2)} - \sqrt{(R_j^2 - r_i^2)}] \tag{6.36}$$

We may interpret this formula as follows: the frequency of profiles found in a given size class is determined by the *sum* of the products of the *frequency of spheres* with the measure of the *probability to obtain a profile of this size class,* whereby summation is over all spheres whose diameter is larger than or equal to the profile diameter considered.

Now remember that we had defined the sizes of spheres and profiles as multiples of the class interval:

$$R_j = j \cdot \Delta r \tag{6.37}$$

$$r_i = i \cdot \Delta r \tag{6.38}$$

If we substitute these values into eq. (6.36) we arrive at the formula

$$f(i) = \frac{1}{\bar{j}} \cdot \sum_{j=i}^{m} F(i) \cdot [\sqrt{(j^2 - (i-1)^2)} - \sqrt{(j^2 - i^2)}] \tag{6.39}$$

which gives the frequencies in each size class as a function of the class numbers j and i respectively. This is a form chosen by Saltykov (1958), to which we shall return later.

We have so far operated with a population of spheres made of discrete size classes R_j, and have sorted the profiles into likewise discrete profile size classes. This simplifying assumption was convenient, but not necessary. It was dictated by the use of $\Pr\{r \mid R\}$ to describe profile frequencies. In eq. (6.13) we have indicated that $\Pr\{r \mid R\}$ is the integral of the fundamental continuous function $\phi(r \mid R)$ between the lower and upper class limits. For analytical purposes it is convenient to allow both the sphere size distribution $F(R)$ and the profile size distribution $f(r)$ to be continuous by letting both dR and dr become infinitesimally small, i.e. $dR \to 0$, $dr \to 0$ at the limit. If we now replace the summation in eq. (6.36) by an integral and substitute, from eqs (6.12) and (6.13)

$$\sqrt{(R^2 - (r - dr)^2)} - \sqrt{(R^2 - r^2)} = \frac{r \cdot dr}{\sqrt{(R^2 - r^2)}} \tag{6.40}$$

we obtain as continuous distribution

$$f(r)\, dr = \frac{1}{\bar{R}} \int_r^{R_{\max}} F(R) \cdot \frac{r \cdot dr}{\sqrt{(R^2 - r^2)}} \cdot dR \tag{6.41}$$

If we further divide both sides by dr we obtain the fundamental probability density function for profile size

$$f(r) = \frac{r}{\bar{R}} \int_r^{R_{\max}} F(R) \cdot \frac{dR}{\sqrt{(R^2 - r^2)}} \tag{6.42}$$

which was derived for the first time by Wicksell in 1925. Further derivations and discussions of this formula are found in Kendall and Moran (1963), Reid (1955), Scheil (1931), Krumbein (1935), Santaló (1955, 1976) and others. This equation presents the relation between the size distribution of spheres, $F(R)$, and that of profiles, $f(r)$. However, $f(r)$ is the observable value, and the distribution we would like to find, namely $F(R)$, is in the integral. It is evident that this must be so, if we verbally retrace the essential steps in the derivation we have here presented: remember that each sphere of radius $R > r$ will produce some profiles with radius r; the relative number of profiles of radius r, measured by $f(r)$, must therefore be given by the sum of the profiles of this size contributed by all spheres with radius larger than r, hence the integration between the limits r and R_{max}. The quantity to be integrated must be the product of the probability of having a profile of size r from spheres of size R, $\phi(r \mid R)$, with the frequency of the spheres of size R, $F(R)$. The final equation is therefore of the following general form‡

$$f(r) = \int_r^\infty F(R) \cdot \phi(r \mid R)\, dR \tag{6.43}$$

This is called an integral equation, more specifically a Volterra integral equation of the first kind. There exist solutions for extracting $F(R)$ from these integral equations, which can be found in any textbook on higher mathematics (e.g. Courant and Hilbert, 1953).

A direct solution of eq. (6.42) can be obtained if a small transformation is made (Kendall and Moran, 1963, p. 87). Let

$$F(R) = R \cdot F_1(R^2) = R \cdot F_1(z) \tag{6.44}$$

Equation (6.42) then becomes

$$f(r) = \frac{r}{2\bar{R}} \int_{r^2}^\infty \frac{F_1(z)}{\surd(z - r^2)} \tag{6.45}$$

which is an integral equation of Abel's type (Courant and Hilbert, 1953) whose solution is

$$F_1(z) = \frac{2}{\pi} \int_r^\infty \frac{\bar{R}}{\surd(r^2 - z)} \cdot \frac{d}{dx}\left(\frac{f(r)}{r}\right) r\, dr \tag{6.46}$$

from which we obtain, by resubstituting $z = R^2$ and $R \cdot F_1(z) = F(R)$,

$$F(R) = \frac{2R \cdot \bar{R}}{\pi} \int_{r^2}^\infty \frac{1}{\surd(r^2 - R^2)} \cdot \frac{d}{dr}\left(\frac{f(r)}{r}\right) dr \tag{6.47}$$

‡ Note that we have changed the upper limit of integration to ∞ instead of R_{max} to allow for continuous distributions of sphere size.

Wicksell (1925) arrived at the same result by using a slightly different approach (p. 88). We should also point to the elegant derivation of these equations presented by Reid (1955), but it would exceed the scope of this presentation to repeat it here.

6.3. Derivation of sphere size distribution from a measured distribution of profile sizes

In the previous section we have dealt with the problem of generating a profile size distribution from a known distribution of sphere sizes. The practical problem however is usually the inverse: we have obtained by measurement a profile size distribution and would like to derive the sphere size distribution from which it had been generated by random sectioning. The problem is complicated by the fact that $F(R)$ is contained in the integral of Abel's integral equation. We have indicated a general solution of this integral equation but should now proceed to present more practical solutions.

6.3.1. INTRODUCTION TO THE PROBLEM

In attempting to derive a size frequency distribution of spheres, $F(R)$, from a size frequency distribution of profiles, $f(r)$, measured on random sections one can use two fundamentally different approaches. In the first instance one can start with a histogram of profile sizes and attempt to derive a histogram of sphere sizes which could have produced the profile histogram; this approach we shall call *reconstruction* of the sphere size distribution. The second approach is more analytical and aims at determining the *moments* of the sphere size distribution. In this second approach one assumes that the measured histogram of profile sizes is an approximation of a continuous profile size distribution and that the sphere size distribution can also be described by a continuous function. In this section we shall deal with the reconstruction methods; the estimation of moments will form the topic of Section 6.5.

Let us first demonstrate the principle of reconstruction methods by studying the example of a profile size distribution which we had constructed on the basis of the model of Fig. 6.6; the resulting histogram is shown in Fig. 6.7. We must first give a clear description of this type of graph and how the measurements are represented.

The data have been grouped, in this particular case, into 12 classes; these are numbered 1, 2, ..., i, ... 12, i being the symbol denoting any of these classes. The width of each class is Δr. The radius characterizing each class, r_i, corresponds to the upper limit of the class; the limits of the class are therefore

$$\{r_i, r_i - \Delta r\} \tag{6.48}$$

The classes are so arranged that

$$r_i = i \cdot \Delta r \tag{6.49}$$

The height of the columns can be expressed in two ways:
(a) either we indicate the actual number of profiles counted $N(r)$ or the numerical profile density $N_A(r)$ on the section area in which the profiles were counted (Fig. 6.8), or
(b) we plot the relative frequencies (Fig. 6.7)

$$f(r) = N(r)/N = N_A(r)/N_A \tag{6.50}$$

where N and N_A are total number of profiles measured or total numerical density respectively. When expressing the column height with $N(r)$ or $N_A(r)$, the area under the histogram must be equal to N or N_A, because, by definition

$$N = \sum_r N(r) \tag{6.51}$$

$$N_A = \sum_r N_A(r) \tag{6.52}$$

In using relative frequencies the area must be

$$\sum_r f(r) = 1 \tag{6.53}$$

What do we know about this profile size distribution? We know (1) that it should correspond to the general formula (eq. 6.35) for discrete size distributions; (2) that each size class of spheres should have contributed profiles to all classes smaller or equal to the sphere size; (3) that the largest profile class is made of profiles derived exclusively from the largest sphere class, and that its numerical density of profiles should be given, according to eq. (6.18) by

$$N_A(r_{max}) = 2N_V(R_{max}) \cdot [\sqrt{(R_{max}^2 - (r_{max} - \Delta r)^2)} - \sqrt{(R_{max}^2 - r_{max}^2)}] \tag{6.54}$$

or, in terms of class numbers (m for largest class of spheres and profiles):

$$N_A(m) = 2N_V(m) \cdot [\sqrt{(m^2 - (m-1)^2)} - \sqrt{(m^2 - m^2)}]\,\Delta r \tag{6.55}$$

which reduces to

$$N_A(m) = 2 \cdot N_V(m) \cdot [\sqrt{(2m-1)}]\,\Delta r \tag{6.56}$$

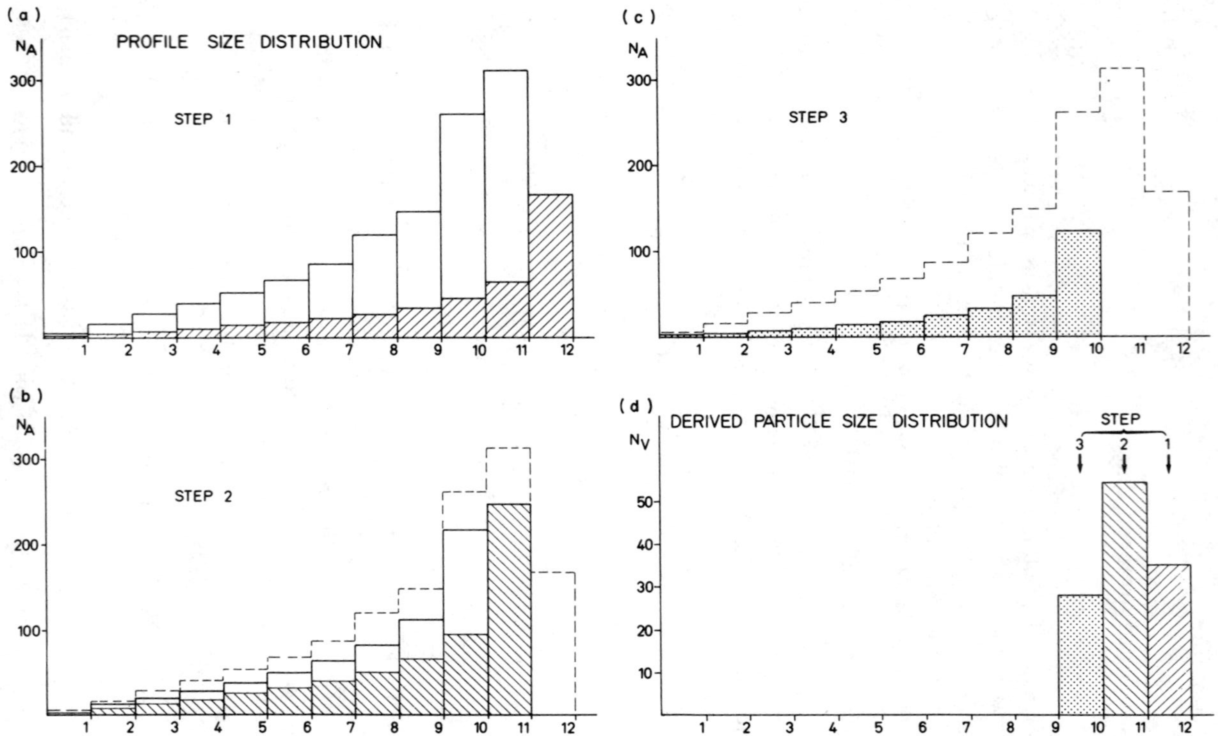

FIG. 6.8: Principle of deriving sphere size distribution from profile size distribution by sequential subtraction method.

From this we evidently derive immediately the numerical density of the largest spheres as

$$N_V(m) = N_A(m)/[2\sqrt{(2m-1)}\,\Delta r] \qquad (6.57)$$

This can be plotted as a column in a new histogram for sphere size distribution (Fig. 6.8d).

Knowing $N_V(R_{max})$ we can now calculate the contribution of these largest spheres to all profile size classes according to eq. (6.18); these can be subtracted from the original histogram, leaving a residual histogram of profiles with the largest profiles in class $(m-1)$ (Fig. 6.8b). We can evidently proceed the same way as before to determine $N_V(m-1)$ for the second largest class of spheres, then subtracting their contribution to the lower profile classes, and so on, until the profile size histogram is exhausted. In Fig. 6.8 this was the case after three such operations, evidently, since the histogram had been constructed from three sphere size classes (Fig. 6.6).

6.3.2. THE METHOD OF SALTYKOV (1958)

The procedure we have just described corresponds to that proposed by Saltykov (1958); it was based on a method first developed by Scheil (1931) and then modified by Schwartz (1934). A very lucid description of these methods is given by Underwood in his chapter contributed to "Quantitative Microscopy" of De Hoff and Rhines (1968) as well as in his own book on stereology (Underwood, 1970). We shall therefore only present a few highlights borrowing heavily on Underwood's presentation of the practical method.

Saltykov assumes a discrete distribution made of up to $m = 15$ classes.‡ The diameter of the largest spheres is D_m and the class width is introduced as

$$\Delta = D_m/m \qquad (6.58)$$

Note that Δ is related to diameter, whereas above we referred to radii; hence

$$\Delta = 2 \cdot \Delta r \qquad (6.59)$$

Some of Saltykov's formulae may therefore differ by a factor of 2 from those presented before and referring to radii. Table 6.1 taken from Underwood (1968) shows that the sphere diameters in each class are taken as integer multiples of Δ; however, the section diameters are

‡ Underwood and Saltykov used k to symbolize maximal size class.

Table 6.1

Definition of terms used in Saltykov method (modified after Underwood, 1970)

Class number	Diameter	Numerical density
	Particles	
1	Δ	$N_V(1)$
2	2Δ	$N_V(2)$
3	3Δ	$N_V(3)$
j	$j\Delta$	$N_V(j)$
m	$m\Delta$	$N_V(m)$
	Profiles	
1	0 to Δ	$N_A(1)$
2	Δ to 2Δ	$N_A(2)$
3	2Δ to 3Δ	$N_A(3)$
i	$(i-1)\Delta$ to $i\Delta$	$N_A(i)$
m	$(m-1)\Delta$ to $m\Delta$	$N_A(m)$

considered to assume a continuous distribution which is then broken up into classes. This amounts to integrating $\phi(R, r)$ between the upper and lower class limit (eq. 6.13).

Saltykov's basic equation is derived in the following straight-forward manner which is worth repeating as an alternative to the approach given above. In accordance with eq. (6.25) the numerical density of profiles in class i is

$$N_A(i) = \sum_{j=1}^{m} N_A(i, j) \tag{6.60}$$

To find the dependence between the number of spheres in class j, $N_V(j)$, and the number of profiles that these spheres contribute to profile class i, $N_A(i, j)$, Saltykov argued in the following way (Fig. 6.9). Consider a sphere of diameter $j \cdot \Delta$. A section plane will generate a profile of diameter $i\Delta$ to $(i-1)\Delta$, the range of profile class i, if the sphere centre is at a distance between h_i and h_{i-1} from the section plane. Accordingly, all spheres of size class j having their centre in a fictitious "slice" of thickness $(h_{i-1} - h_i)$ to either side of the section plane will contribute profiles to class i. Hence

$$N_A(i, j) = N_V(j) \cdot 2(h_{i-1} - h_i) \tag{6.61}$$

the factor 2 entering because each sphere has two regions from which

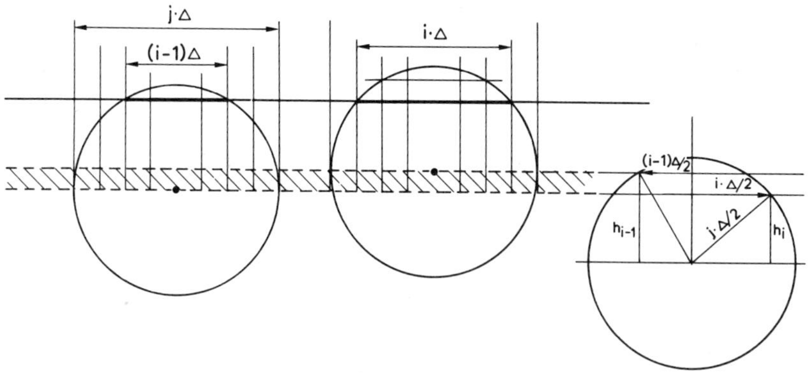

FIG. 6.9: Saltykov's model for deriving profile size frequency in class i.

profiles of this size can be cut. Referring to eq. (6.1) we find that

$$h_i = \sqrt{(R_j^2 - r_i^2)} \tag{6.62}$$

so that, setting $2R = j \cdot \Delta$ and $2r = i \cdot \Delta$, we obtain

$$\begin{aligned} N_A(i, j) &= N_V(j) \cdot \Delta[\sqrt{(j^2 - (i-1)^2)} - \sqrt{(j^2 - i^2)}] \\ &= N_V(j) \cdot \Delta \cdot k_{ij} \end{aligned} \tag{6.63}$$

This is evidently a familiar formula; it is easily obtained from eq. (6.36) by setting $R_j = j \cdot \Delta/2$ and $r_i = i \cdot \Delta/2$ and considering that $\Delta = 2\Delta r$. The term in brackets depends only on class numbers, so that a matrix of coefficients can be readily calculated by substituting integer values for i and j (Table 6.2).

Now, introducing eq. (6.63) into eq. (6.30) we find the observed number of profiles in each size class as

$$\begin{aligned} N_A(i) &= \sum_{j=1}^{m} N_A(i, j) \\ &= \Delta \sum_{j=1}^{m} k_{ij} \cdot N_V(j) \end{aligned} \tag{6.64}$$

Table 6.2

Coefficients k_{ij} for Saltykov method

Profile size class i	Particle size class j														
	$N_V(1)$	$N_V(2)$	$N_V(3)$	$N_V(4)$	$N_V(5)$	$N_V(6)$	$N_V(7)$	$N_V(8)$	$N_V(9)$	$N_V(10)$	$N_V(11)$	$N_V(12)$	$N_V(13)$	$N_V(14)$	$N_V(15)$
$N_A(1)$	1.0000	0.2679	0.1716	0.1270	0.1010	0.0839	0.0718	0.0627	0.0557	0.0501	0.0455	0.0417	0.0385	0.0358	0.0334
$N_A(2)$		1.7321	0.5924	0.4089	0.3164	0.2592	0.2200	0.1913	0.1693	0.1519	0.1378	0.1261	0.1162	0.1078	0.1006
$N_A(3)$			2.2361	0.8184	0.5826	0.4607	0.3836	0.3298	0.2897	0.2586	0.2336	0.2132	0.1961	0.1816	0.1691
$N_A(4)$				2.6458	1.0000	0.7240	0.5800	0.4880	0.4230	0.3742	0.3361	0.3052	0.2798	0.2584	0.2401
$N_A(5)$					3.0000	1.1555	0.8456	0.6832	0.5789	0.5049	0.4490	0.4050	0.3693	0.3397	0.3147
$N_A(6)$						3.3166	1.2934	0.9535	0.7751	0.6603	0.5784	0.5164	0.4674	0.4276	0.3944
$N_A(7)$							3.6056	1.4185	1.0513	0.8586	0.7343	0.6455	0.5781	0.5248	0.4812
$N_A(8)$								3.8730	1.5337	1.1414	0.9354	0.8025	0.7075	0.6352	0.5779
$N_A(9)$									4.1231	1.6411	1.2253	1.0070	0.8661	0.7653	0.6886
$N_A(10)$										4.3589	1.7420	1.3040	1.0742	0.9258	0.8197
$N_A(11)$											4.5826	1.8374	1.3784	1.1377	0.9823
$N_A(12)$												4.7958	1.9282	1.4492	1.1980
$N_A(13)$													5.0000	2.0150	1.5167
$N_A(14)$														5.1962	2.0981
$N_A(15)$															5.3852

This results in a set of linear equations such as

$$\begin{aligned}
N_A(1) &= [k_{11}\cdot N_V(1)+k_{12}\cdot N_V(2)+k_{13}\cdot N_V(3)+\ldots+k_{1m}\cdot N_V(m)]\cdot\Delta\\
N_A(2) &= [k_{22}\cdot N_V(2)+k_{23}\cdot N_V(3)+\ldots+k_{2m}\cdot N_V(m)]\cdot\Delta\\
N_A(3) &= [k_{33}\cdot N_V(3)+\ldots+k_{3m}\cdot N_V(m)]\cdot\Delta\\
&\;\;\vdots\\
N_A(m) &= [k_{mm}\cdot N_V(m)]\cdot\Delta
\end{aligned} \tag{6.65}$$

These are m linear equations with m unknowns ($N_V(j)$) and m experimentally measured quantities ($N_A(i)$); if they are solved for the unknowns we end up with a similar set of linear equations:

$$\begin{aligned}
N_V(1) &= [\alpha_{11}\cdot N_A(1)-\alpha_{21}\cdot N_A(2)-\alpha_{31}\cdot N_A(3)-\ldots-\alpha_{m1}\cdot N_A(m)]\cdot(1/\Delta)\\
N_V(2) &= [\alpha_{22}\cdot N_A(2)-\alpha_{32}\cdot N_A(3)-\ldots-\alpha_{m2}\cdot N_A(m)]\cdot(1/\Delta)\\
N_V(3) &= [\alpha_{33}\cdot N_A(3)-\ldots-\alpha_{m3}\cdot N_A(m)]\cdot(1/\Delta)\\
&\;\;\vdots\\
N_V(m) &= [\alpha_{mm}\cdot N_A(m)]\cdot(1/\Delta)
\end{aligned} \tag{6.66}$$

or, in general form,

$$N_V(j)=\frac{1}{\Delta}\sum_{i=j}^{m}\alpha_{ij}\cdot N_A(i) \tag{6.67}$$

The new coefficients α_{ij} are related to the values of k_{ij} through the solution of the linear equation, but not in a simple fashion, except for the cases where $j=i$ where

$$\alpha_{jj}=1/k_{jj} \tag{6.68}$$

The coefficients α_{ij} have been calculated by Saltykov (1958); they are shown in Table 6.3. They are applicable to any number of classes up to 15, but it seems advisable to work with at least 7 classes (Underwood, 1968, 1970).

The bottom row of coefficients in Table 6.3 is the last value of α, i.e. α_{jj} minus the other values of α in the same column; if these coefficients α_i

Table 6.3

Matrix of coefficients α_{ij} for unfolding particle size distribution according to method of Saltykov (1958) by eq. (6.67). Last row are coefficients α_i for calculating N_V by eq. (6.69). After Saltykov (1958) and Underwood (1970)

Particle size class j	Profile size class i														
	$N_A(1)$	$N_A(2)$	$N_A(3)$	$N_A(4)$	$N_A(5)$	$N_A(6)$	$N_A(7)$	$N_A(8)$	$N_A(9)$	$N_A(10)$	$N_A(11)$	$N_A(12)$	$N_A(13)$	$N_A(14)$	$N_A(15)$
$N_V(1)$	+1.0000	−0.1547	−0.0360	−0.0130	−0.0061	−0.0033	−0.0020	−0.0013	−0.0009	−0.0006	−0.0005	−0.0004	−0.0003	−0.0002	−0.0001
$N_V(2)$		+0.5774	−0.1529	−0.0420	−0.0171	−0.0087	−0.0051	−0.0031	−0.0021	−0.0015	−0.0010	−0.0009	−0.0006	−0.0006	−0.0004
$N_V(3)$			+0.4472	−0.1382	−0.0408	−0.0178	−0.0093	−0.0057	−0.0037	−0.0026	−0.0018	−0.0013	−0.0010	−0.0007	−0.0007
$N_V(4)$				+0.3779	−0.1260	−0.0386	−0.0174	−0.0095	−0.0058	−0.0038	−0.0027	−0.0020	−0.0016	−0.0012	−0.0009
$N_V(5)$					+0.3333	−0.1161	−0.0366	−0.0168	−0.0094	−0.0059	−0.0040	−0.0028	−0.0021	−0.0016	−0.0013
$N_V(6)$						+0.3015	−0.1081	−0.0346	−0.0163	−0.0091	−0.0058	−0.0041	−0.0028	−0.0022	−0.0016
$N_V(7)$							+0.2773	−0.1016	−0.0329	−0.0155	−0.0090	−0.0057	−0.0040	−0.0029	−0.0022
$N_V(8)$								+0.2582	−0.0961	−0.0319	−0.0151	−0.0088	−0.0056	−0.0039	−0.0028
$N_V(9)$									+0.2425	−0.0913	−0.0301	−0.0146	−0.0085	−0.0055	−0.0039
$N_V(10)$										+0.2294	−0.0872	−0.0290	−0.0140	−0.0083	−0.0054
$N_V(11)$											+0.2182	−0.0836	−0.0280	−0.0136	−0.0080
$N_V(12)$												+0.2085	−0.0804	−0.0270	−0.0132
$N_V(13)$													+0.2000	−0.0776	−0.0261
$N_V(14)$														+0.1925	−0.0750
$N_V(15)$															+0.1857
N_V	+1.0000	+0.4227	+0.2583	+0.1847	+0.1433	+0.1170	+0.0988	+0.0856	+0.0753	+0.0672	+0.0610	+0.0553	+0.0511	+0.0472	+0.0441

are multiplied with $N_A(i)$ the sum of these products yields the total numerical density of particles:

$$N_V = \sum_{i=1}^{m} \alpha_i \cdot N_A(i) \tag{6.69}$$

A simple example on the use of the Saltykov method is presented in Volume 1, Section 5.2.9.1.

6.3.3. THE TRANSFORMATION OF WICKSELL (1925)

The first reconstructive procedure of this kind had been proposed by S. D. Wicksell as early as 1925. His paper being published in a biometrical journal it remained unknown to workers in non-biological fields. The fundamental treatise of Wicksell is one of the classical masterpieces of stereology.

Wicksell assumed continuous distributions of both sphere and profile size distributions; for practical purposes, however, they had to be presented in histogram form. He chose to define his classes to have an interval Δr and to be characterized by the value of r or R at the class *mid-point*; the range of class i hence was

$$(i-\tfrac{1}{2})\,\Delta r < r(\text{or } R) < (i+\tfrac{1}{2})\,\Delta r \tag{6.70}$$

This evidently deviates from the pattern used so far and, particularly, in Saltykov's method: the height of the column is obtained by integrating between the limits $(r-\Delta r/2)$ and $(r+\Delta r/2)$, as demonstrated in Fig. 6.10. Furthermore, this design left, near the origin, a half-class of profiles in the range $\{0<r<\Delta r/2\}$; in Fig. 6.11 the data of Fig. 6.7 have been replotted in the form compatible with Wicksell's model. With a sufficiently large class number the differences are minor. According to Wicksell, the relative number of profiles falling into class i is then

$$f_i = k \cdot \bar{R} \int_{\Delta r(i-\frac{1}{2})}^{\Delta r(i+\frac{1}{2})} f(r)\, dr \tag{6.71}$$

whereby $f(r)$ is given by eq. (6.42):

$$f_i = k \int_{\Delta r(i-\frac{1}{2})}^{\Delta r(i+\frac{1}{2})} r\, dr \int_{r}^{R_{\max}} F(R) \frac{dr}{\sqrt{(R^2-r^2)}} \tag{6.72}$$

or, changing the order of integration and integrating over the variable r

$$f_i = k\int_{\Delta r(i-\frac{1}{2})}^{\Delta r(i+\frac{1}{2})} dR \cdot F(R)\,\sqrt{(R^2-\Delta r^2(i-\tfrac{1}{2})^2)}$$
$$+k\int_{\Delta r(i+\frac{1}{2})}^{R_m} dR \cdot F(R)[\sqrt{(R^2-\Delta r^2(i-\tfrac{1}{2})^2)}-\sqrt{(R^2-\Delta r^2(i+\tfrac{1}{2})^2)}] \tag{6.73}$$

The first term is the contribution to f_i of spheres with radii in the range of the class

$$(i-\tfrac{1}{2})\Delta r < R < (i+\tfrac{1}{2})\Delta r \tag{6.74}$$

whereas the second term measures the contribution of larger spheres; note that this second integral is

$$\int_{r+\Delta r/2}^{R_{\max}} F(R)\cdot \phi(r \mid R)\, dR \tag{6.75}$$

a term we have dealt with before (eq. 6.43).

Wicksell then defined a measure of the frequency of spheres in the class $(j\pm\frac{1}{2})$ as

$$P_j = \int_{\Delta r(j-\frac{1}{2})}^{\Delta r(j+\frac{1}{2})} F(R)\, dR \tag{6.76}$$

which amounts to transforming the continuous distribution $F(R)$ into a histogram of class interval $\Delta r(j\pm\frac{1}{2})$. With this discrete size distribution of spheres one can now set

$$f_i = \sum_{j=1}^{m} P_j \cdot k'_{ij} \tag{6.77}$$

where m means the largest size class. It is now evident that, if f_i is measured in terms of numerical profile densities, N_{Ai}, the summand is

$$P_j \cdot k'_{ij} = N_{Aij} \tag{6.78}$$

From eqs (6.73) and (6.77) one can work out that the coefficients k'_{ij} are approximately given by

$$k'_{ij} = \sqrt{(j^2-(i-\tfrac{1}{2})^2)}-\sqrt{(j^2-(i+\tfrac{1}{2})^2)} \tag{6.79}$$

This is evidently very similar to the coefficients k_{ij} of Saltykov (eq. 6.63), differing only by the way the classes had been defined. The table of coefficients k'_{ij} worked out by Wicksell in 1925 for 15 size classes is given in Table 6.4. Note that the lowest class of the histogram extends from 0

Table 6.4

Matrix of coefficients k_{ij}' (eq. 6.79) for Wicksell unfolding method. From Wicksell (1925)

Profile size class i	Particle size class j															
	$N_V(\frac{1}{4})$	$N_V(1)$	$N_V(2)$	$N_V(3)$	$N_V(4)$	$N_V(5)$	$N_V(6)$	$N_V(7)$	$N_V(8)$	$N_V(9)$	$N_V(10)$	$N_V(11)$	$N_V(12)$	$N_V(13)$	$N_V(14)$	$N_V(15)$
$N_A(\frac{1}{4})$	0.2500	+0.1340	+0.0635	+0.0420	+0.0314	+0.0251	+0.0209	+0.0179	+0.0156	+0.0139	+0.0125	+0.0114	+0.0104	+0.0096	+0.0089	+0.0084
$N_A(1)$		0.8660	+0.6136	+0.3599	+0.2605	+0.2052	+0.1696	+0.1447	+0.1263	+0.1120	+0.1006	+0.0914	+0.0838	+0.0772	+0.0717	+0.0668
$N_A(2)$			1.3229	+0.9398	+0.5856	+0.4396	+0.3551	+0.2991	+0.2588	+0.2283	+0.2044	+0.1851	+0.1691	+0.1588	+0.1444	+0.1346
$N_A(3)$				1.6583	+1.1860	+0.7594	+0.5810	+0.4761	+0.4056	+0.3542	+0.3150	+0.2838	+0.2585	+0.2374	+0.2197	+0.2042
$N_A(4)$					1.9365	+1.3913	+0.9048	+0.7003	+0.5793	+0.4974	+0.4372	+0.3909	+0.3539	+0.3237	+0.2982	+0.2769
$N_A(5)$						2.1794	+1.5707	+1.0318	+0.8049	+0.6703	+0.5787	+0.5111	+0.4589	+0.4171	+0.3827	+0.3538
$N_A(6)$							2.3979	+1.7320	+1.1458	+0.8990	+0.7523	+0.6522	+0.5783	+0.5209	+0.4748	+0.4368
$N_A(7)$								2.5981	+1.8798	+1.2500	+0.9849	+0.8274	+0.7196	+0.6399	+0.5780	+0.5281
$N_A(8)$									2.7839	+2.0169	+1.3466	+1.0646	+0.8970	+0.7822	+0.6973	+0.6312
$N_A(9)$										2.9580	+2.1453	+1.4368	+1.1391	+0.9621	+0.8408	+0.7510
$N_A(10)$											3.1225	+2.2666	+1.5219	+1.2092	+1.0234	+0.8961
$N_A(11)$												3.2787	+2.3817	+1.6027	+1.2757	+1.0814
$N_A(12)$													3.4278	+2.4915	+1.6796	+1.3391
$N_A(13)$														3.5707	+2.5967	+1.7533
$N_A(14)$															3.7081	+2.6977
$N_A(15)$																3.8406

to $\frac{1}{2}$, as indicated above, so that its index is $i = 1/4$; this gives a value of k'_{ij} of 0.25.

Since in eq. (6.77) P_j is the unknown and f_i the known variable it is necessary to invert the matrix of k'_{ij} and find a matrix of coefficients which give

$$P_j = \sum_{i=j}^{m} \alpha'_{ij} \cdot f_i \tag{6.80}$$

These are given in Table 6.5 for the fifteen classes of Wicksell. Note that the majority of the coefficients are negative. Fundamentally, α'_{ij} is again equivalent to Saltykov's coefficients α_{ij}. Finally we obtain

$$F_j = \frac{P_j}{\sum_j P_j} \tag{6.81}$$

as the relative frequency of spheres of class j, having radius $R_j = \Delta R(j \pm \frac{1}{2})$.

6.3.4. REVIEW OF THE SALTYKOV AND WICKSELL METHODS

The two methods of Saltykov and Wicksell for unfolding a profile size into a particle size distribution are in essence identical. They differ only with respect to the design of the histogram (Fig. 6.10). Wicksell (1925) considered both the particle and the profile size distributions to be continuous and constructed histograms that approximate these continuous distributions, taking the median of each class as the measure for particle and profile size (Fig. 6.10). The logic of Saltykov's approach was that no profile diameter can be larger than the particle diameter (Fig. 6.3); hence the profile histogram columns are constructed with their upper limits marking the particle and the profile size of each class. The particle size distribution is discrete.

Cruz-Orive (1977) evaluated the performance of the Saltykov procedure on a synthetic example. He found that it did not yield faithful results, particularly in the smaller and larger size classes (Table 6.6). It turned out that this was due to an unfortunate choice of class characteristics. Good results were obtained by constructing histograms according to the Saltykov procedure, i.e. with class limits as multiples of Δ, but choosing, as in the Wicksell procedure, the class mid-point as a measure for the size class (Fig. 6.10c); in a way this is a compromise between the approaches of Saltykov and Wicksell. Since the limits of class i are $i \cdot \Delta$ and $(i-1)\Delta$ and the class parameter is $x_i = (i - 1/2)\Delta$ the matrix of coefficients α''_{ij} is

Table 6.5

Matrix of coefficients α'_{ij} for unfolding particle size distribution according to method of Wicksell (1925) (eq. 6.80). Last row of coefficients are α_i for calculating total N_V by eq (6.69). After Wicksell (1925)

Particle size class j	Profile size class i															
	$N_A(\frac{1}{4})$	$N_A(1)$	$N_A(2)$	$N_A(3)$	$N_A(4)$	$N_A(5)$	$N_A(6)$	$N_A(7)$	$N_A(8)$	$N_A(9)$	$N_A(10)$	$N_A(11)$	$N_A(12)$	$N_A(13)$	$N_A(14)$	$N_A(15)$
$N_V(\frac{1}{4})$	4.0000	−0.6189	+0.0951	−0.0209	+0.0024	−0.0012	−0.0002	−0.0002	−0.0001	−0.0001	−0.0001	−0.0001	−0.0000	−0.0000	−0.0000	−0.0000
$N_V(1)$		1.1547	−0.5356	+0.0529	−0.0258	−0.0027	−0.0037	−0.0019	−0.0014	−0.0009	−0.0007	−0.0005	−0.0004	−0.0003	−0.0003	−0.0002
$N_V(2)$			0.7559	−0.4284	+0.0338	−0.0248	−0.0047	−0.0047	−0.0026	−0.0019	−0.0014	−0.0010	−0.0007	−0.0006	−0.0005	−0.0004
$N_V(3)$				0.6030	−0.3693	+0.0256	−0.0236	−0.0054	−0.0051	−0.0030	−0.0022	−0.0016	−0.0012	−0.0009	−0.0008	−0.0006
$N_V(4)$					0.5164	−0.3297	+0.0211	−0.0223	−0.0057	−0.0052	−0.0032	−0.0024	−0.0017	−0.0014	−0.0010	−0.0009
$N_V(5)$						0.4588	−0.3006	+0.0181	−0.0212	−0.0058	−0.0052	−0.0033	−0.0025	−0.0018	−0.0014	−0.0011
$N_V(6)$							0.4170	−0.2780	+0.0161	−0.0202	−0.0058	−0.0051	−0.0033	−0.0025	−0.0019	−0.0015
$N_V(7)$								0.3849	−0.2600	+0.0146	−0.0193	−0.0058	−0.0050	−0.0033	−0.0025	−0.0019
$N_V(8)$									0.3592	−0.2449	+0.0134	−0.0185	−0.0057	−0.0050	−0.0033	−0.0025
$N_V(9)$										0.3381	−0.2323	+0.0124	−0.0179	−0.0056	−0.0049	−0.0032
$N_V(10)$											0.3203	−0.2214	+0.0116	−0.0172	−0.0054	−0.0048
$N_V(11)$												0.3050	−0.2119	+0.0110	−0.0166	−0.0053
$N_V(12)$													0.2917	−0.2036	+0.0104	−0.0161
$N_V(13)$														0.2801	−0.1961	+0.0099
$N_V(14)$															0.2697	−0.1894
$N_V(15)$																0.2604
N_V	4.0000	0.5358	0.7974	0.2066	0.1575	0.1260	0.1053	0.0905	0.0792	0.0707	0.0635	0.0577	0.0530	0.0489	0.0454	0.0424

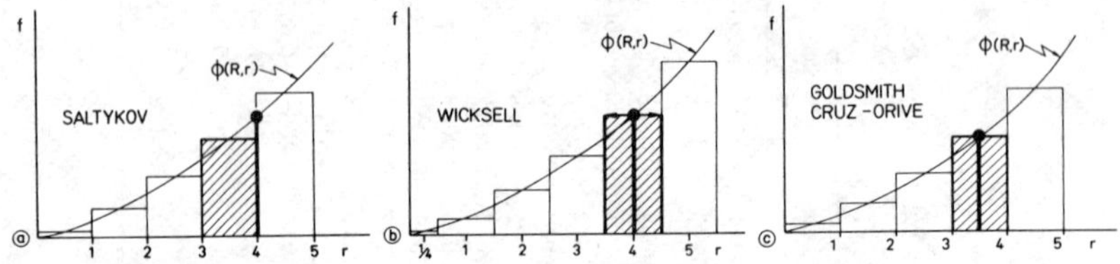

FIG. 6.10: Difference in histogram class definition in (a) Saltykov and (b) Wicksell methods compared to (c) the method proposed by Cruz-Orive (1977) and Goldsmith (1967). Heavy dot notes location of class characteristic for profile and sphere radius.

obtained by inversion of a matrix of coefficients

$$k''_{ij} = \sqrt{[(j-1/2)^2-(i-1)^2]} - \sqrt{[(j-1/2)^2-i^2]} \tag{6.82}$$

for $i = 1, 2, 3, \ldots, (j-1)$

and

$$k''_{jj} = \sqrt{[j-3/4]} \tag{6.83}$$

for $i = j$.

These coefficients and the inverted matrix of coefficients α''_{ij} are given in Tables 6.7 and 6.8 respectively. Table 6.6 shows that these coefficients

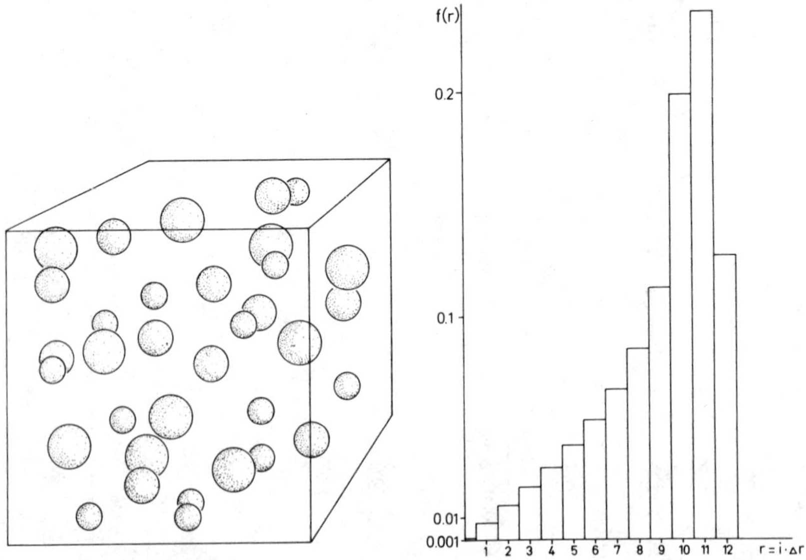

FIG. 6.11. Histogram of Fig. 6.7 replotted in accordance with Wicksell method.

Table 6.6

Comparison of unfolding methods of Saltykov (1958) and Cruz-Orive (1977) on synthetic example of size distribution of spherical particles

Granule diameter (cm or μm) $D(j)$	Numerical densities, $N_V(j) \cdot 10^{-3}$ mm^{-3}		
	True data	Unfolding after Saltykov	Unfolding after Cruz-Orive
(0, 0.2)	341	901	247
(0.2, 0.4)	1703	3268	1405
(0.4, 0.6)	5827	8548	5787
(0.6, 0.8)	13647	16327	14345
(0.8, 1.0)	21893	22307	22958
(1.0, 1.2)	24063	21630	24848
(1.2, 1.4)	18122	14504	17829
(1.4, 1.6)	9349	6841	8858
(1.6, 1.8)	3303	2161	2888
(1.8, 2.0)	799	459	658
N_V(total)	99047	96946	99823
Mean diameter $\bar{D}$ μm	1.05	1.0728	1.0496
Diameter variance S_x^2	0.1024	0.1109	0.0995

yield rather faithful estimates of the particle size distribution, as well as of numerical density, mean diameter and variance of the particles. The practical application of this method is discussed in Volume 1, Section 5.2.9. It turns out that these coefficients had already been derived by Goldsmith (1967) who, starting from the Wicksell theory, had also constructed his histograms according to Fig. 6.10c, taking the class mid-point as the class characteristic. Since this is also the case in Wicksell's method, its performance is equally good. The general recommendation is then not to use the Saltykov method but rather the Wicksell or the Goldsmith–Cruz-Orive method of classification and the corresponding matrix of coefficients.

A large number of papers have dealt with the problem of unfolding particle size distributions under various conditions, partly by assuming certain specific particle size distributions. A sample of these are Fullman (1953), Nicholson (1970, 1976), Tallis (1970), Watson (1971), Keiding *et al.* (1972), Anderssen and Jakeman (1975), Jakeman and Anderssen (1975), Anderssen (1976), to quote but a few.

Table 6.7

Coefficients k''_{ij} by eq. (5.57) after Goldsmith (1967) and Cruz-Orive (1978a). These coefficients correspond also to the coefficients $p_{\alpha i}$ in eq. (8.21)

Particle size class i	Particle size class j														
	$N_V(1)$	$N_V(2)$	$N_V(3)$	$N_V(4)$	$N_V(5)$	$N_V(6)$	$N_V(7)$	$N_V(8)$	$N_V(9)$	$N_V(10)$	$N_V(11)$	$N_V(12)$	$N_V(13)$	$N_V(14)$	$N_V(15)$
$N_A(1)$	0.5000	0.3820	0.2087	0.1459	0.1125	0.0917	0.0774	0.0670	0.0590	0.0528	0.0477	0.0436	0.0401	0.0371	0.0345
$N_A(2)$		1.1180	0.7913	0.4818	0.3564	0.2849	0.2380	0.2046	0.1796	0.1601	0.1445	0.1317	0.1210	0.1119	0.1041
$N_A(3)$			1.5000	1.0695	0.6770	0.5137	0.4184	0.3546	0.3084	0.2732	0.2455	0.2229	0.2043	0.1886	0.1751
$N_A(4)$				1.8028	1.2925	0.8349	0.6428	0.5296	0.4530	0.3970	0.3541	0.3199	0.2919	0.2686	0.2489
$N_A(5)$					2.0616	1.4836	0.9702	0.7541	0.6261	0.5391	0.4752	0.4258	0.3863	0.3539	0.3267
$N_A(6)$						2.2913	1.6533	1.0902	0.8531	0.7123	0.6162	0.5454	0.4906	0.4465	0.4103
$N_A(7)$							2.5000	1.8074	1.1990	0.9428	0.7906	0.6866	0.6097	0.5500	0.5020
$N_A(8)$								2.6925	1.9495	1.2991	1.0255	0.8628	0.7515	0.6691	0.6050
$N_A(9)$									2.8723	2.0821	1.3924	1.1024	0.9300	0.8120	0.7246
$N_A(10)$										3.0414	2.2068	1.4800	1.1747	0.9931	0.8688
$N_A(11)$											3.2016	2.3248	1.5628	1.2429	1.0528
$N_A(12)$												3.3541	2.4372	1.6416	1.3078
$N_A(13)$													3.5000	2.5446	1.7168
$N_A(14)$														3.6401	2.6477
$N_A(15)$															3.7749

Table 6.8

Matrix of coefficients α''_{ij} used in calculation of $N_V(j)$ as derived from eqs (6.82) and (6.83) according to Goldsmith (1967) and Cruz-Orive (1979). Note that some coefficients are negative. The bottom line serves the calculation of the overall particle number per unit volume. These coefficients also correspond to $p^{i\alpha}$ in eq. (8.24)

Particle size	Profile size class i														
class j	$N_A(1)$	$N_A(2)$	$N_A(3)$	$N_A(4)$	$N_A(5)$	$N_A(6)$	$N_A(7)$	$N_A(8)$	$N_A(9)$	$N_A(10)$	$N_A(11)$	$N_A(12)$	$N_A(13)$	$N_A(14)$	$N_A(15)$
$N_V(1)$	+2.00000	−0.68328	+0.08217	−0.02799	−0.00048	−0.00298	−0.00126	−0.00094	−0.00062	−0.00045	−0.00034	−0.00026	−0.00020	−0.00016	−0.00013
$N_V(2)$		+0.89443	−0.47183	+0.04087	−0.02528	−0.00393	−0.00427	−0.00234	−0.00167	−0.00117	−0.00087	−0.00066	−0.00051	−0.00040	−0.00033
$N_V(3)$			+0.66667	−0.39550	+0.02903	−0.02416	−0.00516	−0.00488	−0.00288	−0.00208	−0.00150	−0.00113	−0.00087	−0.00069	−0.00055
$N_V(4)$				+0.55470	−0.34778	+0.02308	−0.02293	−0.00565	−0.00512	−0.00315	−0.00230	−0.00169	−0.00129	−0.00101	−0.00080
$N_V(5)$					+0.48507	−0.31409	+0.01947	−0.02176	−0.00582	−0.00518	−0.00327	−0.00242	−0.00180	−0.00139	−0.00109
$N_V(6)$						+0.43644	−0.28863	+0.01704	−0.02071	−0.00584	−0.00516	−0.00332	−0.00248	−0.00186	−0.00145
$N_V(7)$							+0.40000	−0.26850	+0.01527	−0.01977	−0.00579	−0.00509	−0.00332	−0.00250	−0.00189
$N_V(8)$								+0.37139	−0.25208	+0.01393	−0.01893	−0.00571	−0.00501	−0.00330	−0.00250
$N_V(9)$									+0.34816	−0.23834	+0.01287	−0.01818	−0.00560	−0.00491	−0.00327
$N_V(10)$										+0.32880	−0.22663	+0.01200	−0.01751	−0.00549	−0.00481
$N_V(11)$											+0.31235	−0.21649	+0.01128	−0.01691	−0.00538
$N_V(12)$												+0.29814	−0.20761	+0.01067	−0.01636
$N_V(13)$													+0.28571	−0.19973	+0.01015
$N_V(14)$														+0.27472	−0.19269
$N_V(15)$															+0.26491
N_V	+2.00000	+0.21115	+0.27700	+0.17208	+0.14056	+0.11435	+0.09722	+0.08436	+0.07453	+0.06674	+0.06042	+0.05519	+0.05080	+0.04705	+0.04382

6.4. Size distribution of profiles when spheres are cut with a section of finite thickness

All that has been said so far on the relationship between sphere and profile size distributions assumed that section thickness was negligible and that the spheres were cut by a true plane. In practice this is often not the case and we must now examine the effect that finite section thickness has on the profile size distribution and then seek ways to account for this when we try to derive a sphere size distribution from that of observed profiles.

6.4.1. EFFECT OF SECTION THICKNESS ON THE PROFILE SIZE DISTRIBUTION

We have seen above that a finite section thickness t yields an overestimate of the number of profiles seen on a section as compared to an infinitely thin section, such that

$$N_A = N_V(D+t) = N_V \cdot D + N_V \cdot t \tag{6.84}$$

If we now look at one given size class of particles we note that all the additional $N_V \cdot t$ profiles are in the largest profile size class. The reason is easily seen from Fig. 6.12: if t increases the number of profiles for which the particle's great circle is projected increases whereas the number of profiles of particles with their centre outside the slice, but cut by one of the slice surfaces, remains unchanged. This becomes immediately clear if we construct the size distribution of profiles by the graphical procedure shown in Fig. 6.13, in analogy to that used previously in Fig. 6.4. It becomes evident that the probability of finding a profile in the largest size

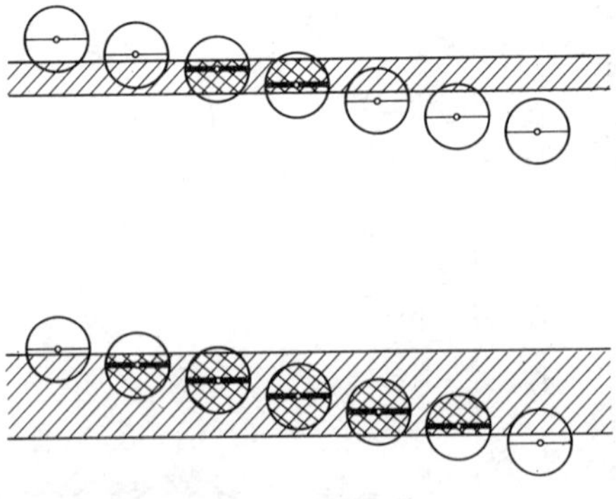

FIG. 6.12: Effect of section thickness is to increase the number of spheres whose main circle is seen on the projection.

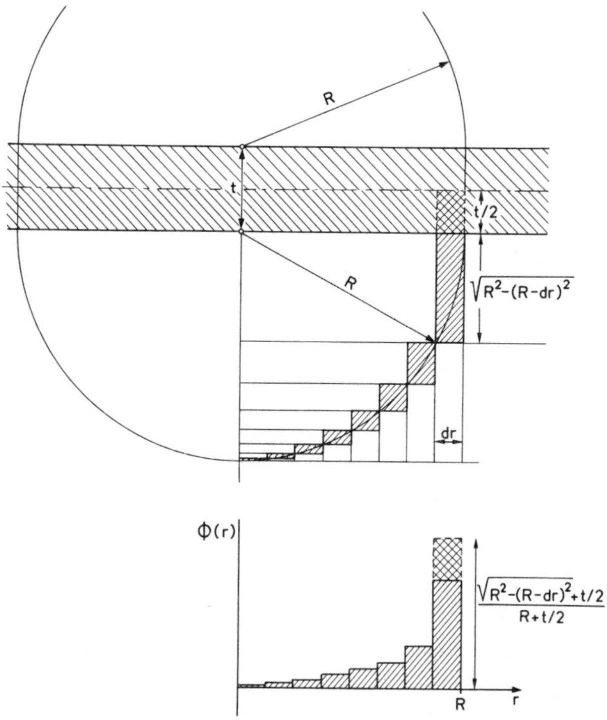

FIG. 6.13: Modification of graphical derivation of profile size distribution (see Fig. 6.4) to allow for section thickness t.

class increases by $t/2$ for the hemisphere or by t for the entire sphere. The frequency of profiles in this largest size class is

$$\phi(r_{\max}) = \frac{\sqrt{[R^2-(R-dr)^2]}+t/2}{R+t/2} \tag{6.85}$$

instead of

$$\phi(r_{\max}) = \frac{\sqrt{[R^2-(R-dr)^2]}}{R} \tag{6.86}$$

as found before. The number of profiles in the other size classes is unaffected by t, but their relative frequency will change.

The effect that finite section thickness has on the profile size distribution derived from a population of spheres of varying diameter $2R_j$ is immediately evident: the numerical density of profiles in size class i will contain two types of profiles:

(1) those representing spheres cut by one of the surfaces of the thick section or slice; they will obey the laws described in the preceding paragraphs, i.e. the spheres from which they are derived must be in the size range $\{R_j \geqslant r_i\}$;
(2) projections of great circles of spheres of size $R_j = r_i$; the contribution of these spheres must be proportional to $t/(2R_j)$.

Hence, the numerical density of profiles observed on a thick section is, considering eq. (6.84),

$$N'_A(r_i) = 2 \cdot \sum_{j=i}^{m} N_{Vj}[\sqrt{[R_j^2 - (r_i - \Delta r)^2]} - \sqrt{[R_j^2 - r_i^2]}] + \frac{t}{2R_j} \cdot N_V(R_i) \tag{6.87}$$

Expressing the numerical densities as frequencies, with $f(r)$ applying to profiles and $F(R)$ to spheres, and choosing a form analogous to eq. (6.42), Bach (1959, 1967) has written this as

$$f(r)\,\Delta r = 2r\,\Delta r \sum \frac{F(R)\,dR}{\sqrt{[R^2 - r^2]}} + t \cdot F(r)\,\Delta r \tag{6.88}$$

or, in integral form

$$f(r) = 2r \int_0^{\infty} \frac{F(R)\,dR}{\sqrt{[R^2 - r^2]}} + t \cdot F(r) \tag{6.89}$$

where $F(r)$ is the frequency of spheres of radius $R = r$.

In order to show this effect we have assumed in Fig. 6.14 a symmetrical size distribution of spheres with mean 10.5. We have then derived, by a Saltykov procedure, the profile size distribution resulting from cutting with a plane (open histogram) and have added to this the additional profiles observed because of a finite section thickness $t = 2$; they are shown as hatched bars. There are evidently no additional profiles in all those classes that do not contain spheres. The practical problem now is to find a method for unfolding the profile size distribution by accounting for the section thickness effect.

6.4.2. THE METHOD OF BACH (1967) FOR DERIVING SPHERE SIZE DISTRIBUTION FROM PROFILE SIZE DISTRIBUTIONS OBSERVED ON THICK SECTIONS

Bach (1967) has devised a method for deriving a sphere size distribution from an observed size distribution of profiles in a manner analogous to

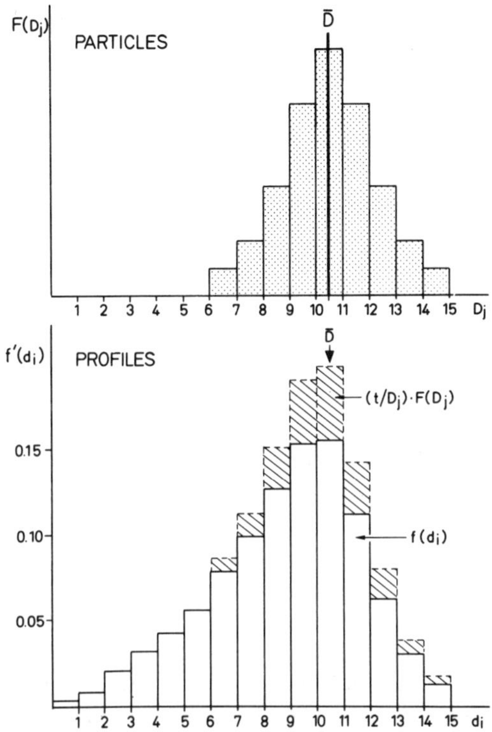

FIG. 6.14: Section thickness adds profiles in those size classes containing particles.

the methods of Wicksell (1925) and Saltykov (1958), but allowing for finite section thickness. He starts with eq. (6.89) and forms classes of profile (and sphere) size by integrating between the lower (r_-) and upper (r_+) class limits; the profile frequency in class i is:

$$p_i = \int_{r_-}^{r_+} f(r)\, dr = t \cdot P_i + 2 \int_{r_-}^{r_+} r \int_0^{\infty} \frac{F(R)\, dr}{\sqrt{[R^2 - r^2]}}\, dr \qquad (6.90)$$

where P_i is the frequency of spheres in size class i. Note that the frequencies are

$$p_i = \frac{N_A(i)}{N_A}; \qquad P_i = \frac{N_V(i)}{N_V} \qquad (6.91)$$

Without going through the entire derivation we state his final result, where he finds, with Δr as the class interval, for the frequency of profiles in class i

$$p_i = (t + 2 \cdot \Delta r \cdot k_{ii}) P_i + 2\Delta r \sum_{j=i}^{m} k_{ji} \cdot P_j \qquad (6.92)$$

(Note: this has been rewritten to have the classes range from $i = 1, 2, \ldots, m$ in conformity with the preceding text, whereas Bach used $i = 0, 1, 2, \ldots, n$ thus indexing the lower class limit.)
The coefficients k_{ji} are the same as those used in Saltykov's procedure (see Section 6.3.2); they are shown in Table 6.2. Here again it is preferable to use the coefficients k''_{ij} modified by Cruz-Orive (1977) given in Table 6.7 in order to eliminate bias (see Section 6.3.4). From this the sphere frequency becomes

$$P_i = \frac{p_i}{(t + 2\Delta r \cdot k_{ii})} - \frac{2\Delta r}{(t + 2\Delta r \cdot k_{ii})} \cdot \sum_{j=i}^{m} k_{ji} \cdot P_j \qquad (6.93)$$

One difference to the methods of Saltykov and Wicksell must be noted: it is not possible to calculate once and for all a matrix of coefficients α_{ij}, as given in Tables 6.3, 6.5 and 6.8 because this matrix depends on t.

These formulae apply for discrete sphere size distributions. Bach (1967) has derived a similar set of equations applying to continuous distributions; the only thing that changes is the set of coefficients k_{ji}. However, Bach also showed that the difference between the two methods is minor, provided a sufficiently large number of classes is used.

We should finally also mention that Goldsmith (1967) has worked out a number of relations pertaining to the unfolding of particle size distributions when these are studied on thick slices whose thickness is of the order of the particle diameter.

6.5. Relationship between moments of particle and profile size distribution

The rather elaborate procedures for reconstructing a sphere size distribution from one of observed profiles by either one of the methods of Wicksell (1925), Saltykov (1958) or Bach (1967) very often yield much more information than is actually required. In many instances one wants to know the number of particles, their mean diameter and the standard deviation of the distribution, perhaps also mean volume and surface. These parameters are, however, directly accessible without going through a reconstruction procedure, this because Wicksell (1925) and Bach (1959, 1967) have derived some important relationships between the moments of particle size distribution and those of the profile size distribution.

Expressed in terms of particle and profile radii, R and r respectively, the kth moments of their distributions are defined as

$$M_k = \int_0^\infty R^k \cdot F(R)\, dR \tag{6.94}$$

$$m_k = \int_0^\infty r^k \cdot f(r)\, dr \tag{6.95}$$

where we are using m_k for profiles and M_k for particles. If $F(R)$ and $f(r)$ are normalized or relative frequencies of particles and profiles respectively, it is evident that the 0th moments, M_0 and m_0 respectively, are just unity. If they are, however, expressed as real numbers, or as numerical densities, then the 0th moments are the overall numerical densities of all particles or profiles in all size classes. With either definition of frequencies, we find that

$$\frac{m_1}{m_0} = \frac{\int_0^\infty r \cdot f(r)\, dr}{\int_0^\infty f(r)\, dr} = \bar{r} \tag{6.96}$$

and

$$\frac{M_1}{M_0} = \frac{\int_0^\infty R \cdot F(R)\, dR}{\int_0^\infty F(R)\, dR} = \bar{R} \tag{6.97}$$

are the mean profile and particle radius respectively; the second moment

$$\frac{M_2}{M_0} = \frac{\int_0^\infty R^2 \cdot F(R)\, dR}{\int_0^\infty F(R) \cdot dR} = g \cdot \bar{s} \tag{6.98}$$

is related to the particle surface with $g = 1/(4\pi)$ as a geometric factor for spheres, and so on.

The relationship between m_k and M_k is found, according to Bach (1959, 1967) by multiplying eq. (6.89) with r^k and integrating over r from 0 to ∞:

$$m_k = \int_0^\infty r^k \cdot f(r)\, dr = 2\int_0^\infty F(R) \int_0^R \frac{r^{k+1}}{\sqrt{[R^2 - r^2]}}\, dr\, dR + t \cdot M_k \tag{6.99}$$

By integrating this double integral Bach (1967) found

$$m_k = \sqrt{\pi} \cdot \frac{\Gamma\left(\frac{k+2}{2}\right)}{\Gamma\left(\frac{k+3}{2}\right)} \cdot M_{k+1} + t \cdot M_k \tag{6.100}$$

$\Gamma(x)$ is the Gamma function whose values are easily found in statistical tables; for integer values of x we have $\Gamma(1+x) = x!$. The measurement of profile size distribution allows estimation of m_k by

$$m_k = \sum_{i=0}^{n} p_i \cdot r_i^k \tag{6.101}$$

where p_i is the measured frequency of profiles in class i.

The simplest relationship derived from eq. (6.100) is that between the 0th moments

$$\begin{aligned} m_0 &= 2 \cdot M_1 + t \cdot M_0 \\ &= M_0(2\bar{R} + t) \end{aligned} \tag{6.102}$$

or, defining $F(R)$ and $f(r)$ as numerical densities,

$$N_A = N_V(2\bar{R} + t) \tag{6.103}$$

the well-known relationship between profile and particle numbers derived in eq. (5.85). From this we immediately obtain the mean particle radius

$$\bar{R} = \frac{M_1}{M_0} = \frac{1}{2}\left(\frac{N_A}{N_V} - t\right) \tag{6.104}$$

The variance $\sigma^2(R)$ is

$$\sigma^2(R) = M_2/M_0 - (M_1/M_0)^2 \tag{6.105}$$

which is found to be

$$\sigma^2(R) = \frac{2}{\pi}\left(\frac{N_A}{N_V} \cdot \bar{r} - t \cdot \bar{R}\right) - (\bar{R})^2 \tag{6.106}$$

By the same approach it is also possible to derive the mean particle surface, $\bar{s}$, and volume, $\bar{v}$:

$$\bar{s} = 4\pi \frac{M_2}{M_0} = 8 \cdot \left(\frac{N_A}{N_V} \cdot \bar{r} - t \cdot \bar{R}\right) \tag{6.107}$$

$$\bar{v} = \frac{4\pi}{3} \cdot \frac{M_3}{M_0} = 2t^2 \cdot \bar{R} + (2\bar{R} + t) \cdot [\pi(\sigma^2(r) + \bar{r}^2) - 2t \cdot \bar{r}] \tag{6.108}$$

where the new parameter introduced, $\sigma^2(r)$, is the variance of the profile radii

$$\sigma^2(r) = m_2/m_0 - \bar{r}^2 \tag{6.109}$$

which can be calculated from the profile size distribution. Note also that evidently

$$\bar{r} = m_1/m_0 \tag{6.110}$$

The practical value of this approach is that one can easily calculate the moments of the profile size distributions by

$$m_k = \sum_{i=0}^{m} r_i^k \cdot f(i) \tag{6.111}$$

where summation is over all classes. Secondly, Bach (1967) has developed a formula by which one can directly obtain N_V from the profile size

Table 6.9

Coefficients $g(r_i/t)$ for calculating N_V by eq. (6.112) according to method of Bach (1967). From Bach (1967)

$r_i < t$		$r_i \geq t$			
r_i/t	$g(r_i/t)$	$r_i/t)$	$g(r_i/t)$	r_i/t	$g(r_i/t)$
0.00	1.2533	1.0	0.3535	3.0	0.1307
0.05	1.1372	1.1	0.3270	3.2	0.1228
0.10	1.0373	1.2	0.3039	3.4	0.1158
0.15	0.9509	1.3	0.2837	3.6	0.1095
0.20	0.8756	1.4	0.2659	3.8	0.1039
0.25	0.8097	1.5	0.2502	4.0	0.0988
0.30	0.7517	1.6	0.2361	5	0.0793
0.35	0.7004	1.7	0.2234	6	0.0662
0.40	0.6548	1.8	0.2120	7	0.0568
0.45	0.6140	1.9	0.2017	8	0.0497
0.50	0.5775	2.0	0.1923	9	0.0442
0.55	0.5446	2.1	0.1838	10	0.0398
0.60	0.5149	2.2	0.1759	11	0.0362
0.65	0.4880	2.3	0.1686	12	0.0332
0.70	0.4634	2.4	0.1620		
0.75	0.4410	2.5	0.1558	for larger values:	
0.80	0.4205	2.6	0.1501		
0.85	0.4017	2.7	0.1447		
0.90	0.3844	2.8	0.1397		
0.95	0.3683	2.9	0.1351	$g(r_i/t) \approx t/\{r_i \cdot \sqrt{(2\pi)}\}$	

distribution; it reads

$$N_V = \frac{1}{t}\sqrt{\left(\frac{2}{\pi}\right)} \cdot \sum_{i=1}^{m} f(i) \cdot g(r_i/t) \tag{6.112}$$

where $g(r_i/t)$ is a complex function whose values can be read from Table 6.9.

It is evident that with N_V calculated by this formula, and $N_A = m_0$, $\bar{r} = m_1/m_0$ etc., one can now directly obtain the mean and variance of the particle diameter distribution, as well as the mean particle volume and surface area. For many practical applications this is really all the information one needs.

CHAPTER 7

Estimation of particle size distribution from measurements of intercept length on a random line

A number of methods have been developed which establish a relationship between the distribution of particle diameters and the distribution of intercept lengths formed between the particles and long random test lines (Spektor, 1950; Lord and Willis, 1951; Cahn and Fullman, 1956). These methods have been presented in detail by Underwood (1968, 1970).

7.1. Spheres of uniform size

If a random line crosses a sphere, a certain chord length l_i will be contained within the sphere, whereby l_i evidently depends on the sphere diameter $D = 2R$ and on the distance, d_i, between the sphere centre and the line (Fig. 7.1):

$$l_i = 2\sqrt{(R^2 - d_i^2)} \tag{7.1}$$

It is also evident that all chords lying on a cylindrical surface of radius d_i around the sphere centre will have the same length; furthermore, all chords lying within this cylinder will have chord length lying between l_i and $2R$, the sphere diameter, whereas those lying outside will be shorter than l_i. From this we can define the probability that chords of a certain range of lengths are obtained.

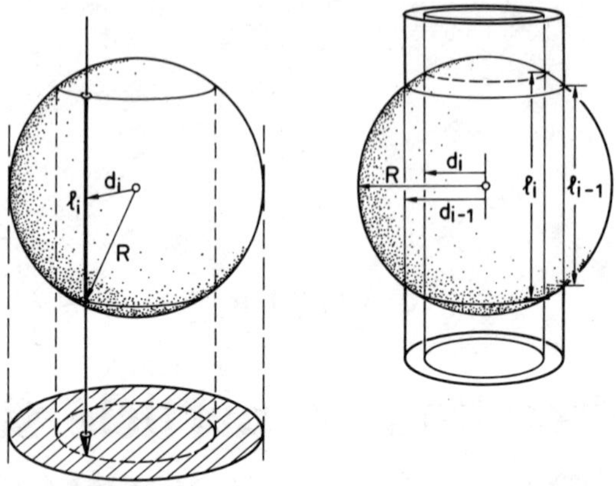

FIG. 7.1: Intercept length l of test line with sphere of radius R depends on distance d from centre of sphere.

The probability that a random line hits the sphere is, according to Section 2.2, proportional to its projection area πR^2. Likewise the probability that this line falls within the cylinder of radius d_i is proportional to its cross-section πd_i^2. Hence the probability to find an intercept $l \geq l_i$

$$\Pr\{l \geq l_i\} = \frac{\pi d_i^2}{\pi R^2} = \frac{d_i^2}{R^2} \tag{7.2}$$

Expressing the cylinder diameter in terms of l_i, by eq. (7.1), this becomes

$$\Pr\{l \geq l_i\} = \frac{R^2 - \frac{1}{4} l_i^2}{R^2} = \frac{D^2 - l_i^2}{D^2} \tag{7.3}$$

Likewise the probability that $l \geq l_{i-1}$ is

$$\Pr\{l \geq l_{i-1}\} = \frac{D^2 - l_{i-1}^2}{D^2} \tag{7.4}$$

It is now easy to see that intercepts that have length in the range $\{l_{i-1} < l < l_i\}$ must hit the sphere within a hollow cylindrical shell of outer and inner radius d_{i-1} and d_i respectively, (Fig. 7.1) so that the probability of obtaining such intercepts is

$$\Pr\{l_{i-1} < l < l_i\} = \Pr\{l \geq l_i\} - \Pr\{l \geq l_{i-1}\}$$

$$= \frac{l_i^2 - l_{i-1}^2}{D^2} \tag{7.5}$$

If l is measured in constant class intervals dl this may be rewritten as

$$\Pr\{l\}\,dl = \frac{l^2-(l-dl)^2}{D^2}$$

$$= \frac{2dl}{D^2}\cdot l - \frac{dl^2}{D^2} \tag{7.6}$$

This result is interesting since it shows that $\Pr\{l\}$ is linearly proportional to l; indeed if $dl\to 0$ the probability density function is

$$f(l) = \begin{cases}(2/D^2)\cdot l & \text{for} \quad l\leqslant D\\ 0 & \text{for} \quad l>D\end{cases} \tag{7.7}$$

The obvious result of this relationship is that the frequency distribution of intercepts is a straight line from the origin to the sphere diameter D but falls to 0 at $l=D$ (Fig. 7.2a).

Alternatively, retaining a classified measurement of intercept lengths, we may express both the sphere diameter D and l as multiples of dl such that

$$l = i\cdot dl, \qquad D = j\cdot dl \tag{7.8}$$

From eq. (7.6) we then obtain the frequency distribution

$$f(i) = (2i-1)/j^2 \tag{7.9}$$

As shown in Fig. 7.2b this yields a histogram which approximates the straight line function shown in relation to eq. (7.7). Note that i marks the upper limit of each size class.

Having so far elaborated the relative frequencies of intercept lengths, it now remains to work out the number of chords we may expect if a long test line of length L traverses a randomly distributed population of

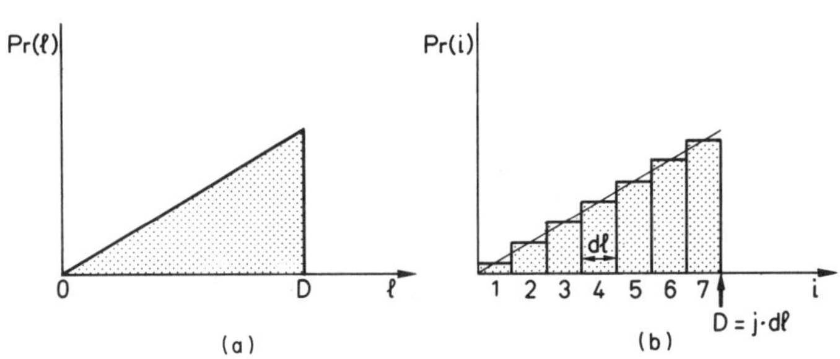

Fig. 7.2: Probability density functions for intercept lengths obtained on sphere of diameter D.

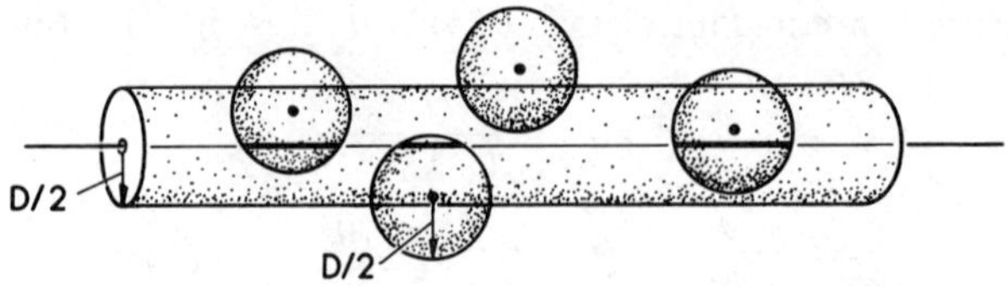

FIG. 7.3: Probability of hitting spheres with test line.

spheres of equal diameter D. We will express this as the number of chords per unit test line length,

$$N_L = N/L_T \tag{7.10}$$

in relation to the numerical density of spheres, N_V, defined as the number of sphere centres per unit volume.

It is quite evident from Fig. 7.3 that all spheres which have their centre within a cylinder of diameter D around the test line will be traversed by the line and will hence form a chord. It immediately follows that the number of chords is

$$N = N_V \cdot (\pi/4)D^2 \cdot L_T \tag{7.11}$$

or

$$N_L = N_V(\pi/4)D^2 \tag{7.12}$$

N_L is the total number of intercepts per unit test line length expected from N_V spheres of diameter D. Their length will vary from 0 to D. The number of intercepts of length l to $(l + dl)$ is obtained by combining eqs (7.6) and (7.12):

$$\begin{aligned} N_L(l)\, dl &= N_L \cdot \Pr(l)\, dl \\ &= N_V \cdot (\pi/4) \cdot (2 \cdot dl \cdot l - dl^2) \end{aligned} \tag{7.13}$$

If l is measured in discrete size classes, such that $l = i \cdot dl$, this becomes

$$N_L(i) = N_V \cdot (\pi/4) \cdot (2i - 1)\, dl^2 \tag{7.14}$$

Alternatively, if l is measured as a continuous intercept length distribution we find by eqs (7.7) and (7.12)

$$N_L(l) = N_L \cdot f(l) \tag{7.15}$$

For a population of spheres of equal diameter D the numerical frequency of intercept lengths l is again a straight line function between 0 and D.

7.2. Spheres of varying size

If a population of spheres of varying diameter D is considered, the distribution of intercept lengths will be affected in two ways: (a) the probability of obtaining an intercept is going to depend on the sphere's projection area which is proportional to D^2; (b) the intercept lengths are a function of D which is variable, and of the position of the intercept with respect to the sphere centre. The practical problem will be to unfold the sphere size distribution from the measured distribution of intercepts. This can be done either analytically or graphically.

7.2.1. GRAPHIC SOLUTION

Because of the simple straight line function obtained for the probability density of intercept lengths from spheres of uniform size (Fig. 7.2), a graphic solution is particularly easy. Each size class of spheres will yield a triangular distribution of intercept lengths extending from 0 to D; the overall intercept length distribution is simply the sum of these triangles (Fig. 7.4). Now the area of each triangle, $a(D)$, is proportional to the total number of intercepts derived from spheres of size D, hence, by eq. (7.12),

$$\begin{aligned} a(D) &= k \cdot N_L(D) \\ &= k \cdot (\pi/4) \cdot D^2 \cdot N_V(D) \end{aligned} \tag{7.16}$$

The graphical solution proposed by Lord and Willis (1951) allows $N_V(D)$ to be directly read off the graph if the triangles are extended towards an auxiliary vertical line at distance A from the origin. It is easily demonstrated as follows.

The area of the triangle is

$$a(D) = \tfrac{1}{2} D \cdot h \tag{7.17}$$

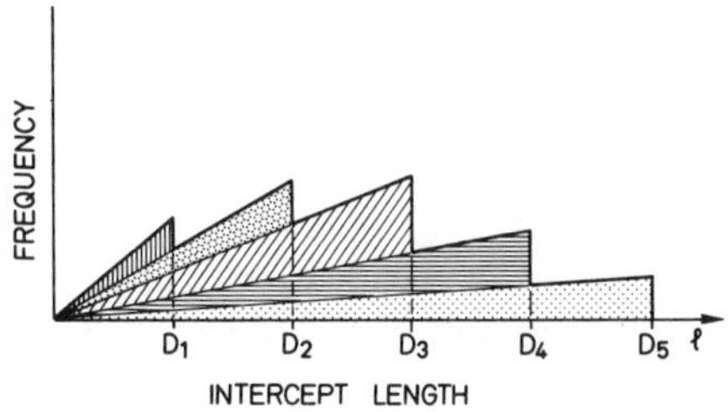

FIG. 7.4: Combined intercept length frequency distribution if spheres of varying diameter D are cut by line.

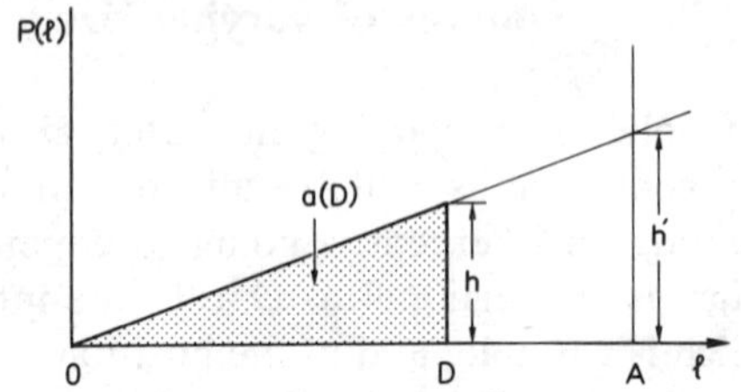

FIG. 7.5: Intercept length distribution from sphere of one size class D introducing graphical method.

where h is the height of the triangle. Since (Fig. 7.5)

$$\frac{D}{A}=\frac{h}{h'} \tag{7.18}$$

this becomes

$$a(D)=\tfrac{1}{2}D^2\cdot\frac{h'}{A} \tag{7.19}$$

and by substitution into eq. (7.16) and rearrangement we derive

$$N_V(D)=\left(\frac{2}{\pi\cdot k\cdot A}\right)\cdot h' \tag{7.20}$$

The term in brackets is a constant. Since the relative numerical frequency is

$$N_N(D)=\frac{N_V(D)}{\sum_D N_V(D)} \tag{7.21}$$

one immediately obtains

$$N_N(D)=\frac{h'(D)}{\sum_D h'(D)} \tag{7.22}$$

The very simple graphical procedure for estimating $N_N(D)$ is now immediately developed (Fig. 7.6):

(1) Plot the histogram of the observed intercept length distribution: note that D is at the upper limit of the classes.
(2) Draw straight lines from the origin through the centre of the columns (median) to the right hand limit of the class to generate the family of triangles, as in Fig. 7.6a.
(3) Extend these lines to the auxiliary line at A, and measure the line segments $h'(D)$. Equation (7.22) then allows immediate calculation of

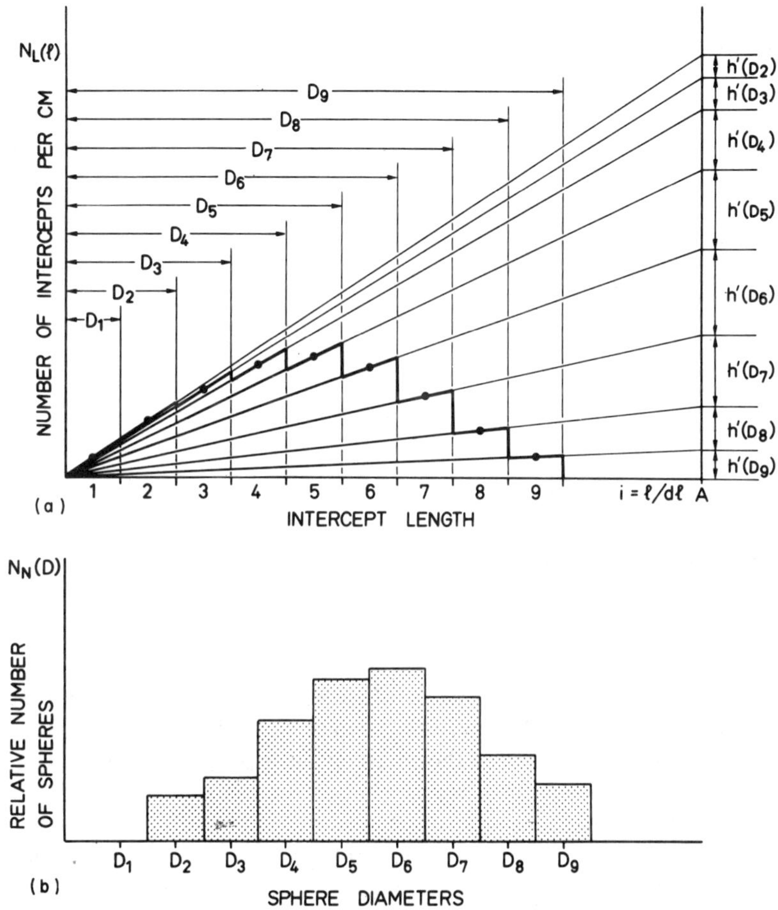

FIG. 7.6: Graphical method of Lord and Willis (1951) for estimating sphere size distribution from distribution of intercept lengths.

$N_N(D)$ and the plotting of a histogram of sphere size distribution (Fig. 7.6b). The practical procedures are further discussed in Volume 1, Section 5.3.

7.2.2. ANALYTICAL SOLUTION

Besides the original methods of Spektor (1950) and Lord and Willis (1951), a particularly compact development has been given by Cahn and Fullman (1956). We shall not go through the entire derivation, since it was lucidly reviewed by Underwood (1968, 1970).

The number of intercepts derived from spheres of the size class D to $(D+dD)$ is, by eq. (7.12),

$$N_L(D) = N_V(D) \cdot (\pi/4) \cdot D^2 \tag{7.23}$$

and the number of intercepts of length l to $(l+dl)$ derived from these spheres is

$$N_L(l, D)\, dl = N_V(D) \cdot \tfrac{1}{2}\pi l \cdot dl \tag{7.24}$$

Integrating over all sphere sizes that may contribute to l (note that $D \geqslant l$), we obtain

$$N_L(l)\, dl = \frac{\pi}{2}\, l \cdot dl \int_l^{\infty} N_V(D)\, dD \tag{7.25}$$

Solving eq. (7.25) for $N_V(D)$ yields

$$N_V(D=l) = -\frac{2}{\pi} \cdot \frac{d}{dl}\left[\frac{N_L(l)}{l}\right] \tag{7.26}$$

which by differentiation becomes

$$N_V(D=l) = \frac{2}{\pi l^2} \cdot N_L(l) - \frac{2}{\pi l} \cdot \left[\frac{dN_L(l)}{dl}\right] \tag{7.27}$$

The term in brackets is the slope of the intercept length distribution at l. Evidently, the sphere size distribution can now be derived by measuring this slope at various points of the intercept length distribution.

An important practical question is whether an intercept length distribution measured on random section profiles of a population of spheres is suitable for this analysis, i.e. whether it has the same probability density function as that of intercepts generated by a random test line in space. This is the case if the following conditions are observed: (1) the population of spheres is sampled with the section plane in proportion to their diameters, i.e. large spheres are more likely to be cut than small ones; (2) the population of profiles is sampled with a test line in proportion to the profile diameter. These two conditions are natural if isotropic uniform random sampling of both the structure and the section is performed.

CHAPTER 8

Estimation of size distribution of non-spherical particles from information on sections

Much has now been said about the ways and means by which one can arrive at an estimate of size distribution of spherical particles. But the sphere is not really an ubiquitous, not even a frequently occurring shape of natural objects. The constraints of the "environment" often impose a particular shape on cells and their constituents: thus cells which are basically spherical objects become polyhedral in shape when they are densely packed; and the nuclei simulate the shape of the cell, becoming flattened or oblate ellipsoids in flattened cells, or prolate ellipsoids in columnar or elongated cells.

In Section 5.5 we discussed extensively how the mean caliper diameter of convex solids is related to their shape and dimensions; in Volume 1, Section 5.1 we discussed the limitations of approximating the shape of a non-spherical particle with a model sphere. The conclusion was that there is a certain error introduced if one substitutes a "volume-equivalent" sphere for non-spherical particles, but that this error is predictable. For higher order polyhedra it is rather small, but it is appreciable for shapes such as cubes, octahedra or cylinders. With ellipsoids the error evidently increases as their axial ratio deviates from 1.

If we wish to consider the relationship between size distribution of non-spherical particles and their profiles we evidently meet with greater problems than in the preceding Chapters 6 and 7. If we allow the size and the shape of the particles to vary, we will have to establish a relationship between a particle size and shape distribution and the corresponding profile size and shape distribution resulting from random sectioning. This

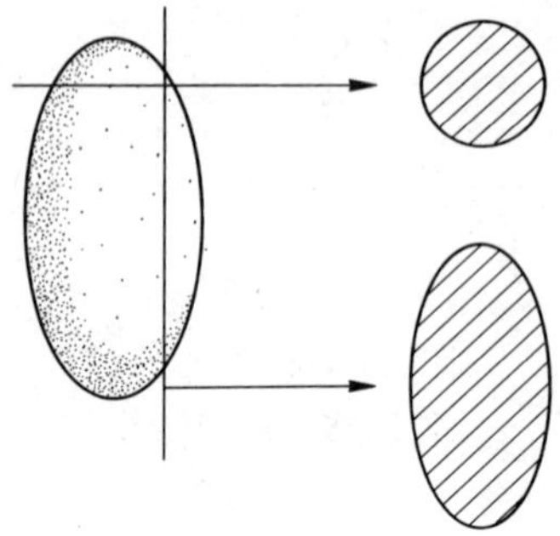

FIG. 8.1: Transverse and longitudinal section of spheroid. Axial ratio of spheroid is reflected in axial ratio of longitudinal section.

is not a simple straight-forward matter, because even if the particles should all have the same shape, the profiles will not: if, for example, a prolate ellipsoid of axial ratio 2:1 is randomly cut the profiles will range from circles (axial ratio 1:1) to ellipses of axial ratio 2:1 depending on the spatial orientation between the particle and the section plane (Fig. 8.1). The situation is hence rather complicated.

8.1. The method of Wicksell for ellipsoids cut by a plane

In his second classical memoir on the corpuscle problem Wicksell (1926) dealt with the case of ellipsoidal particles.

Wicksell chose the geometric means of the diameters of triaxial ellipsoids and of their profiles as the characteristic measure or "diameter". He then found first that the frequency of profile "diameters" is related to the "diameter" of the central ellipse of the particle by the same fundamental Abel-type integral equation as he had found for spheres (eq. 6.47). For prolate spheroids (ellipsoids of revolution) this was also the case if he chose the *minor* axis of both ellipsoids and section ellipses, or if for oblate spheroids he used the *major* axis. Without stating any of his mathematical reasoning—worth reading in the original—let us just quote his conclusion in section 11 (p. 166):

> It is thus seen that we have exactly the same relations between the functions F (frequency of the central diameter of the ellipsoid) and ϕ (relative frequency of diameters of section ellipses) as in the case of spherical corpuscles. Hence applying the reductions (or transformations as used here) deduced for the case of spherical corpuscles to the distribution of the *minor* diameters in section ellipses we obtain the distribution of the *minor* diameters of the corpuscles, if they are all *prolate* ellipsoids of revolution. ... It is

evident that the same result will hold if the corpuscles are all *oblate* ellipsoids of revolution, if we then apply the spherical reductions to the distribution of the *major* diameters in the section ellipses. The result obtained will be the distribution of the *major* diameters of the corpuscles.

With this conclusion we now have a method at hand for estimating the size distribution of ellipsoids of revolution. If we know that they are prolate spheroids we measure on the section the *minor* diameter of the elliptic profiles; if they are oblate we measure the *major* diameter. We then use any of the methods proposed in Sections 6.3 and 6.4 and in Volume 1, Section 5.2 to obtain the size distribution of minor or major axes respectively for the prolate or oblate spheroids. If we have an estimate of the axial ratio of these spheroids we can then also calculate other parameters, by the formulae given in Section 5.5 and Table 5.2, such as the mean caliper diameter (required for estimating the particle number), the volume or surface distributions, etc.

The difficulty with Wicksell's analysis is that it is not valid for a general population of ellipsoids because it was based on the rather strong assumption that size and shape were statistically independent. The general problem, which involves bivariate distributions of size and shape, has been solved by Cruz-Orive (1976, 1978a), as shown in the next section.

8.2. Size-shape distributions of ellipsoidal particles derived from plane sections‡

8.2.1. THE PROBLEM

Spherical particles, or particles modelled by geometrical bodies of a constant shape (e.g. ellipsoids with constant axial ratios, cylinders with a constant height-to-diameter ratio, etc.), can be described by a single size parameter as proposed by Wicksell (1926). The particulate phase is then described by a univariate size distribution, and the stereological problem consists simply in identifying this distribution from the univariate size distribution describing the set of profiles produced by a random plane section through the aggregate of particles. In practice, the problem consists in estimating the particle size histogram from the corresponding profile size histogram.

Particles exhibiting variation about a given type of shape as well as size variation, however, cannot be described by a single parameter. Such is the case of, say, ellipsoidal particles of variable size and excentricities. Thus a complete description of a variable triaxial ellipsoid requires three

‡ I am most grateful to Dr Luis M. Cruz-Orive for having contributed this section.

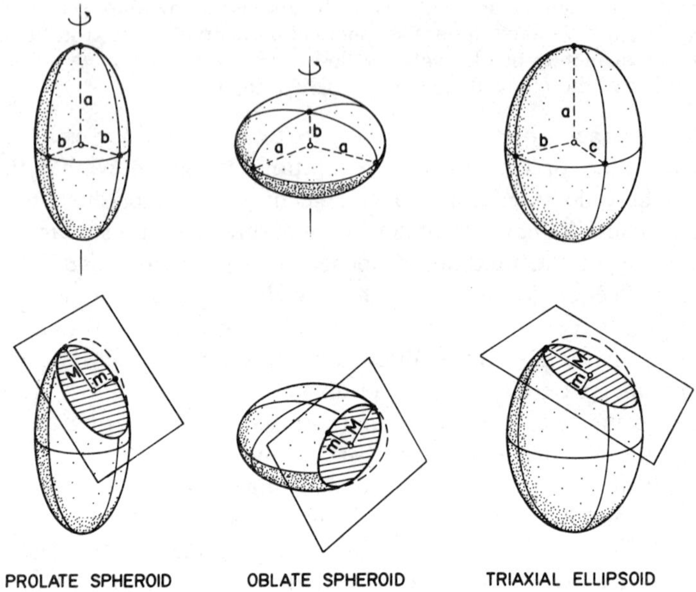

FIG. 8.2: Characterization of ellipsoids by three axial half diameters (a, b, c) and of ellipses of section by two half diameters (m, M).

parameters, whereas that of a variable ellipsoid of revolution (= *spheroid*) requires two parameters (Fig. 8.2). The corresponding distribution functions will be three- and bi-dimensional, respectively. It seems intuitively clear that a p-dimensional particle distribution can be identified from the corresponding profile distribution only if the latter has a dimension greater than, or equal to p. For example, the problem of identifying a three-variate distribution describing variable triaxial ellipsoids from plane sections would be indeterminate, because all the resulting profiles are ellipses (Fig. 8.2), which can be fully described by a bivariate distribution only (e.g. that governing their major and minor principal axes). This fact was conjectured by Moran (1972) and proved by Cruz-Orive (1976). The reader is referred to the introduction of the latter paper for a brief history of the problem, which starts with Wicksell (1926). Solutions for special cases have been given. Thus Tallis (1970) considers oriented triaxial ellipsoids cut either by a transversal plane, or by a vertical plane.

On the other hand, the problem of identifying a bivariate distribution describing a phase of variable spheroids from the corresponding bivariate profile distribution is determinate *provided* the spheroids are all of the same type (Fig. 8.3), namely, either all prolate (= generated by ellipses revolving around their *major* principal axis), or all oblate (= generated by

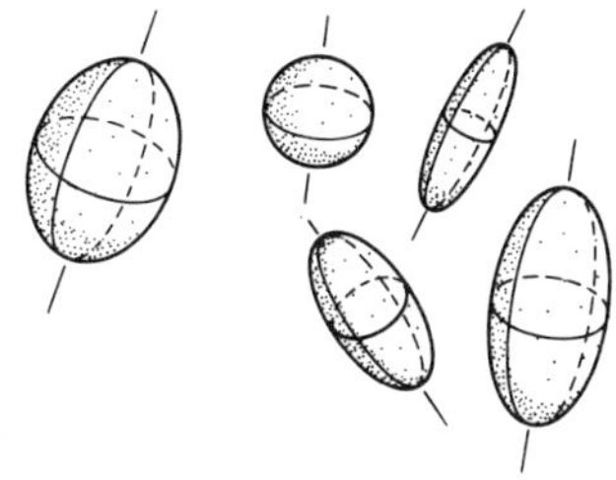

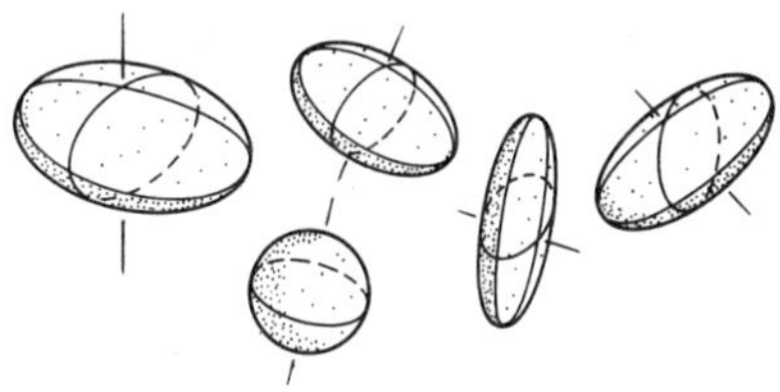

FIG. 8.3: Shape variation by variation of axial ratio within the types of spheroids (prolate or oblate).

ellipses revolving around their *minor* principal axis). Otherwise, the relative proportion of each type of spheroids in the mixture introduces an extra parameter (note that a given elliptical profile does not provide information regarding the parent spheroid type).

Modelling a phase of particles of variable size and shape by variable spheroids of the same type cut by a sectioning plane seems appropriate for three main reasons. (1) Often, the model will be much more realistic than a sphere's model, or a model consisting of spheroids of identical shape (= "similar"). (2) The profiles are always of the same type, namely ellipses, thus making the problem manageable in theory and practice. (This property is not shared by bodies like polyhedra, cylinders or cones.) (3) The stereological problem is determinate. (If the probes used are lines instead of planes, the problem becomes indeterminate, because intercept length distributions are essentially univariate.)

In setting up the model the spheroid centres are assumed to be uniformly scattered within the specimen; the spheroids are isotropically and independently orientated about their centres and, finally, the size and shape of a randomly chosen spheroid are independent from its position within the specimen. All the foregoing assumptions will be fulfilled only approximately, because the spheroids cannot overlap.

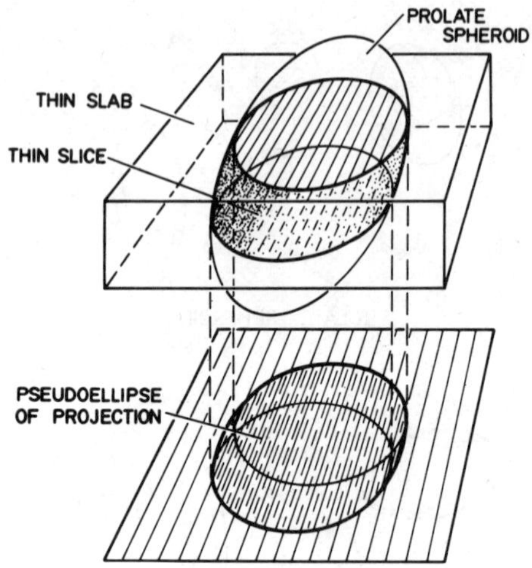

FIG. 8.4: Pseudoellipse seen on the projected image of a slice of a spheroid.

The probe used is a true plane of section. Should section thickness be introduced in the model, insurmountable analytical problems would arise, because a slice of a spheroid may not appear as an ellipse when projected on to the observation plane (Fig. 8.4).

In the following, both the continuous and the discrete approaches to the problem are briefly examined. The former assumes a continuous variation in the size and shape of spheroids and profiles. However unrealistic this approach may look, it brings the problem into a deep perspective, and allows the derivation of useful moment relationships. The discrete approach yields practical algorithms for estimating ("unfolding") the bivariate frequency histogram of the particles from that of the profiles. For completeness, the reader is referred to the papers by Cruz-Orive (1976, 1978a, 1978b).

8.2.2. CONTINUOUS SIZE-SHAPE SPHEROID MODELS

Either a prolate, or an oblate spheroid is fully determined by giving the lengths a and b of its major and minor principal axes, respectively (Fig. 8.2). In our model, both a and b are variable. The derivations become much simpler, however, if a prolate spheroid is described by the variables (b, x^2) and an oblate spheroid by (a, x^2). The size component b (or a) may vary between 0 and ∞, whereas the shape component $x^2 = 1-(b/a)^2$ (which is the square excentricity parameter) may vary between 0 and 1. (The value $x^2=0$ corresponds to a sphere; for $x^2=1$, a prolate spheroid becomes a line segment, whereas an oblate spheroid becomes a circular platelet.) The respective size-shape density functions are denoted by $g(b, x^2)$ and by $g(a, x^2)$.

Likewise, an ellipse on section is determined by the lengths M and m of its major and minor principal axes. However, the variables adopted are (m, y^2) if the parent spheroid is prolate, and (M, y^2) if it is oblate. Both m and M may vary between 0 and ∞, whereas the square excentricity parameter $y^2 = 1-(m/M)^2$ always varies between 0 and 1. The profile size-shape density functions are denoted by $f(m, y^2)$, $f(M, y^2)$ in the respective cases.

The integral equations expressing the profile densities in terms of the particle densities are shown to be as follows. For prolate spheroids,

$$f(m, y^2) = (2\bar{H})^{-1} m(1-y^2)^{\frac{1}{2}} \int_m^\infty \frac{db}{\surd(b^2-m^2)} \int_{y^2}^1 \frac{g(b, x^2)\, dx^2}{x\surd(1-x^2)\surd(x^2-y^2)} \tag{8.1}$$

valid for $(0<m<\infty,\ 0\leq y^2<1)$.

For oblate spheroids,

$$f(M, y^2) = (2\bar{H})^{-1} M(1-y^2)^{-2} \int_M^\infty \frac{da}{\surd(a^2-M^2)} \int_{y^2}^1 \frac{(1-x^2)^{\frac{3}{2}} g(a, x^2)}{x\surd(x^2-y^2)}\, dx^2 \tag{8.2}$$

valid for $(0<M<\infty,\ 0\leq y^2<1)$.

In both cases, $\bar{H}$ denotes the overall mean caliper diameter of the spheroids, namely,

$$\bar{H} = \int_0^\infty \int_0^1 H(b, x^2) g(b, x^2)\, db\, dx^2 \tag{8.3}$$

where $H(b, x^2)$ is the mean caliper diameter of a fixed spheroid of parameters (b, x^2). If the spheroid is prolate,

$$H(b, x^2) = \tfrac{1}{2} b(1-x^2)^{-\frac{1}{2}} Q_p(x) \tag{8.4}$$

where $Q_p(x) = 1 + x^{-1}(1-x^2)\tanh^{-1} x$. If the spheroid is oblate,

$$H(a, x^2) = \tfrac{1}{2} a Q_0(x) \tag{8.5}$$

where $Q_0(x) = (1-x^2)^{\frac{1}{2}} + x^{-1} \sin^{-1} x$. As x^2 tends to 0, both $Q_p(x)$ and $Q_0(x)$ tend to 2.

Both eqs (8.1) and (8.2) are double integral equations of Abel-type. They constitute a generalization of the simple integral equation derived in Section 6.3 for spheres. Indeed, both equations collapse into the latter for $x^2=0$. Note that the first integral depends exclusively on particle and profile size, whereas the second integral contains both the bivariate size-shape distribution and the shape parameters.

The inversion of these two equations yields the following solutions. For *prolate spheroids* and provided that $f(\infty, y^2)=0$ and $(1-y^2)^{-\frac{1}{2}}f(m, y^2)\to 0$ as $y^2\to 1$, we have,

$$g(b, x^2)=4\pi^{-2}\bar{H}bx\sqrt{(1-x^2)}\int_b^\infty \frac{dm}{\sqrt{(m^2-b^2)}}\int_{x^2}^1 (y^2-x^2)^{-\frac{1}{2}} \times\frac{\partial^2}{\partial m\,\partial y^2}\left[\frac{f(m, y^2)}{m\sqrt{(1-y^2)}}\right]dy^2 \quad (8.6)$$

valid for $(0<b<\infty,\ 0\leqslant x^2<1)$.

For *oblate spheroids*, and provided that $f(\infty, y^2)=0$, $(1-y^2)^2 f(M, y^2)\to 0$ as $y^2\to 1$, we have

$$g(a, x^2)=4\pi^{-2}\bar{H}ax(1-x^2)^{-\frac{3}{2}}\int_a^\infty \frac{dM}{\sqrt{(M^2-a^2)}}\int_{x^2}^1 (y^2-x^2)^{-\frac{1}{2}} \times\frac{\partial^2}{\partial M\,\partial y^2}\left[\frac{(1-y^2)^2 f(M, y^2)}{M}\right]dy^2 \quad (8.7)$$

valid for $(0<a<\infty,\ 0\leqslant x^2<1)$.

On the basis of these basic relations we can now derive some relations between moments of the particle and the profile distributions. The expected value of a function of the form

$$u(b, x^2)=b^p x^r(1-x^2)^s\{Q(x)\}^k \quad (8.8)$$

becomes available from the two preceding relationships in terms of profile measurements alone. These relationships are useful because, by suitable choices of the constants p, r, s and k, a large number of interesting parameters are obtained. For example, let the spheroids be prolate. Putting $p=1$ and $r=s=k=0$, one has $u=b$, the minor axis length; $p=r=k=0$ and $s=\frac{1}{2}$ gives $u=b/a$, the axial ratio, and so forth. For prolate spheroids, the relationship mentioned has the following form,

$$E_g\{b^p x^r(1-x^2)^s[Q(x)]^k\}=\bar{H}c_p^{-1}E_f\{m^{p-1}q(y^2\mid r, s, k)\} \quad (8.9)$$

For oblate spheroids,

$$E_g\{a^p x^r(1-x^2)^s[Q(x)]^k\}=\bar{H}c_p^{-1}E_f\{M^{p-1}(1-y^2)^{\frac{5}{2}}q(y^2\mid r, s-2, k)\} \quad (8.10)$$

In both cases, $Q(x)$ denotes an arbitrary shape function which is differentiable with respect to x^2. E_g and E_f denote expectations with respect to the densities g and f, respectively, whereas

$$c_p=\frac{\sqrt{\pi}}{2}\Gamma\left(\frac{p+1}{2}\right)\Big/\Gamma\left(\frac{p+2}{2}\right), \qquad p>-1 \quad (8.11)$$

where $\Gamma(n)$ is the gamma function, having the recurrence relation $\Gamma(n) = (n-1)\,\Gamma(n-1)$ and $\Gamma(\frac{1}{2}) = \sqrt{\pi}$, $\Gamma(1) = 1$. Thus $c_0 = \pi/2$, $c_1 = 1$, $c_2 = \pi/4$, $c_3 = \frac{2}{3}$, etc.

The function $q(y^2 \mid r, s, k)$ is a shape factor having the following expression

$$q(y \mid r, s, k) = (2/\pi) y^r (1-y^2)^{-\frac{1}{2}} \int_0^{\pi/2} (\sin\theta)^r (1 - y^2 \sin^2\theta)^{s-\frac{1}{2}} \{Q(y \sin\theta)\}^{k-1}$$
$$\cdot \{(r+1-(r+2s+2)y^2 \sin^2\theta) Q(y\sin\theta)$$
$$+ 2ky^2 \sin^2\theta (1 - y^2 \sin^2\theta) Q'(y \sin\theta)\}\, d\theta \quad (8.12)$$

where $Q'(y \sin\theta) = [dQ(x)/dx^2]_{x=y\sin\theta}$.
For $k = 0$,

$$q(y \mid r, s, 0) = (2/\pi) y^r (1-y^2)^{-\frac{1}{2}} \int_0^{\pi/2} (\sin\theta)^r (1 - y^2 \sin^2\theta)^{s-\frac{1}{2}}$$
$$\times \{r+1-(r+2s+2)y^2 \sin^2\theta\}\, d\theta \quad (8.13)$$

A number of interesting $q(y^2 \mid r, s, k)$ coefficients appear in Table 8.1. It should be pointed out that the integrals leading to these coefficients

Table 8.1

Auxiliary coefficients $q(y^2 \mid r, s, k)$ for estimating size shape parameters of spheroid distribution (eqs 8.12–8.13). Reproduced with permission from Cruz-Orive (1978b)

y^2	$q(y \mid 0, 0, 0)$	$(1-y^2)^{\frac{5}{2}} \cdot q(y \mid 0, -1, 0)$	$(1-y^2)^{\frac{5}{2}} \cdot q(y \mid 0, -2, 0)$	$(1-y^2)^{\frac{3}{2}} \cdot q_1(y \mid 0, -1, 2)$	$(1-y^2)^{\frac{5}{2}} \cdot q_2(y \mid 0, -2, 2)$
0.00	1.0000	1.0000	1.0000	4.0000	4.0000
0.05	0.9869	0.9380	1.0125	3.7529	3.9504
0.10	0.9724	0.8771	1.0252	3.5117	3.9015
0.15	0.9564	0.8172	1.0379	3.2765	3.8533
0.20	0.9384	0.7584	1.0507	3.0475	3.8059
0.25	0.9183	0.7007	1.0635	2.8250	3.7592
0.30	0.8955	0.6441	1.0765	2.6091	3.7133
0.35	0.8694	0.5887	1.0896	2.4001	3.6681
0.40	0.8394	0.5345	1.1028	2.1982	3.6236
0.45	0.8043	0.4816	1.1161	2.0038	3.5799
0.50	0.7628	0.4299	1.1295	1.8173	3.5369
0.55	0.7128	0.3796	1.1431	1.6390	3.4947
0.60	0.6516	0.3306	1.1568	1.4695	3.4532
0.65	0.5744	0.2831	1.1706	1.3093	3.4124
0.70	0.4742	0.2371	1.1846	1.1594	3.3723
0.75	0.3382	0.1928	1.1987	1.0206	3.3328
0.80	0.1420	0.1501	1.2131	0.8944	3.2939
0.85	−0.1680	0.1092	1.2277	0.7831	3.2556
0.90	−0.7419	0.0703	1.2425	0.6902	3.2176
0.95	−2.2417	0.0338	1.2577	0.6242	3.1799
0.99	−10.5911	0.0065	1.2692	0.6122	3.1494

converge rapidly, for most of the interesting values of p, r, s, k and y^2, by a standard numerical quadrature (e.g. Simpson's); their evaluation presents no problem with the aid of a programmable calculator.

Putting $p = r = s = k = 0$ in each of the moment relationships (8.9) and (8.10), the key quantity $\bar{H}$ comes out. Thus, for prolate spheroids,

$$\bar{H} = \tfrac{1}{2}\pi / E_f\{m^{-1}q(y^2 \mid 0, 0, 0)\} \tag{8.14}$$

and for oblate spheroids,

$$\bar{H} = \tfrac{1}{2}\pi / E_f\{M^{-1}(1-y^2)^{\frac{5}{2}}q(y^2 \mid 0, -2, 0)\} \tag{8.15}$$

Surprisingly, rather beautiful relations (because of their independence from q-coefficients) hold for the mean axial ratios. For prolate spheroids,

$$E_g(a/b) = \frac{2}{\pi}\bar{H}E_f(M/m^2) \tag{8.16}$$

For oblate spheroids,

$$E_g(b/a) = \frac{\pi}{2}\bar{H}E_f(m/M^2) \tag{8.17}$$

To conclude this section let us look at the relationship between the bivariate size-shape distributions of particles and profiles on a random section plane (Fig. 8.5). Cruz-Orive (1976) constructed a synthetic example, assuming that the particles were an aggregate of prolate spheroids

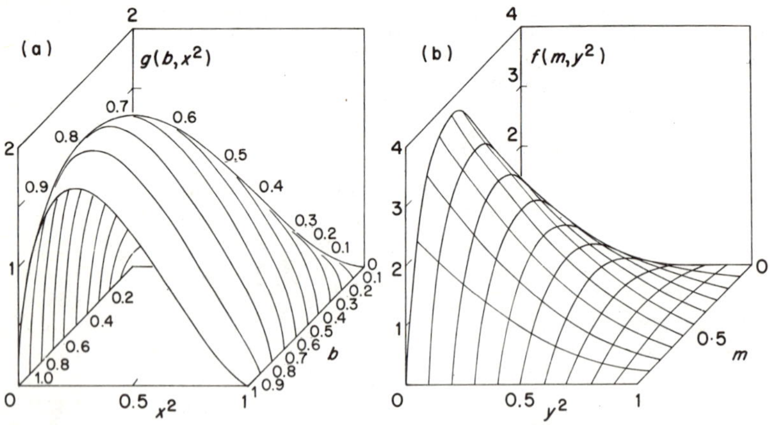

FIG. 8.5: Three-dimensional representation of size-shape probability density functions for spheroids and their sections. The surface in (a) represents an hypothetical population of variable prolate spheroids (eq. 8.18). The surface in (b) corresponds to the variable elliptic profiles generated from the spheroids in (a) by random sectioning (eq. 8.19). Reproduced with permission from Cruz-Orive (1976).

whose size and shape was governed by a joint probability density function of the form

$$g(b, x^2) = \frac{16}{\pi} b \cdot x\sqrt{(1-x^2)}[(1-b^2)+(1-x^2)] \tag{8.18}$$

for the ranges $\{0<b\leq 1\}$ and $\{0\leq x^2<1\}$. This function describes the surface shown in Fig. 8.5a. The joint size-shape probability density function describing the random ellipses of section is obtained through eq. (8.1), with g given by eq. (8.18). The result is

$$f(m, y^2) = \tfrac{90}{19} m\sqrt{(1-m^2)}(1-y^2)(2-m^2-y^2) \tag{8.19}$$

for the ranges $\{0\leq m\leq 1\}$ and $\{0\leq y^2\leq 1\}$. This is shown in Fig. 8.5b. Cruz-Orive (1976) has shown that, by using the relationship of eq. (8.6), one can also derive the distribution of Fig. 8.5a if that of Fig. 8.5b (eq. 8.19) is the measured distribution. For further details see the original article (Cruz-Orive, 1976).

8.2.3. DISCRETE SIZE-SHAPE SPHEROID MODELS: UNFOLDING PROCEDURE

The continuous bivariate distributions discussed in the preceding section are not very realistic from a practical point of view, because distributions of profile size and shape are usually obtained by distributing the measurements into size and shape classes. The distributions resulting are discrete (Fig. 8.6). The practical procedures for unfolding a spherical particle size distribution from a distribution of circular profile sizes described in Section 6.3 all worked with discrete distributions; in the following we shall proceed in a very similar way in order to derive unfolding procedures for size-shape distributions of spheroids (Cruz-Orive, 1978a).

For either spheroid type, prolate or oblate, the size variable, b or a, respectively, is assumed to vary in the range $(0, B)$, where B is a constant bigger than, or equal to the largest value of b (or of a). This range is divided into s classes of equal width $\Delta = B/s$, $(s>1)$. The shape component, x^2, varies in the range $(0, 1)$, which is divided into k classes of equal width $1/k$. Thus, the rectangular domain of variation of (b, x^2), or of (a, x^2), is divided into a grid comprising $s\times k$ classes, each class being represented by a rectangle of sides Δ and $1/k$ (Fig. 8.6b). Each spheroid belongs to exactly one of such classes. A spheroid belonging to the ijth class is called an ij-spheroid; it must satisfy the inequalities $(i-1)\Delta<b$ (or $a)\leq i\Delta$ and $(j-1)/k<x^2\leq j/k$, where $i=1, 2, \ldots, s$ and $j=1, 2, \ldots, k$. The number of ij-spheroids per unit volume of specimen is

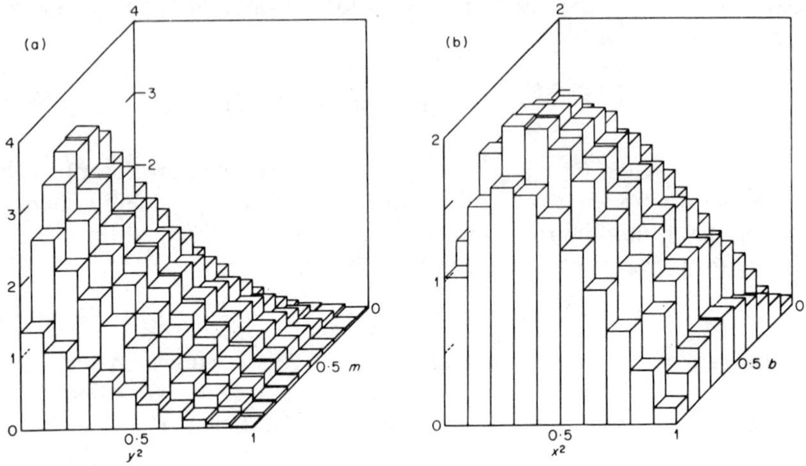

FIG. 8.6: Discrete size-shape distributions of prolate spheroids (b) and their sections (a) derived from the probability density functions shown in Fig. 8.5. Each size-shape class, (m, y^2) or (b, x^2), is represented by a column whose height indicates frequency. The size-shape distribution of section profiles shown in (a) is used to derive the particle size-shape distribution shown in (b). Reproduced with permission from Cruz-Orive (1978a).

denoted by $N_V(i, j)$, so that $\sum_{i=1}^{s} \sum_{j=1}^{k} N_V(i, j) = N_V$, which is the overall numerical density of spheroids.

The elliptical profiles are classified by means of the same $s \times k$ size-shape grid used for the spheroids (Fig. 8.6a). Thus, an $\alpha\beta$-ellipse satisfies the inequalities $(\alpha-1)\Delta < m$ (or $M) \leqslant \alpha\Delta$ and $(\beta-1)/k < y^2 \leqslant \beta/k$, for all $\alpha = 1, 2, \ldots, s$, and $\beta = 1, 2 \ldots, k$. The number of $\alpha\beta$-ellipses observed per unit area of section is denoted by $N_A(\alpha, \beta)$. Clearly, $\sum_{\alpha=1}^{s} \sum_{\beta=1}^{k} N_A(\alpha, \beta) = N_A$, is the total number of profiles per unit area of section. The ellipse number densities are related to the spheroid ones as follows,

$$N_A(\alpha, \beta) = \Delta \sum_{i=\alpha}^{s} \sum_{j=\beta}^{k} p_{\alpha i} N_V(i, j) q_{j\beta} \tag{8.20}$$

for each $\alpha = 1, 2, \ldots, s$; $\beta = 1, 2, \ldots, k$. This relationship is the discrete version of eqs (8.6) and (8.7), and it constitutes a generalization of that obtained for spheres in Section 6.3. The coefficients $p_{\alpha i}$ are size-correctors; they are independent from Δ—as in the sphere model—and from spheroid type. Their form is as follows,

$$p_{\alpha i} = \begin{cases} \sqrt{((i-\frac{1}{2})^2 - (\alpha-1)^2)} - \sqrt{((i-\frac{1}{2})^2 - \alpha^2)}, & \text{for} \quad \alpha < i \\ \sqrt{(i-\frac{3}{4})}, & \text{for} \quad \alpha = i \\ 0, & \text{for} \quad \alpha > i \end{cases} \tag{8.21}$$

Note that these size-correctors are identical to the coefficients k''_{ij} given for spheres in eqs (6.82) and (6.83), with some changes in symbols: i now standing for particle size class (j in the sphere case), α for elliptical profile size (i in the sphere case). The values for these correctors can evidently be read from Table 6.7.

The coefficients $q_{j\beta}$ are shape-correctors, and do depend on k and on the spheroid type. For prolate spheroids,

$$q_{j\beta} = \begin{cases} \sqrt{(t_1^2-1)}\{f(t_\beta)-f(t_{\beta+1})\}, & \text{for} \quad \beta<j \\ \sqrt{(t_1^2-1)}f(t_j), & \text{for} \quad \beta=j \\ 0, & \text{for} \quad \beta>j \end{cases} \tag{8.22}$$

where $f(t)=t/(t^2-1)+\text{arc tanh}\, t$, $t_\beta=\{(2k-2\beta+2)/(2j-2\beta+1)\}^{\frac{1}{2}}$. For oblate spheroids,

$$q_{j\beta} = \begin{cases} \sqrt{(1+t_1^{-2})}\{f(t_\beta)-f(t_{\beta+1})\}, & \text{for} \quad \beta<j \\ \sqrt{(1+t_1^{-2})}f(t_j), & \text{for} \quad \beta=j \\ 0, & \text{for} \quad \beta>j \end{cases} \tag{8.23}$$

where $f(t)=t/(1+t^2)+\text{arc tan}\, t$, $t_\beta=\{(2j-2\beta+1)/(2k-2j+1)\}^{\frac{1}{2}}$

The inverse relation to eq. (8.20) gives us the required frequencies in terms of the known ones. It reads as follows,

$$N_V(i,j)=\Delta^{-1}\sum_{\alpha=i}^{s}\sum_{\beta=j}^{k} p^{i\alpha}N_A(\alpha,\beta)q^{\beta j} \tag{8.24}$$

for all $i=1,2,\ldots,s$; $j=1,2,\ldots,k$. The coefficient $p^{i\alpha}$ is the $i\alpha$th element of the inverse of a matrix whose αith element is just $p_{\alpha i}$; likewise, $q^{\beta j}$ is the βjth element of the inverse of a matrix having $q_{j\beta}$ as $j\beta$th element. All these coefficients can be obtained quickly by means of a programmable calculator, preferably equipped with matrix operations hardware. A 15×15 matrix $\{p^{i\alpha}\}$ can be seen in Table 6.8, a 10×10 matrix $\{q^{\beta j}\}$ for prolate spheroids, in Table 8.2, and one for oblate spheroids, in Table 8.3.

The result of this kind of unfolding of bivariate discrete distributions is shown for a synthetic example in Fig. 8.6; it has been based on the same probability density functions as the continuous distributions of Fig. 8.5.

Once the $N_V(i,j)$'s have been unfolded, N_V is quickly estimated by the sum of all of them. Similarly, the expected value of a function of the form $u=w(b)\cdot z(x^2)$, where $w(\cdot)$ depends only on size and $z(\cdot)$ on shape, is estimated as follows,

$$\bar{u}=\frac{1}{N_V}\sum_{i=1}^{s}\sum_{j=1}^{k} w(b_i)\cdot N_V(i,j)\cdot z(x_j^2) \tag{8.25}$$

Table 8.2

A 10×10 matrix $\{q^{\beta j}\}$ of shape correctors for unfolding *prolate* spheroid size-shape distributions (eq. 8.24)

β \ j	1	2	3	4	5	6	7	8	9	10
1	0.49569									
2	−0.37899	0.85769								
3	0.01355	−0.65931	1.10586							
4	−0.03731	0.02286	−0.85599	1.30631						
5	−0.01593	−0.06611	0.02844	−1.02049	1.47790					
6	−0.01502	−0.02897	−0.08801	0.03185	−1.16948	1.62863				
7	−0.01244	−0.02785	−0.04004	−0.10881	0.03300	−1.31379	1.76168			
8	−0.01177	−0.02406	−0.03978	−0.05250	−0.13216	0.03058	−1.46729	1.87572		
9	−0.01230	−0.02467	−0.03737	−0.05595	−0.07105	−0.16552	0.01978	−1.66537	1.95767	
10	−0.01677	−0.03279	−0.04871	−0.06722	−0.09480	−0.11892	−0.24814	−0.02771	−2.11959	1.89881

Table 8.3

A 10 × 10 matrix $\{q^{\beta i}\}$ of shape correctors for unfolding *oblate* spheroid size-shape distributions (eq. 8.24)

β \ j	1	2	3	4	5	6	7	8	9	10
1	0.50424									
2	−0.31071	0.82933								
3	0.02091	−0.50032	1.01062							
4	−0.01743	0.03482	−0.59337	1.12028						
5	−0.00309	−0.02632	0.04290	−0.63445	1.17847					
6	−0.00241	−0.00400	−0.02862	0.04788	−0.63470	1.19279				
7	−0.00096	−0.00303	−0.00342	−0.02704	0.05017	−0.59600	1.16492			
8	−0.00041	−0.00099	−0.00248	−0.00209	−0.02234	0.04910	−0.51396	1.09179		
9	−0.00012	−0.00031	−0.00054	−0.00141	−0.00052	−0.01483	0.04213	−0.37525	0.96377	
10	−0.00001	−0.00003	−0.00006	−0.00008	−0.00031	0.00034	−0.00490	0.02058	−0.14607	0.75827

where $b_i = (i - \frac{1}{2})\Delta$ ($= a_i$ for oblate spheroids), whereas $x_j^2 = (j - \frac{1}{2})/k$, i.e. the mid-point class values.

The main drawback of the preceding approach is that it might require very large profile samples. If the latter are not available, direct moment estimates may be sought via the relationships of eqs (8.9) and (8.10). For example,

$$N_V = N_A/\bar{H}$$

where

$$\bar{H} = (\pi/2)n \Big/ \sum_{i=1}^{n} \{m_i^{-1} q(y_i^2 \mid 0, 0, 0)\} \tag{8.26}$$

for prolate spheroids, and

$$\bar{H} = (\pi/2)n \Big/ \sum_{j=1}^{n} \{M_i^{-1}(1 - y_i^2)^{\frac{5}{2}} q(y_i^2 \mid 0, -2, 0)\} \tag{8.27}$$

for oblate spheroids. n is the number of profiles recorded, and (m_i, y_i^2), or (M_i, y_i^2) are the data recorded from the ith profile. Unfortunately, this alternative approach is not free from drawbacks either. Being functions of shape-weighted harmonic means of m_i (or of M_i), the present estimates are bound to have large error variances.

8.3. Size distribution of triaxial ellipsoids from intercept length distribution

We showed in Chapter 7 that one can derive a size distribution of spheres from the distribution of intercept lengths measured along a random test line system. It is plausible that this approach might be advantageous with respect to some ellipsoids because the measured distribution will be simply one in size with no variation of shape.

The principles set out in Chapter 7 have been extended by De Hoff and Bousquet (1970) to an application to triaxial ellipsoids of *constant shape*; for derivation of the following formulae the reader is referred to the original article.

Consider a triaxial ellipsoid (Fig. 8.2) with a as its longest axis and

$$b = q_1 \cdot a, \qquad c = q_2 \cdot a$$

as the two other axes. Note that $q_1 = q_2 = 1$ describes a sphere, and $q_1 = q_2$ an ellipsoid generated by revolution of an ellipse, i.e. a spheroid.

A shape factor

$$k_3 = \frac{\pi}{4} \cdot q_1 \cdot q_2$$

is introduced.

It is shown that

$$N_V(D) = -\frac{1}{2k_3} \cdot \frac{d}{dl}\left[\frac{N_L(l)}{l}\right] \tag{8.28}$$

which is evidently identical to eq. (7.26) except for the shape factor k_3. But note that if $q_1 = q_2 = 1$ as in the spheres $k_3 = \pi/4$ so that eq. (8.28) becomes eq. (7.26). Here again the particle size distribution is related to the slope of $N_L(l)$, for by differentiation we obtain, in analogy to eq. (7.27),

$$N_V(D = l) = \frac{1}{2k_3}\left[\frac{N_L(l)}{l} - \frac{1}{l} \cdot \frac{dN_L(l)}{dl}\right] \tag{8.29}$$

De Hoff and Bousquet have shown that estimation of the fractional frequency distribution, rather than the actual numerical densities, requires no knowledge of the shape factor. If one considers the number of intercepts in the *smallest* size class, l_0, then the cumulative fractional frequency distribution is simply

$$F(l) = 1 - \frac{N_L(l)}{N_L(l_0)} \cdot \frac{l_0}{l} \tag{8.30}$$

The problem with this nice formula is that the smallest size class of intercepts is often fraught with considerable experimental error, because many of the very small profiles, which contribute mainly to this class, cannot be recognized on a section.

8.4. Polyhedral particles

As set out in Volume 1, Section 5.1 it will often be acceptable to adopt a spherical or ellipsoidal model for higher order polyhedra; this particularly because cuts of the corners are often not recognized—an artifact which tends to "smooth" the polyhedral surfaces!

Problems arise however with particles that resembles cubes. The shape of the profiles will be anything between triangles and hexagons so that it does not make sense to define any particular "diameter". Saltykov (1967) has proposed to measure the area a of the profiles and to express it as a

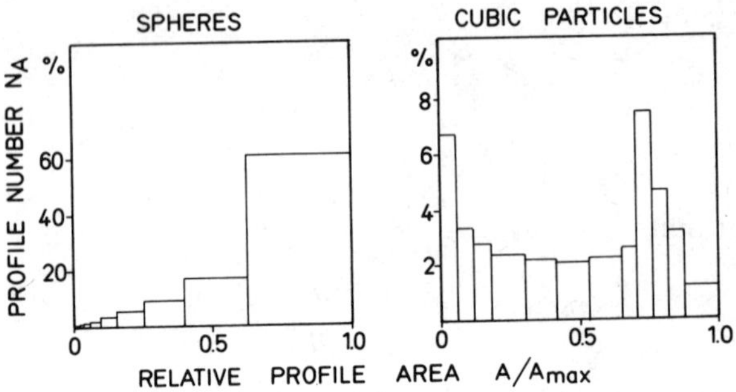

FIG. 8.7: Profile size distribution of spheres and cubic particles expressed in terms of relative profile area. Reproduced with permission from Saltykov (1967).

fraction of maximal possible profile area $a_{\max}$. Figure 8.7 compares the frequency distributions of profile areas for cubes with that for spheres, in both cases for equal size. It is evident that considerable problems may arise in the case of cubes, particularly because of the two peaks and because one of the peaks is in the lowest size class, that which is generally missed altogether. Underwood (1970) has described this method in detail. An alternative method was proposed by Myers (1967).

Hull and Houk (1953) have studied on models of regular polyhedra the shapes and size distributions of profiles obtained by sectioning with a plane. Some rather general solutions may be forth-coming through the work of Coleman (1969, 1973, 1978), Warren (1976) and Warren and Naumovich (1977), who consider the length distribution of random intercepts of convex polyhedral particles.

8.5. Determination of moments of particle size distribution from intercept measurements

Hilliard (1968) has proposed an approach by which one can obtain directly the moments of a size distribution of particles of any shape from measurements of intercept lengths. The assumptions are (1) that the shape of the particles is known and, if more than one shape is present, the frequency of occurrence of the different shapes is known; (2) that profiles are observed on true sections of zero thickness; (3) that the intercept lengths are measured by random test lines applied to random sections. The other conditions are such that the method should apply to any convex solid, including polyhedra.

In this method the particle size is measured by its *maximum intercept length* D; this deviates from preceding considerations where D generally designated a caliper diameter, but for most convex solids the maximum intercept length is equal to the maximum caliper diameter. The shape of the particles is considered through some shape coefficients. We need to be concerned about two of them:

(1) $$\alpha = \overline{a_p}/D^2 = s/(4D^2) \tag{8.31}$$

is the ratio of mean projected area a_p to D^2, which by Cauchy's formula (see eq. 4.5) is related to the particle surface s.

(2) $$f(\lambda) = f(l/D) \tag{8.32}$$

is the frequency distribution of "reduced" intercept length l divided by the particle "diameter"; from $f(\lambda)$ moments of λ are derived which are shape characteristics of the particle.

The measurements to be performed are determinations of intercept lengths l and their numerical density on the random test line, $N_L(l)$. From that one obtains the relative frequencies by

$$f(l) = N_L(l)/N_L \tag{8.33}$$

where N_L is the total number of intercepts. The kth moment $M_k(x)$ of a variable x is here defined as

$$M_k(x) = \int_0^\infty x^k \cdot f(x)\, dx \tag{8.34}$$

and since $f(x)$ is a normalized frequency, $M_0 = 1$.

Let us just state Hilliard's main result without going through the derivations. He finds the following dependence for the moments

$$M_k(l) = \left(\frac{\alpha \cdot N_V}{N_L}\right) \cdot M_{k+2}(D) \cdot M_k(\lambda) \tag{8.35}$$

or

$$M_k(D) = \left(\frac{N_L}{\alpha \cdot N_V}\right) \cdot M_{k-2}(l)/M_{k-2}(\lambda) \tag{8.36}$$

In these relations α and $M_k(\lambda)$ are shape dependent factors. The problem is that N_V, the total number of particles, is contained in this formula, but evidently this is an unknown. It can, however, be determined from the measurements of intercept lengths if the number of intercept lengths in the smallest size class is considered and eq. (5.104) is applied. Unfortunately, the intercept counts in the smallest size class will be fraught with considerable experimental error because we will miss a great number of them.

Another way of handling this problem, as proposed by Hilliard (1968), is to eliminate N_V from the above relations. Indeed, setting $K=2$ in eq. (8.36) we find

$$M_2(D) = \frac{N_L}{\alpha \cdot N_V} \tag{8.37}$$

because we had said that $M_0(x) = 1$. If one now divides eq. (8.36) by eq. (8.37) we end up with

$$\frac{M_k(D)}{M_2(D)} = \frac{M_{k-2}(l)}{M_{k-2}(\lambda)} \tag{8.38}$$

With this one can now determine the ratio of any moment $M_k(D)$ to the second moment of the particle size distribution, which can often yield interesting information. The main problem is that the shape factor $M_{k-2}(\lambda)$ cannot yet be determined for most shapes, although it seems theoretically possible to find a general formula. Hilliard (1968) has indicated that two of these moments can be expressed as ratios of volume v, surface s, and "diameter" D of the particles:

$$M_1(\lambda) = 4v/(s \cdot D) \tag{8.39}$$

$$M_4(\lambda) = 12v^2/(\pi s \cdot D^4) \tag{8.40}$$

Making use of the formulae for v and s for various shapes given in Table 5.2 one could estimate these moments; but notice that D as here defined is the maximum intercept length or maximum diameter and *not* the mean caliper diameter as given in Table 5.2. In addition, the ratio $M_2(\lambda)/[M_1(\lambda)]^2$ has been calculated to be 1.583 for the regular tetrahedron, and 1.333 and 1.487 for infinite cylinders of circular or square cross section respectively.

The special case of the sphere is simple because $f(\lambda) = 2\lambda$ (Cahn and Fullman, 1956) so that

$$M_k(\lambda) = 2/(k+2) \tag{8.41}$$

Consequently we find

$$\frac{M_k(D)}{M_2(D)} = \frac{k}{2} \cdot M_{k-2}(l) \tag{8.42}$$

It is evident that this approach is a promising one, and one must hope that it becomes practical one day if better information on $f(\lambda)$ for the general convex solid becomes available.

CHAPTER 9

Thickness of tissue sheets or barriers

9.1. Definition of measures of barriers

The organization of living matter into compartments requires the establishment of "*barriers*" between them. Thus blood in the capillaries is separated from the cells it supplies by the capillary endothelium and by some interstitial tissue; in the lung (Fig. 9.1) blood and air are separated by a tissue barrier made of three layers (Weibel, 1963, 1973). Hence the task of estimating the thickness of such barriers will occur frequently.

We can define a "barrier" as a mass of tissue bounded by two surfaces, Y and Y' (Fig. 9.2); its thickness is much less than its extension along the bounding surfaces. The thickness of a natural tissue barrier is, in general, variable; a "local thickness" $\tau(u)$ at a given point u on one of the two bounding surfaces, say Y, is defined as the shortest distance from u to the other surface Y'. To appreciate the overall barrier thickness, we subdivide the reference surface Y into very small elements dY of equal area $s(dY)$ so that the barrier related to this unit area can be represented by a little "disc" of constant thickness $\tau(dY)$ (Fig. 9.2). With this "model barrier" we can define the following barrier properties.

(1) The *distribution of thickness* is given by the relative number of unit discs of a certain thickness τ_i. With $S(Y)$ as the area of the reference surface

$$N(dY) = S(Y)/s(dY) \tag{9.1}$$

is the total number of discs formed by the process of subdivision. If $N_i(dY)$

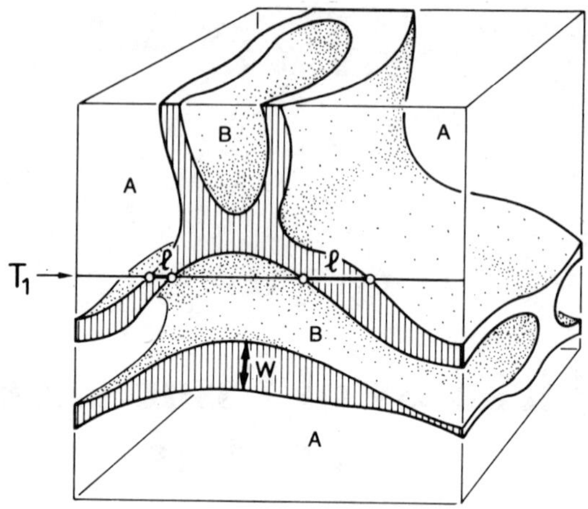

FIG. 9.1: Diagrammatic representation of barrier such as is found, for example, in the lung, separating air (A) from blood (B). On a section the apparent width (w) is the shortest distance from the air to the blood surface; the intercept length (l) is the distance between intersection points with air and blood surface on a random test line (T_1). Compare Volume 1, Figs 6.1, 6.4 and 8.10.3.

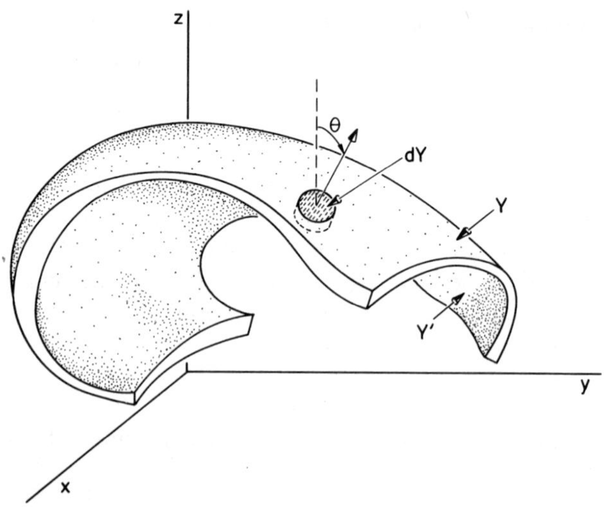

FIG. 9.2: Model barrier with reference surface Y and opposing surface Y', showing barrier element dY.

discs have thickness τ_i the frequency of τ_i is

$$F(\tau_i) = \frac{N_i(dY)}{N(dY)} = \frac{S(\tau_i)}{S(Y)} = S_S(\tau_i) \tag{9.2}$$

where $S(\tau_i) = N_i(dY) \cdot s(dY)$ is the area of the reference surface above which the barrier has thickness τ_i. It is hence evident that the barrier thickness distribution is expressed as a fraction of the reference surface, $S_S(\tau_i)$.

(2) The *average barrier thickness* $\bar{\tau}$ is

$$\bar{\tau} = \int_a^b F(\tau_i) \cdot \tau_i \cdot d\tau \tag{9.3}$$

where a and b are the minimal and maximal values of τ respectively. Introducing eq. (9.2) we find

$$\bar{\tau} = \frac{1}{S(Y)} \int_a^b S(\tau_i) \cdot \tau_i \cdot d\tau \tag{9.4}$$

It is evident that the integral of the product of barrier thickness with the reference surface represents the volume of the tissue barrier, V_t, so that

$$\bar{\tau} = V_t/S(Y) \tag{9.5}$$

Note that the mean barrier thickness is again expressed with respect to a reference surface; this can either be one of the two natural bounding surfaces, or, alternatively, the average of the two natural surfaces (Weibel and Knight, 1964).

(3) The effective thickness of the barrier with respect to passive transport is related to the reciprocal barrier thickness $1/\tau(u)$ at each point (Weibel and Knight, 1964). The average of these reciprocal thicknesses is the *harmonic mean barrier thickness* τ_h which is defined as follows:

$$\frac{1}{\tau_h} = \int_a^b F(\tau_i) \cdot \frac{1}{\tau_i} \cdot d\tau \tag{9.6}$$

It should be noted that $\bar{\tau}$ and τ_h are only equal if the barrier has equal thickness throughout. If the barrier thickness varies

$$\bar{\tau} > \tau_h \tag{9.7}$$

because the harmonic mean is weighted in favour of the thin areas. It has been shown that this favours the diffusion properties of natural barriers; in the lung the diffusion resistance of the irregular air-blood barrier is about one-third that of a barrier of equal mass and even thickness (Weibel and Knight, 1964; Weibel, 1973).

9.2. Apparent width of barrier on random section

It must first be noted that the apparent width w of a barrier profile on a random section (Fig. 9.1) is

$$w \geq \tau$$

and $w = \tau$ only if the barrier is cut exactly perpendicularly. We shall now demonstrate the relation between τ and the average barrier profile width $\bar{w}$.

Let us look first at one barrier element dY of thickness τ defined in relation to the reference surface Y; it is suspended at random within a specimen X (Fig. 9.3). This specimen is cut by an isotropic random plane T_2. The probability that T_2 intersects dY is,

$$\Pr\{T_2 \uparrow dY\} = \Pr\{T_2 \uparrow dY \mid \theta\} \cdot \Pr\{\theta\} \tag{9.8}$$

We know from Section 2.3 that

$$\Pr\{\theta\} = \sin\theta \tag{9.9}$$

and that the probability of cutting dY is the ratio between the caliper diameters of dY and of the specimen X:

$$\Pr\{T_2 \uparrow dY \mid \theta\} = H(dY \mid \theta)/H(X) \tag{9.10}$$

With $D(dY)$ as the diameter of the element we have

$$H(dY \mid \theta) = D(dY) \cdot \sin\theta \tag{9.11}$$

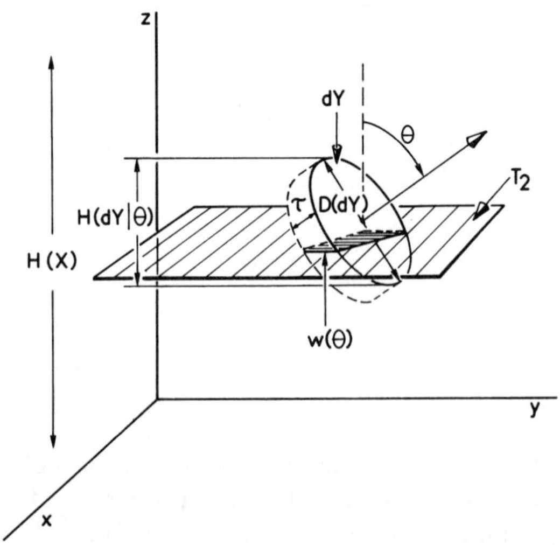

FIG. 9.3: Barrier element dY of diameter $D(dY)$ and thickness τ cut by random plane section T_2.

Note that we consider only the diameter of the element at the reference surface. Substituting eqs (9.9–9.11) into eq. (9.10) we find

$$\Pr\{T_2 \uparrow dY\} = [D(dY)/H(X)] \cdot \sin^2 \theta \tag{9.12}$$

The width of the band resulting from the intersection between the disc and T_2 is

$$w(T_2 \cap dY \mid \theta) = \tau/\sin \theta \tag{9.13}$$

Equations (9.12) and (9.13) now allow the derivation of average width of the barrier profile $E\{w\}$, allowing θ to vary in the range $\{0 \leqslant \theta \leqslant \pi/2\}$:

$$E\{w\} = \frac{\int_0^{\pi/2} \Pr\{T_2 \uparrow dY\} \cdot w(T_2 \cap dY \mid \theta) \cdot d\theta}{\int_0^{\pi/2} \Pr\{T_2 \uparrow dY\} \cdot d\theta}$$

$$= \frac{\tau \cdot \int_0^{\pi/2} \sin \theta \, d\theta}{\int_0^{\pi/2} \sin^2 \theta \, d\theta} \tag{9.14}$$

Performing the integrations we find

$$E\{w\} = \bar{w} = \frac{4}{\pi} \cdot \tau \tag{9.15}$$

for the relation between average width of the barrier profile and the thickness τ of a barrier of constant thickness.

Let us now consider the case of a barrier of variable thickness. The measure of the frequency of a given thickness τ is the surface area $S(Y \mid \tau)$ of the reference surface Y associated with τ; if the barrier is suspended in a specimen of volume $V(X)$ this can be expressed as the surface density of Y associated with τ:

$$S_V(\tau) = S(Y \mid \tau)/V(X) \tag{9.16}$$

The length of the trace of Y per unit area of a random section plane T_2 is, by eq. (3.76),

$$B_A = \int B(Y \cap T_2) / \int A(X \cap T_2) = \frac{\pi}{4} \cdot S(Y)/V(X) \tag{9.17}$$

where $B(Y \cap T_2)$ is, as in eq. (3.67), the integral of the traces of $(dY \cap T_2)$ and $A(X \cap T_2)$ is the area of the specimen profile on T_2; the integrals in eq. (9.17) are taken over the entire sample of sections. The width of the barrier profile $w(\tau)$ is measured along this trace, and we can hence express the relative frequency of $w(\tau)$ in relation to the length density $B_A(w)$ associated with width w. Cruz-Orive (1979) has shown that

$$B_A(w) = \frac{\pi}{4} \int_{R(w)} f(w \mid \tau) \cdot S_V(\tau)\, d\tau \tag{9.18}$$

where $R(w) = \{0 < \tau \leq w\}$ is the range over which τ is to be integrated, and $f(w \mid \tau)$ is the frequency of w to be expected if the barrier has local thickness τ. This frequency is (Cruz-Orive, 1979):

$$f(w \mid \tau) = \tau^3 / [w^3 \sqrt{(w^2 - \tau^2)}] \tag{9.19}$$

Inserting the right-hand side of eq. (9.19) into eq. (9.18) we obtain

$$B_A(w) = \int_{R(w)} \frac{\tau^3}{w^3 \sqrt{(w^2 - \tau^2)}} \cdot S_V(\tau)\, d\tau \tag{9.20}$$

This is an Abel-type integral equation, similar to the one found in Section 6.2, where the parameter to be derived, $S_V(\tau)$, occurs within the integral. Its solution is (Sneddon, 1966; Cruz-Orive, 1979)

$$S_V(\tau) = \frac{2}{\pi \tau^2} \int_0^{\tau} \frac{1}{\sqrt{(\tau^2 - w^2)}} \cdot \frac{d}{dw} \{w^3 \cdot B_A(w)\}\, dw \tag{9.21}$$

From this we can first calculate the relationships between the moments of τ and w, considering relation (9.13). In particular we find the average of w to be related to the mean barrier thickness by

$$E\{\tau\} = (\pi/4) \cdot E\{w\} \tag{9.22}$$

which is identical to eq. (9.15); hence the arithmetic mean barrier thickness is directly related to the mean width of profiles by

$$\bar{\tau} = \frac{\pi}{4} \cdot \bar{w} \tag{9.23}$$

Cruz-Orive (1979) has derived a general equation relating various moments of the thickness distribution to the corresponding moments of the measured profile width distribution; for $k < 3$ he finds

$$E\{\tau^k\} = (\sqrt{\pi}/2) \cdot \frac{\Gamma\{(4-k)/2\}}{\Gamma\{(3-k)/2\}} \cdot E\{w^k\} \tag{9.24}$$

where $\Gamma(\cdot)$ is the complete gamma function. As a result of special interest the harmonic mean barrier thickness τ_h is obtained by letting $k = -1$:

$$E\{1/\tau\} = (3\pi/8) \cdot E\{1/w\} \tag{9.25}$$

Equation (9.21) also allows a derivation of the actual thickness distribution of τ. Cruz-Orive (1979) has given a procedure for numerical calculation of this distribution in terms of $S_V(\tau)$. It is similar to the one described below (Section 9.3.3) for intercept lengths.

Casley-Smith and Davy (1978) have analysed the procedures for estimating "distances between parallel membranes"—which is identical to barrier thickness—on thick sections. They considered effects of truncation (grazing sections not measurable) and of uneven stain distribution throughout the section thickness, but limited their considerations to barriers of constant thickness.

9.3. Barrier thickness and random linear intercept lengths

A practical way of estimating barrier thickness is to allow long test lines to traverse the barrier at random and to measure the length of the linear intercept, l, between the points at which the line crosses the two boundary surfaces (Fig. 9.1). Evidently $l = \tau$ only if the line traverses the barrier exactly perpendicularly to the reference surface; in all other cases $l > \tau$. We shall now develop the relationship between the distributions of l and τ by following the approach of Weibel and Knight (1964), Gundersen *et al.* (1978) and Cruz-Orive (1979).

9.3.1. THE MODEL AND MOMENT RELATIONSHIPS

Let us first derive the distribution of intercept lengths for a barrier of even thickness τ. Consider, as a model (Fig. 9.4), a barrier element dY of area $s(dY)$ and thickness τ which may assume any orientation in space, and let a line T_1 penetrate the specimen X containing the plate. We would like to establish the relation between τ and the distribution of intercept lengths as T_1 intersects the barrier at random.

Suppose the barrier element oriented at an angle θ measured between the direction of T_1 and the normal to the element's reference surface. Given that T_1 hits dY the length of the intercept is (Fig. 9.4)

$$l(\theta \mid T_1 \uparrow dY) = \tau/\cos\theta \tag{9.26}$$

The probability of having an intercept at orientation θ, given T_1 hits dY, is obtained by

$$\Pr\{\theta \mid T_1 \uparrow dY\} = \Pr\{T_1 \uparrow dY \mid \theta\} \cdot \Pr\{\theta\}/\Pr\{T_1 \uparrow dY\} \tag{9.27}$$

We know $\Pr\{\theta\}$ from eq. (9.9) to be $\sin\theta$. The probability of hitting dY at a given orientation is the ratio of the projected areas of dY and of the

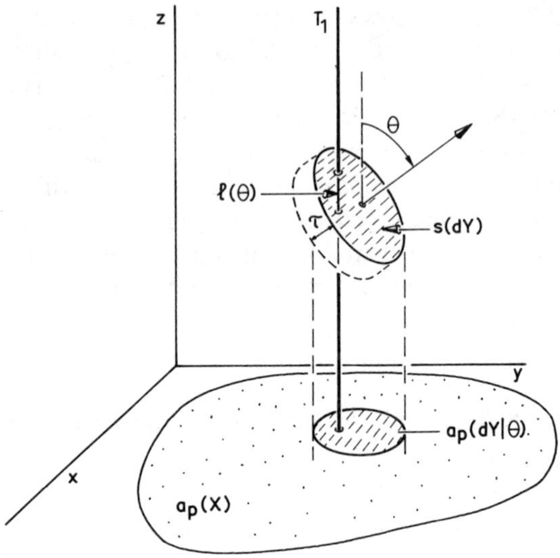

FIG. 9.4: Barrier element dY of surface $s(dY)$ and thickness τ cut by test line T_1.

containing space which we find to be

$$\Pr\{T_1 \uparrow dY \mid \theta\} = [s(dY)/a_p(X)] \cdot \cos\theta \tag{9.28}$$

and from this we find by integrating over all orientations

$$\Pr\{T_1 \uparrow dY\} = \tfrac{1}{2}[s(dY)/a_p(X)] \tag{9.29}$$

Introducing all this into eq. (9.27) we obtain

$$\Pr\{\theta \mid T_1 \uparrow dY\} = 2\cos\theta \sin\theta \, d\theta \tag{9.30}$$

which gives us the probability of having an intercept in the range of angles $\{\theta, \theta + d\theta\}$.

The first result we can derive from eqs (9.26) and (9.30) is the relation between τ and the mean intercept length which is

$$E\{l(T_1 \cap dY)\} = \frac{\int_0^\pi l(\theta \mid T_1 \uparrow dY) \cdot \Pr\{\theta \mid T_1 \uparrow dY\}}{\int_0^\pi \Pr\{\theta \mid T_1 \uparrow dY\}}$$

$$= \tau \cdot \frac{\int_0^\pi \sin\theta \, d\theta}{\int_0^\pi \cos\theta \sin\theta \, d\theta} \tag{9.31}$$

Performing the integrations we obtain

$$E\{l\} = \bar{l} = 2 \cdot \tau \tag{9.32}$$

This relation holds for a barrier of constant thickness τ whose elements are randomly oriented.

Let us now consider barriers of *variable thickness.* As before (eq. 9.16) the thickness distribution is expressed in terms of the area of the reference surface associated with (local) thickness τ or rather its surface density $S_V(\tau)$. In Chapter 3 (eq. 3.103) we had shown that the total number of intercepts per unit length of test line N_L is related to S_V, the total surface density of the reference surface, by

$$N_L = \int N(Y \cap T_1) \Big/ \int L(X \cap T_1) = \tfrac{1}{2} S(Y)/V(X) \tag{9.33}$$

(Note that we have substituted the symbol N for I because each intercept is associated with one intersection with the reference surface.)

It is thus evident that the frequency distribution of intercept lengths can be expressed in terms of $N_L(l) = N_L\{l(Y \cap T_1)\}$ which is related to $S_V(\tau)$ by (Cruz-Orive, 1979)

$$N_L(l) = \tfrac{1}{2} \int_0^l f(l \mid \tau) \cdot S_V(\tau)\, d\tau \tag{9.34}$$

The first problem is to find the frequency of l given τ, $f(l \mid \tau)$. Differentiating eq. (9.26) and rearranging we obtain

$$\sin\theta\, d\theta = \cos^2\theta \cdot dl/\tau \tag{9.35}$$

Substituting into eq. (9.30) from eqs (9.26) and (9.35) and integrating one finds

$$f(l \mid \tau) = 2\tau^2/l^3 \tag{9.36}$$

(Gundersen *et al.*, 1978; Cruz-Orive, 1979). As shown in Fig. 9.5, $f(l \mid \tau)$ is represented by a cubic hyperbola; intercept lengths of unit length (i.e. $l = \tau$) have highest frequency, and 75 per cent of all intercepts have lengths between 1 and 2 units of thickness, but there are also a small number of very long intercepts.

By substituting eq. (9.36) into eq. (9.34) we derive the integral equation

$$N_L(l) = \frac{1}{l^3} \int_0^l \tau^2 \cdot S_V(\tau)\, d\tau \tag{9.37}$$

which can be solved for $S_V(\tau)$ with the result (Gundersen *et al.*, 1978; Cruz-Orive, 1979)

$$S_V(\tau) = 3 \cdot N_L(l) + l \cdot \frac{d}{dl}[N_L(l)] \tag{9.38}$$

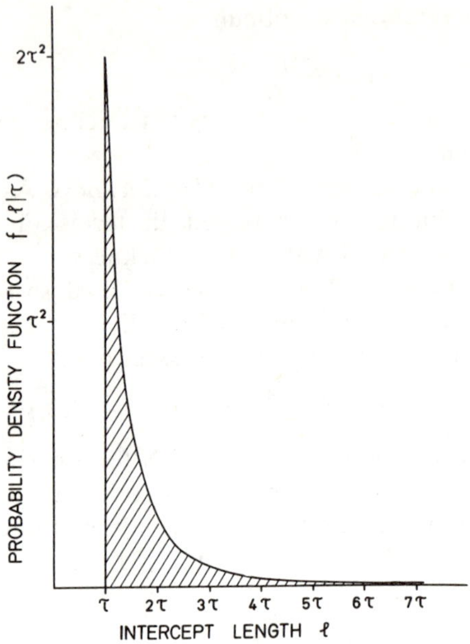

FIG. 9.5: Probability density function for intercept length l for barrier of constant thickness τ.

The area fraction of the barrier having thickness $\tau = l$ is hence related to three times the number of intercepts in the range l to $l + dl$ and to the slope of the empirical distribution in the same range multiplied by the intercept length. Both parameters can be derived from an empirical distribution of intercept lengths, as shown below.

Gundersen *et al.* (1978) and Cruz-Orive (1979) have considered on this basis the relations between powers k of τ and l leading to a derivation of the relationship between moments. The relation corresponding to eq. (9.34) reads

$$N_L(l^k) = (\tfrac{1}{2}) \int_0^{l^k} f(l^k \mid \tau) \cdot S_V(\tau)\, d(\tau) \tag{9.39}$$

In analogy to eq. (9.36) the conditional distribution is

$$f(l^k \mid \tau) = \frac{2}{|k|} \cdot \tau^2 / l^{(k+2)} \tag{9.40}$$

and the solution of the integral equation, similar to eq. (9.38):

$$S_V(\tau^k) = (k+2) N_L(l^k) + k \cdot l^k \frac{d}{dl^k} [N_L(l^k)] \tag{9.41}$$

The moment relationship between τ and l can likewise be obtained (Gundersen *et al.*, 1978; Cruz-Orive, 1979):

$$E\{\tau^k\} = (1 - k/2) \cdot E\{l^k\} \tag{9.42}$$

which holds for $k < 2$. From this we obtain the relation between arithmetic mean thickness and intercept length as

$$E\{\tau\} = (\tfrac{1}{2}) \cdot E\{l\} \tag{9.43}$$

which is the same as eq. (9.32), and for that between harmonic means of τ and l:

$$E\{1/\tau\} = (\tfrac{3}{2}) \cdot E\{1/l\} \tag{9.44}$$

the relation derived by Weibel and Knight (1964).

9.3.2. UNFOLDING OF THICKNESS DISTRIBUTION: GRAPHICAL SOLUTION

Whilst for most practical purposes the estimation of arithmetic or harmonic mean barrier thickness is entirely satisfactory, it may occasionally be useful to assess the thickness distribution as well. Gundersen *et al.* (1978) and Cruz-Orive (1979) have derived procedures by which such distributions can be derived from the distribution of intercept lengths.

The first case relates to the distribution of reciprocal thickness, important for any attempt at characterizing the passive transport properties of barriers. Gundersen *et al.* (1978) have shown that for a barrier of constant thickness τ the reciprocal intercept length, $1/l$, has an interesting probability density function. The probability density function for $(1/l)$ can be directly obtained from eq. (9.40) by letting $k = -1$. This yields

$$g(1/l) = g(l^{-1} \mid \tau^{-1}) = 2\tau^2 \cdot (1/l) \tag{9.45}$$

This turns out to be particularly convenient because it leads to a simple graphical procedure for deriving the distribution of $(1/\tau)$ from that of $(1/l)$. If $g(1/l)$ is plotted against $(1/l)$ it is represented by a triangle (Fig. 9.6), with $g(1/l) = 0$ if $1/l = 0$, and peak frequency at $1/\tau$. Assume now a barrier of varying thickness. For each thickness class τ_i^{-1} we can expect $g(l^{-1} \mid \tau_i^{-1})$ to take the form of such a triangle, when plotted against $1/l$, with peak frequency at $1/\tau_i$; these distributions are superimposed to yield an overall distribution of $G(1/l)$ for all thickness classes (Fig. 9.7). The area of each triangle is proportional to the area of the barrier's reference surface which is associated with τ_i^{-1}.

In trying to derive from an empirical distribution of reciprocal intercept lengths the distribution of reciprocal barrier thicknesses we must try to

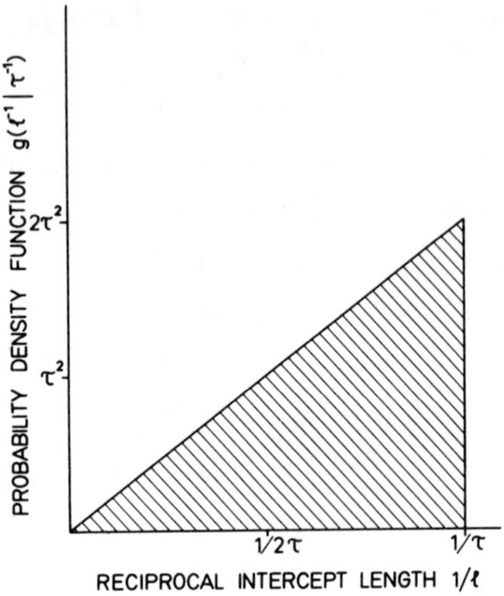

FIG. 9.6: Probability density function for reciprocal intercept length for barrier of constant thickness τ.

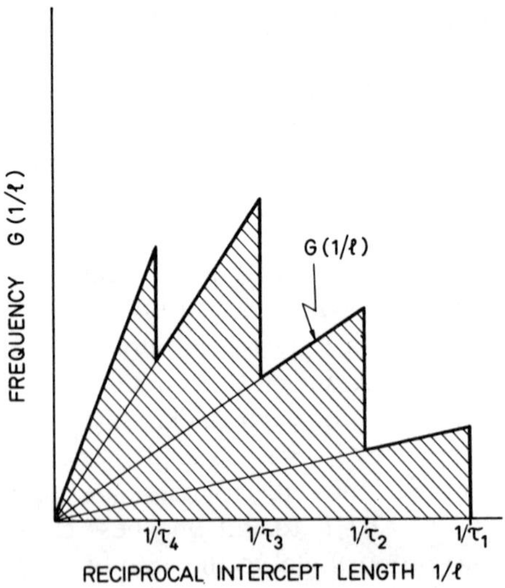

FIG. 9.7: Frequency distribution for reciprocal intercept length for barrier of varying reciprocal thickness.

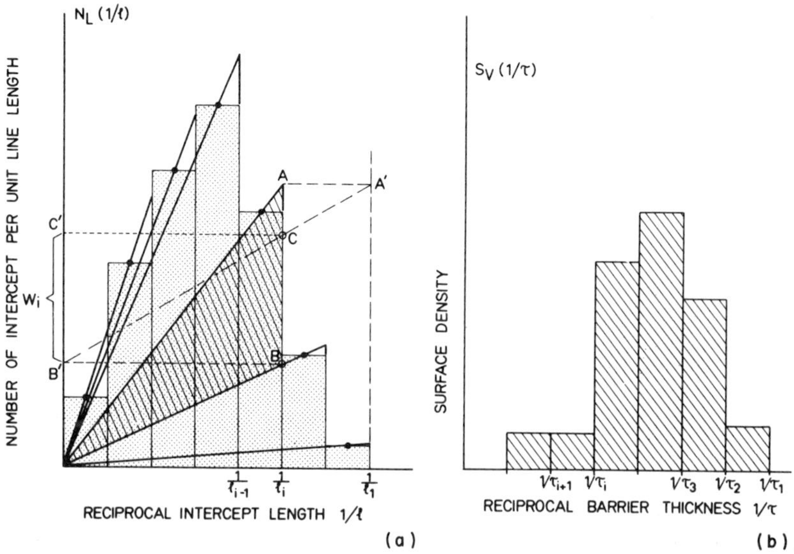

FIG. 9.8: Graphical procedure for unfolding frequency distribution of reciprocal barrier thickness from that of reciprocal intercept length. Reproduced with permission from Gundersen *et al.* (1978).

split the empirical distribution into a set of superimposed triangles and then to estimate their areas, Gundersen *et al.* (1978) have proposed a graphical solution, as shown in Fig. 9.8. The empirical frequency distribution of $1/l$ is first plotted as a histogram, with $N_L(i)$ being the number of intercepts in the ith class. Straight lines are now drawn from the origin through the middle of the top of each column to its right-hand limit. The triangles thus constructed correspond to those shown in Fig. 9.6 for each size class; thus, evidently, the area of the triangle ending at the upper limit of class i is proportional to the barrier surface associated with reciprocal thickness $1/\tau_i$, or, more precisely, with reciprocal thickness in the range $\{1/\tau_{i-1} < 1/\tau < 1/\tau_i\}$.

Figure 9.8a shows a simple graphical way of estimating the relative area of all these triangles; for its proof see Gundersen *et al.* (1978). According to the dashed line, project the peak of the triangle A to the line above the right-hand end point of the distribution (A'), the bottom right-hand point B to the ordinate (B'), and connect A' and B'; the length of the line segment BC above $1/l_i$ is proportional to the area of the triangle and hence to the weight of the barrier thickness class τ_i. If this is done for all classes the distribution of $1/\tau$ as shown in Fig. 9.8b can be derived. The relative weights of each class are obtained by

$$f(i) = W_i \Big/ \sum_{i=1}^{n} W_i \tag{9.46}$$

where n is the total number of classes in $1/l$.

The area a_i of the ith triangle can also be calculated arithmetically by

$$a_i = \frac{N_L(i) \cdot l_m(i) - N_L(i-1) \cdot l_m(i-1)}{2 \cdot l_i^2} \tag{9.47}$$

where $l_m(i)$ is the intecept length corresponding to the midpoint of class i, l_i being the length corresponding to its right-hand limit. The relative weight $f(i)$ for all classes is then obtained, in analogy to eq. (9.46) by

$$f(i) = \frac{a_i}{\sum_{i=1}^{n} a_i} = S_S(1/\tau_i) \tag{9.48}$$

Note that $f(i)$ represents the fraction of the barrier surface associated with reciprocal thickness $1/\tau_i$. Note also the similarity of this method with the method of Lord and Willis (1951) for estimating a size distribution of spheres (Section 7.2); a graphical procedure similar to that of Fig. 7.6 would also be applicable here as an alternative to Fig. 9.8.

The practical estimation of the distribution of τ from that of l is not quite as simple as that for the reciprocals because the probability density function $f(l \mid \tau)$ has the shape of a cubic hyperbola extending from τ to ∞ (eq. 9.36; Fig. 9.5). The solution of the corresponding integral equation is given in eq. (9.38). $S_V(\tau)$ is shown to be related to three times the numerical density (on the unit test line length) of intercepts in the range $\{\tau < l < \tau + d\tau\}$, and to the slope of the empirical distribution multiplied with l. The practical problem is to estimate these two parameters from the measured distribution of intercept lengths.

Gundersen *et al.* (1978) have again developed an easy graphical method for obtaining the relative weights of the barrier thickness distribution (Fig. 9.9a). Note first that the intercept lengths are grouped in classes of constant *relative* width

$$\Delta l_i = f \cdot l_i \tag{9.49}$$

This is achieved by measuring l with a *logarithmic scale* (see Volume 1, Section 6.1, Fig. 6.5). The empirical distribution is now plotted on a double-logarithmic scale and a line of slope 2 is drawn from the *left* limit of the class through the value of $N_L(l)$ plotted at the median of the class. At the median of the ith class the next lower line cuts off a line segment whose length W_i is proportional to $S_V(l_i)$, and can thus be directly used to

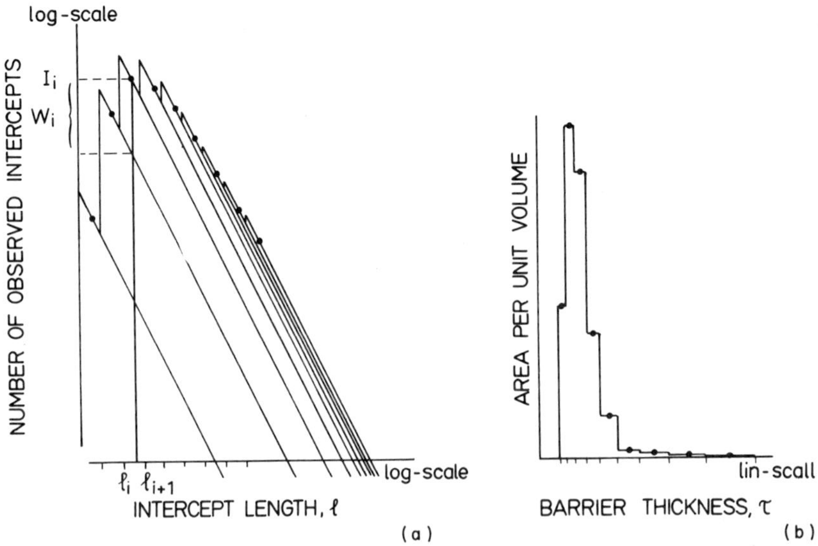

FIG. 9.9: Graphical procedure for unfolding barrier thickness distribution from that of intercept length. Reproduced with permission from Gundersen *et al.* (1978).

calculate the relative weight of thickness τ_i, as in eq. (9.46); the thickness distribution thus derived is plotted in Fig. 9.9b.

9.3.3. NUMERICAL SOLUTION FOR UNFOLDING OF THICKNESS DISTRIBUTION

A numerical solution for unfolding the barrier thickness distribution from a measured distribution of intercept lengths by computational procedures has recently been developed by Cruz-Orive (1979). It turns out to be most convenient to unfold this distribution in terms of some power k of τ and l; for example, Gundersen and Østerby (1973) showed that the reciprocal of the square root of the thickness of the glomerular basement membrane, $\tau^{-\frac{1}{2}}$, was normally distributed, whereas τ showed marked positive skewness.

The numerical solution proposed therefore starts from eqs (9.39) and (9.40) relating the distribution of l^k to that of τ^k. These distributions are continuous; in view of numerical treatment they need to be transformed into discrete distributions (histograms). For that purpose the range of values τ^k is subdivided into a suitable number s of classes of equal width Δ, such that the range $\{(i-1)\Delta < \tau^k \leq i \cdot \Delta\}$ is class i; the classes are numbered $i=1,2,\ldots,s$. Similarly, the values of l^k are divided into the same classes of width Δ, such that $\{(\alpha-1)\Delta < l^k \leqslant \alpha\cdot\Delta\}$ is class α, where $\alpha=1,2,\ldots,s$. Note that the scale on which τ and l are measured is linear only if $k=1$; otherwise its units are proportional to $\tau^{k/s}$, i.e. they are increasing exponentially with τ. Accordingly, the measurements of l should be classified on a convenient logarithmic scale (see Volume 1, Section 6.1, Fig. 6.5 and Gundersen *et al.*, 1978).

Evidently, the frequency of each class is given by $S_V(i)$ for thickness and $N_L(\alpha)$ for intercept length. The continuous distributions connected by eq. (9.34) are now transformed into discrete distributions (histograms) which evidently correspond to the practical situation. The discrete version of eq. (9.39) is

$$N_L(\alpha)=\sum_{i\in R(\alpha)} p_{\alpha i}\cdot S_V(i) \tag{9.49a}$$

The summation is performed for all classes i which occur in the range of α, $R(\alpha)$. Note that $R(\alpha)$ is different depending on whether the exponent k is positive or negative: for $k>0$, $R(\alpha)=\{\alpha,\alpha+1,\ldots\}$; for $k<0$, $R(\alpha)=\{1,2,\ldots\alpha\}$. This inversion of the range becomes clear if we compare Figs 9.5 and 9.6: for $k>0$ (Fig. 9.5) the values of $f(l\mid\tau)$ extend from τ to ∞, whereas for $k<0$ (Fig. 9.6) the values of $f(l^{-1}\mid\tau^{-1})$ range from 0 to τ^{-1}.

The coefficients $p_{\alpha i}$ are obtained by integrating the density functions between the class limits (Cruz-Orive, 1979); depending on k they take the following values

$$p_{\alpha i}=\begin{cases}|F_\alpha-F_{\alpha-1}|, & \text{for } (\alpha<i,\ k<0);\ (\alpha>i,\ k>0)\\ \frac{1}{2}-F_{i-1}, & \text{for } (\alpha=i,\ k<0)\\ \frac{1}{2}-F_i, & \text{for } (\alpha=i,\ k>0)\\ 0, & \text{for } (\alpha>i,\ k<0);\ (\alpha<i,\ k>0)\end{cases} \tag{9.50}$$

where

$$F_\alpha=(\tfrac{1}{2})\cdot\{(i-\tfrac{1}{2})/\alpha\}^{2/k} \tag{9.51}$$

If the ranges of τ^k and l^k are divided into s classes of equal width the values of $p_{\alpha i}$ represent an $s\times s$ matrix. The unfolding algorithm is then given by

$$S_V(i)=\sum_{\alpha\in R(i)} p^{i\alpha}N_L(\alpha) \tag{9.52}$$

where the range of i, $R(i)=\{i=1,2,\ldots,s\}$, and $p^{i\alpha}$ is the (i,α)th

Table 9.1

Matrix $\{p^{i\alpha}\}$ for estimating the surface density histogram of $\tau^{-\frac{1}{2}}$ with equally spaced classes and from linear probes (eq. 9.52), for example, $p^{28} = 0.3153$. Reproduced with permission from Cruz-Orive (1979).

i \ α	1	2	3	4	5	6	7	8	9	10
1	2.0000	−0.4923	0.2335	−0.1418	0.0975	−0.0722	0.0563	−0.0454	0.0376	−0.0319
2		2.4923	−1.6210	0.9843	−0.6769	0.5016	−0.3907	0.3153	−0.2614	0.2213
3			3.3875	−3.1883	2.1923	−1.6245	1.2654	−1.0214	0.8468	−0.7168
4				4.3457	−4.9363	3.6578	−2.8491	2.2998	−1.9067	1.6141
5					5.3233	−6.7720	5.2748	−4.2578	3.5300	−2.9882
6						6.3094	−8.6566	6.9875	−5.7932	4.9041
7							7.2999	−10.5710	8.7643	−7.4191
8								8.2931	−12.5050	10.5860
9									9.2879	−14.4530
10										10.2840

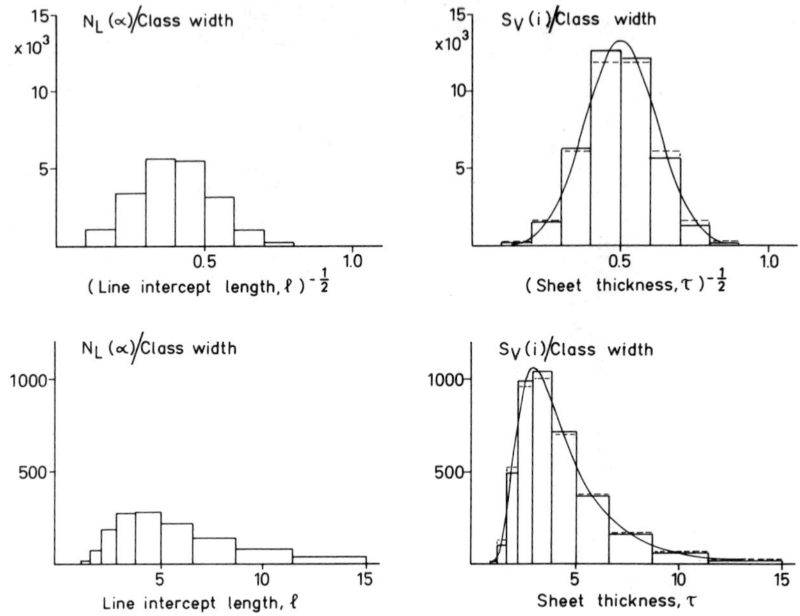

FIG. 9.10: Model distribution of barrier thickness unfolded by numerical procedure. Reproduced with permission from Cruz-Orive (1979).

element of the inverse of the $s \times s$ matrix of $p_{\alpha i}$ given by eq. (9.50). Table 9.1 reproduces the inverse matrix for $p^{i\alpha}$ for $k = -\frac{1}{2}$ and $s = 15$, i.e. for thickness distributions expressed in terms of $\tau^{-\frac{1}{2}}$. Figure 9.10 shows the appearance of histograms for $N_L(\alpha)$ and $S_V(i)$ for a synthetic example where it was assumed that $\tau^{-\frac{1}{2}}$ was normally distributed (Cruz-Orive, 1979).

One must note that in this procedure the values of τ^k are given by the mid-point value of the class Δ, such that $\tau_i^k = (i - \frac{1}{2}) \cdot \Delta$. Note that this choice of class value is similarly motivated as that for the size distribution of spherical particles discussed in Section 6.3.4. In the method of Gundersen *et al.* (1978) exposed in the preceding section, the class value is implicitly taken as the upper class limit $\tau_i^k = i \cdot \Delta$, in analogy to the Saltykov procedure for particles; this introduces some bias.

Cruz-Orive (1979) has shown that the calculation of $S_V(i)$ from $N_L(\alpha)$ is simplified if the working range of τ^k is divided into s classes of exponentially increasing widths. This is justified and also practical because thickness distributions tend to have a long upper tail, i.e. they show positive skewness. Designating by τ_0^k and τ_s^k the lowest and highest values

Table 9.2

Coefficients $q_{|\alpha-i|+1}$ for estimating the surface density histogram of τ with exponentially increasing classes (by a factor $\Delta = 15^{0.1} = 1.31$) and from linear probes, by formula (9.54). Reproduced with permission from Cruz-Orive (1979)

q_1	q_2	q_3	q_4	q_5	q_6	q_7	q_8	q_9	q_{10}
8.4305	−11.3354	8.6463	−6.5951	5.0305	−3.8371	2.9268	−2.2325	1.7028	−1.2989

of τ^k the ith class becomes

$$\tau_s \cdot \Delta^{-s+i-1} < \tau^k \leq \tau_s^k \cdot \Delta^{-s+i} \tag{9.53}$$

where $\Delta = (\tau_s^k/\tau_0^k)^{1/s}$ and $i = 1, 2, \ldots, s$. One proceeds similarly for the classification of the intercept lengths in terms of l^k. This choice of scale modifies the calculation of the matrix $p_{\alpha i}$. It is shown (the reader is referred to the original article by Cruz-Orive, 1979, for details) that the set of coefficients required for calculating $S_V(i)$ from $N_L(\alpha)$ is reduced to s coefficients $q_{|\alpha-i|+1}$ which replace the $s \times s$ coefficients $p^{i\alpha}$ in eq. (9.52). In Table 9.2 these coefficients are given for $k = 1$ and $s = 10$, the working range for τ extending over $\{1 < \tau < 15\}$. The class intervals are hence $\Delta = 15^{0.1} = 1.31$. The alternative algorithm to eq. (9.52) is

$$S_V(i) = \sum_{\alpha \in R(i)} q_{|\alpha-i|+1} \cdot N_L(\alpha) \tag{9.54}$$

where the range $R(i)$ is $\{i = 1, 2, \ldots, s\}$.

9.4. Barrier thickness estimated from point counts

In eq. (9.5) we had indicated that the arithmetic mean barrier thickness $\bar{\tau}$ was equal to the ratio of barrier volume to the area of the reference surface; this can be rewritten as

$$\bar{\tau} = V_V(t)/S_V(r) \tag{9.55}$$

where $V_V(t)$ is the volume density of the barrier and $S_V(r)$ the surface density of the reference surface in the containing space. It is hence evident that $\bar{\tau}$ can be estimated by determining the two densities by a point and intersection count respectively. This is achieved if a test system of lines and points is applied to a section (Weibel and Knight, 1964): if each test point is associated with a test line element of length l_T we find

$$\bar{\tau} = l_T \cdot P_t/(2I_r) \tag{9.56}$$

where P_t is the number of points falling on the barrier and I_r the number of test line intersections with the reference surface. If intersections with the two barrier surfaces, I_a and I_b, are not distinguished I_r can be taken as the average of the two, so that eq. (9.56) becomes

$$\bar{\tau} = l_T \cdot P_t/(I_a + I_b) \tag{9.57}$$

This is a useful approach for which practical methods are presented in Volume 1, Section 6.1.

9.5. Mean free distance between particles

If discrete convex particles are suspended in a matrix, the latter in a way forms a "barrier" between the particles (Fig. 9.11). A useful quantitative descriptor of the separation of particles is the "mean free distance" λ; as defined by Fullman (1953) and by Underwood (1968, p. 82)

$$\lambda = (1 - V_{V\alpha})/N_{L\alpha} \tag{9.58}$$

where $V_{V\alpha}$ is the volume density of the particles and $N_{L\alpha}$ the number of linear intercepts between the particles and random test lines of known length. Noting that each intercept intersects the particle surface twice, so that $I_L = 2N_L$, we find (see also Santaló, 1943)

$$\lambda = 2(1 - V_{V\alpha})/I_{L\alpha} = 4(1 - V_{V\alpha})/S_{V\alpha} \tag{9.59}$$

The mean free distance between the particles is hence four times the ratio of the volume density of the matrix to the surface density of the interface between particles and matrix. It is evident that λ is similarly defined as

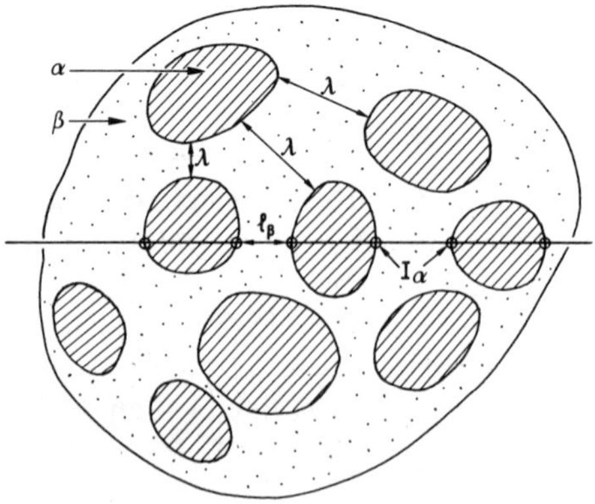

FIG. 9.11: Estimating mean free distance λ between convex particles α from intercept lengths in matrix, l_β.

the mean barrier thickness $\bar{\tau}$ given in eq. (9.55); it can hence also be estimated by a point and intersection count. Using a test system of lines and points, such that a line element of length l_T is associated with each point, we find the mean free distance between particles by

$$\lambda = 2 \cdot l_T \cdot (P_T - P_\alpha)/I_\alpha = 2 \cdot l_T \cdot P_\beta/I_\alpha \tag{9.60}$$

where P_T is the total number of test points used, P_α and P_β point hits with the particles and the matrix respectively, and I_α intersections of the lines with the interface between particles and matrix.

The mean free distance is also related to the mean linear intercept length $\bar{l}_\beta$ in the matrix (Fig. 9.11). Noting that, according to eq. (3.30),

$$1 - V_{V\alpha} = V_{V\beta} = L_{L\beta} \tag{9.61}$$

and, again, that each intercept in the matrix is associated with two intersections with the particle surface, we can easily find that

$$\lambda = 2 \cdot L_{L\beta}/I_{L\alpha} = \bar{l}_\beta \tag{9.62}$$

In other words, the mean free distance between particles is nothing else than the mean linear intercept length of test lines in the matrix.

These procedures hold, strictly, only for convex particles; if the particles are non-convex a bias is introduced which depends on the degree of non-convexity of the particles because some intercepts will connect surface points on the same particle.

The spatial separation of different components can be analysed by the methods of "hit-or-miss transformations" of Serra (1969, 1972, in preparation) and Matheron (1967, 1972, 1975). These transformations require a powerful image analysing system; indeed this a where automatic image analysers are probably strongest. Cruz-Orive (1976b) has recently proposed a circle test procedure by which the "pattern" by which components are assembled can be quantified by rather simple counting procedures.

CHAPTER 10

Anisotropic structures and stereology

10.1. Identification of the problem

All, or almost all, stereological methods discussed in the preceding chapters are valid only under the condition that the sample planes (or lines), the "sections", are *isotropic* with respect to the structure space, i.e. that they may assume any orientation in space with equal likelihood. As the stereological estimators depend on expectations or averages one assumes that these are obtained through a very large number of such isotropic random samples. If that condition is fulfilled there are no restrictions on the properties of the structure, i.e. its components may well be highly ordered, orientated in a preferred direction, or clustered: the estimators are said to be invariant with respect to rotations or translations.

In practice, the requirement of testing a specimen with a large number of isotropic sections can rarely be fulfilled. One is often limited to cutting only a few, mostly only one single section from each specimen; at best one can attempt to produce a section satisfying the condition of isotropy, more generally one will have to be content with an arbitrary section, hoping that the condition of isotropic random sectioning is approximately fulfilled by repeating this "isotropic random" (or arbitrary) sampling on several independent samples of the specimen. The stereological principles are then still valid if one has reason to assume that the elements of structure—cells, membranes, filaments, particles—are isotropic with respect to the structure space, i.e. if they show no preferred orientations.

These assumptions are however very strong and sometimes unjustified, because the processes leading to the assembly of a structure often result in some order among the elements, and this may imply the development of preferred orientations (see Volume 1, Sections 3.6, 6.4, 8.4–8.9). The problem we are therefore facing is that of testing for anisotropy in the structure, and of accounting for it in deriving stereological methods which are applicable to cases with demonstrated anisotropy. This is, unfortunately, a vastly underdeveloped field of stereological theory; the reason is that it is apparently very difficult—if at all possible—to treat this problem with any degree of generality. We shall therefore have to limit ourselves to some indications of the principles to be applied.

10.2. The geometric nature of anisotropic structures

Anisotropic structures are not easy to imagine. Let us therefore begin by describing the type of structure we shall be considering.

On a *plane* there is only one kind of object that can make anisotropic structures: lineal elements dB which can either be line segments or "needles" of finite length (Fig. 10.1a) or infinitesimal elements of a

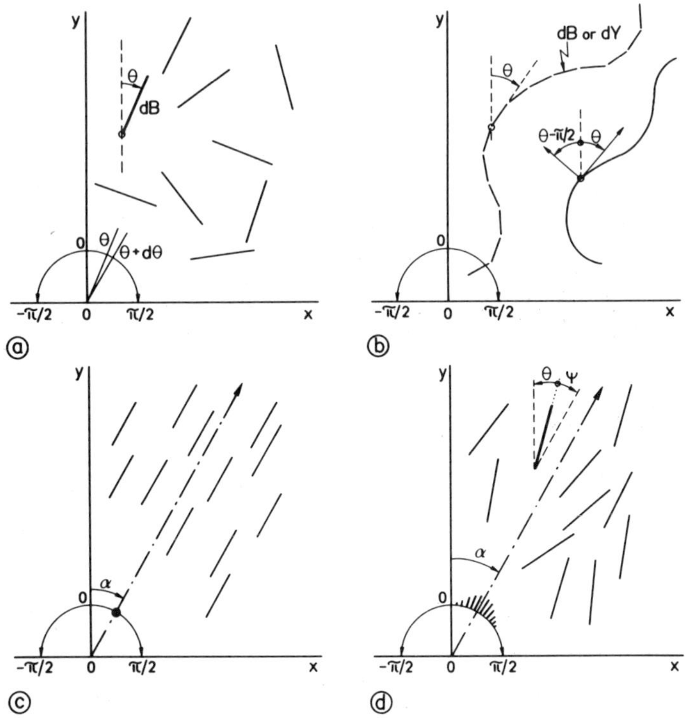

FIG. 10.1: Orientation of a field of needles (a) or line segments (b) on a plane, with total (c) or partial anisotropy (d).

curved line (Fig. 10.1b); their length is $l(dB)$ and their orientation is measured by their angle θ to the y-axis of the coordinate system (Fig. 10.1). Note that, for a smooth curve, the direction at any point of the curve can be measured either by the direction of the tangent (representing the direction of the infinitesimal element dB) or by that of the normal; the two angles simply differ by $\pi/2$ (Fig. 10.1b) and in calculating probabilities as below, this proves immaterial.

A planar structure is *isotropic* if all orientations θ occur with equal likelihood. In terms of line length this means, more precisely, that the length of line found in an orientation interval $\{\theta, \theta+d\theta\}$ is the same for all values of θ in the range $\{-\pi/2 \leq \theta \leq \pi/2\}$. Note that it suffices to consider orientation on half the orientation circle (Fig. 10.1a). If we project the orientation of each needle onto the orientation circle isotropy means that all parts of the circle will show such projections with equal density.

A planar structure is *anisotropic* if a certain orientation is preferred. Consider first a set of line segments of equal length all perfectly aligned in parallel at an orientation $\theta = \alpha$, but otherwise randomly positioned within the x–y plane (Fig. 10.1c). Such a field of needles has perfect anisotropy. If the orientations of the lines are projected onto the orientation circle they all fall onto the same point. Now allow each of these aligned line segments some freedom of orientation in that it may deviate from α by an angle ψ (Fig. 10.1d). It is clear that the degree of anisotropy will depend on the frequency or probability density function for ψ, $f(\psi)$. If $f(\psi)$ favours orientations with respect to α the projections of the orientations θ onto the orientation circle will be inhomogeneously distributed (Fig. 10.1d). Examples for such orientation distributions are shown in Fig. 10.6.

Let us now consider the construction of *anisotropic structures in three-dimensional space*. They can be made of two types of structural elements for which orientation can be defined:

(a) *Line segments dJ* of length $l(dJ)$ which are, for example, elements of space curves or of fibre systems; their orientation is measured by the pair of polar coordinates θ, ϕ in an x, y, z coordinate system (see Fig. 10.8).
(b) *Planar surface elements dS* of area $s(dS)$, mostly elements of (curved) surfaces in space; their orientation is given by the direction of their *normals* measured again by the pair of angles θ, ϕ (Fig. 10.10).

A spatial structure is *isotropic* if all orientations (θ, ϕ) occur with equal likelihood. It is evident that the "frequency" of orientation (θ, ϕ) must be defined to be proportional to the length of curve or the area of surface which shows this orientation. Now we must remember that orientation in space is measured with respect to the surface of an orientation sphere (Fig. 2.16), whereby it suffices to consider one hemisphere. The degree of orientation can be visualized if we project the orientation of each element of equal length or surface onto the orientation sphere where it is marked off as a point at that direction. The structure is isotropic if the sphere surface is homogeneously covered with such projection points. We must remember, however, that the orientation distribution is measured with respect to a surface element of area $\sin\theta\, d\phi\, d\theta$ on the orientation sphere (see Section 2.3).

Let us now consider how we can construct and characterize *anisotropic structures in space.* Assume first a *spatial field of needles or fibres* all perfectly oriented at an angle α to the z-axis of the x, y, z coordinate system, but otherwise randomly positioned in space (Fig. 10.2a); for convenience we choose the coordinate system such that the preferred orientation α lies in the x–z plane so that $\phi = 0$. It is clear that, if we project the orientation of these needles onto the orientation sphere, they will all project onto the same point at α. Such a structure will be called "*fascicular*" because it corresponds to bundles or fascicules of parallel fibres.

We now relax orientation by allowing the needles to be tilted by an angle ψ away from α, but we allow the tilt to occur isotropically in any direction (Fig. 10.2b). If we project the resulting orientations onto the orientation sphere the projection points will lie homogeneously on a circle around α, the radius of the circle evidently being defined by ψ; in fact it is given by $\sin\psi$. The tilt angle ψ can assume any value in the range $\{0 \leq \psi \leq \pi/2\}$. Of particular interest is the case $\psi = \pi/2$, i.e. deflecting all needles isotropically by 90° (Fig. 10.2c). Such a structure may appear isotropic at first glance, but it clearly is not: the orientation of these needles projects onto a main circle in the equatorial plane normal to α (Fig. 10.2c). Alternatively, shifting these needles to meet on a point on the radius α presents them in form of a "flat star".

In designing "realistic" anisotropic structures all we now need to do is to allow needles to assume any orientations ψ with a certain probability density function $f(\psi)$ which is represented on the orientation sphere by zones of changing density covering the hemisphere around α (Fig. 10.2d), but within each zone the density is equal throughout.

Such structures are said to have a single anisotropy axis, namely the preferred orientation α. If one wanted to construct an orientation surface on which the orientation probability density function $f(\psi)$ was uniformly distributed (equal density on each surface element) then this would correspond roughly to a rotation "ellipsoid", obtained by rotating an

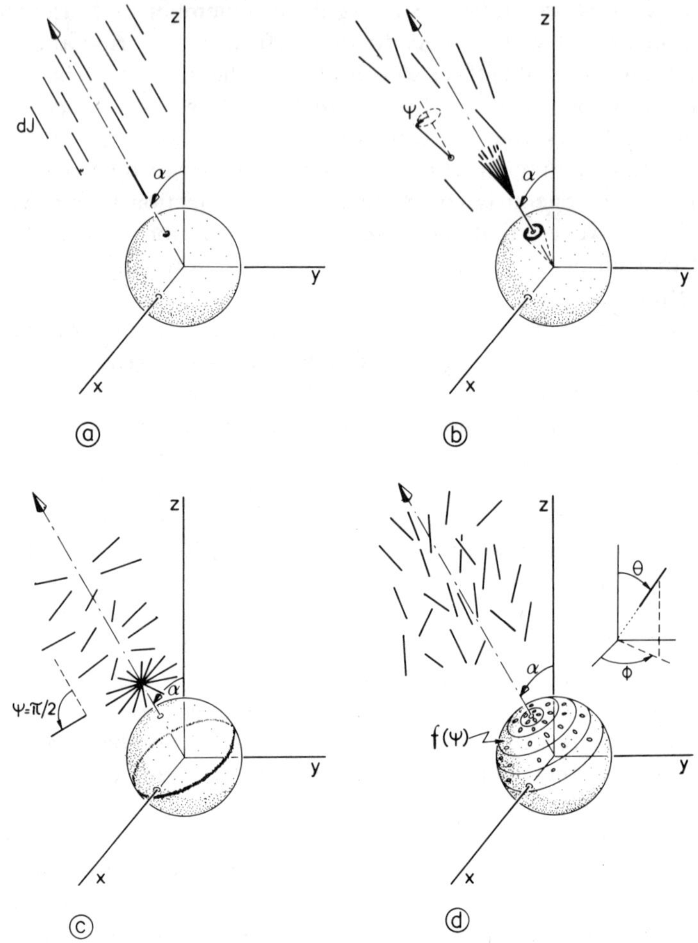

FIG. 10.2: Orientation of needles in space related to orientation sphere: (a) all needles parallel to preferred orientation α, (b) at angle ψ to α, (c) at right angles to α, (d) with varying orientation distribution. Inset: angles defining needle orientation.

"ellipse" around the axis α, but the shape of this "ellipse" would depend on the nature of $f(\psi)$.

The construction of *anisotropic surfaces* follows similar considerations. We start out with a spatial set of little planar surface elements dS (Fig. 10.3) whose orientation is given by that of their normals, so that the preferred orientation α also indicates the direction of the normals of the surface elements.

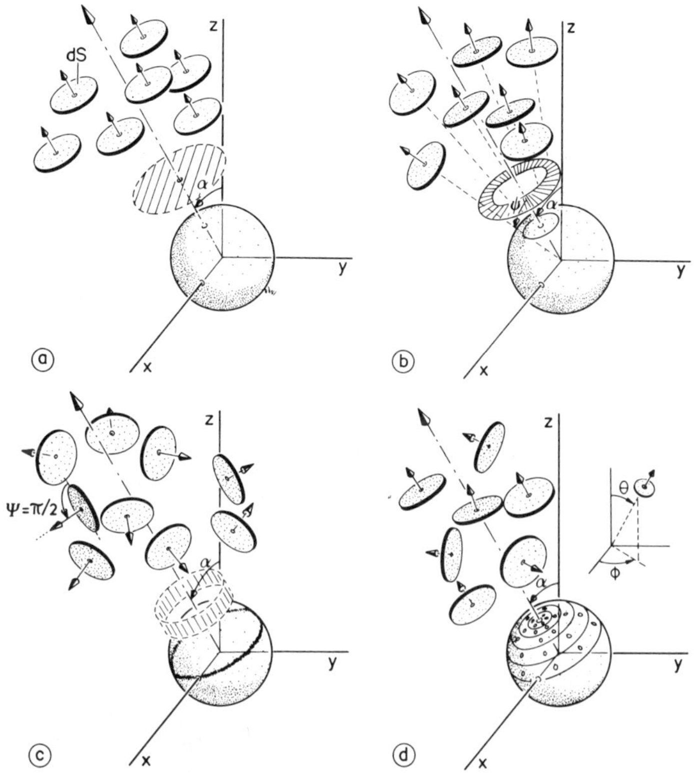

FIG. 10.3: Orientation of planar surface elements in space, measured by direction of normals (inset in (d)): (a) all plates in direction α, (b) at angle ψ to α, (c) at right angles to α, (d) with orientation distribution.

We start again with a set of surfaces whose direction is parallel to α but whose position in space is otherwise random (Fig. 10.3a). If these elements are shifted without changing their orientation they can be made to cover a plane perpendicular to α, and their orientations project onto a single point at α on the orientation sphere. Such a structure we will call "*lamellar*" because it is evidently made of a set of parallel flat lamellae or sheets.

We now relax orientation by tilting each disc by an angle ψ away from α but allowing the deflection to occur in any random direction (Fig. 10.3b). Shifting the discs without changing orientation we can make them cover a narrow strip of a surface cut out of a cone with the normals inclined at ψ to the axis α; and the orientations project onto a circle of radius $\sin \psi$ around α on the orientation sphere.

Again the case $\psi = \pi/2$ is of special interest (Fig. 10.3c) because the surface elements can be made to cover a cylindrical surface by shifting them around. Such structures are therefore represented by, for example, cylindrical tubules orientated in the direction α, so that we shall call them "*tubular*" structures. Note that the orientation normals of the *surface* elements are at right angles to the orientation axis of the tubule which coincides with α!

If we now allow the surface elements dS to assume any orientation ψ with a probability density function $f(\psi)$, we can construct "realistic" anisotropic surfaces. The procedure of shifting the discs around allows us to understand the nature of the surface generated. Thus we will find that the elements can be made to cover a surface corresponding roughly to a rotation "ellipsoid"; if $f(\psi)$ favours small angles ψ then the "ellipsoid" is oblate (it is derived from a plane when $\psi = 0$ for all elements), but if $f(\psi)$ favours large angles the "ellipsoid" is prolate, tending eventually to a long cylinder when all $\psi = \pi/2$. But clearly the true shape of this equivalent surface will depend on the nature of $f(\psi)$. In any case, however, $f(\psi)$ will be represented on the orientation sphere by varying density in zones around α, as shown in Fig. 10.3d. The equivalent surface discussed above corresponds to the shape that an orientation surface should assume to show $f(\psi)$ uniformly distributed, in analogy to what we discussed in relation to needles above.

The anisotropic structures made of filaments or surfaces so far discussed all have a single anisotropy axis α. One can further complicate them by introducing a second anisotropy axis oriented at some angle to α. Such structures clearly exist in nature; with respect to surfaces they could be represented, for example, by cylinders with an elliptic or even rectangular cross-section, or by triaxial ellipsoids. Similar fibrillar structures can be imagined; the orientation surface for homogeneous representation of the orientation projections would be something like a triaxial ellipsoid, for example. Such anisotropies have not yet been treated theoretically; they are very difficult to handle and are probably not accessible to stereological analysis, so that we shall not pursue them further. All subsequent examples of distributions have only a single anisotropy axis.

10.3. Geometric probability in anisotropic systems

10.3.1. BUFFON'S NEEDLE PROBLEM FOR NON-RANDOM DIRECTIONS

Let us first look at the simplest situation where anisotropy can be introduced: a field of needles dB of length $l(dB)$ spread out on a plane.

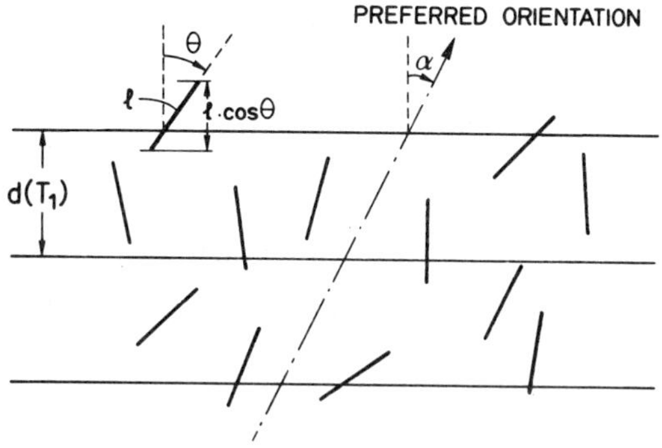

FIG. 10.4: Buffon problem for anisotropic field of needles on plane.

Let a grid of parallel lines T_1 of spacing $d(T_1)$ intersect this field and determine the probability that one needle intersects one of the grid lines (Fig. 10.4). This is Buffon's needle problem, as it was presented in Section 3.2.3, where we noted that the probability of intersection depended on the orientation angle θ between the needle and the normal to the test lines.

In deriving the solution to Buffon's problem we had assumed random directions of the needles, i.e. all orientations to be equally likely. Anisotropy prevails if some directions are preferred by the needles. Designating by $g(\theta)$ the probability density function of θ we can write for the probability of obtaining an intersection

$$\Pr\{dB \uparrow T_1\} = \int_0^{\pi} \Pr\{dB \uparrow T_1 \mid \theta\} \cdot g(\theta)\, d\theta \tag{10.1}$$

This differs from eq. (3.109) only by the fact that we have introduced a variable weight for the different orientations, thus allowing for non-random directions.

In Section 3.2.3 we have found that

$$\Pr\{dB \uparrow T_1 \mid \theta\} = l(dB) \cdot \cos \theta / d(T_1) \tag{10.2}$$

so that eq. (10.1) becomes

$$\Pr\{dB \uparrow T_1\} = \frac{l(dB)}{d(T_1)} \cdot \int_0^{\pi} |\cos \theta| \cdot g(\theta)\, d\theta \tag{10.3}$$

where $|\cos \theta|$ means absolute value of $\cos \theta$ because positive and negative values occur.

It is evident that $\Pr\{dB \uparrow T_1\}$ can only be obtained if $g(\theta)$ is known or if reasonable assumptions about the orientation distribution can be made.

The simplest case is to assume *uniform orientation distribution*, i.e. isotropy, where

$$g(\theta) = 1/\pi \tag{10.4}$$

whereby the solution of eq. (10.3) becomes

$$\Pr\{dB \uparrow T_1\} = (2/\pi) \cdot [l(dB)/d(T_1)] \tag{10.5}$$

which is the probability we derived in Section 3.2.3 (eq. 3.110) for an isotropic field of needles.

A convenient way of dealing with the anisotropic situation is to assume that the needles tend to prefer a certain orientation. Let the orientation of each needle be described by the angle ψ it makes with the preferred orientation (Figs 10.4 and 10.5); we can then describe the orientation distribution in the field of needles by a probability density function $f(\psi)$ which we assume to be symmetrical about the preferred orientation. Note that, if the preferred orientation makes an angle α with the test lines, we find

$$\theta = \alpha + \psi \tag{10.6}$$

as shown in Fig. 10.5.

A good choice for $f(\psi)$ is the von Mises distribution (Mardia, 1972; Streit, 1978) which is in essence similar to a normal distribution of ψ about the preferred orientation so that α, $(\psi = 0)$, is the mean and the mode of the needle orientations. The von Mises probability density function is

$$f(\psi) = [\pi I_0(K)]^{-1} \cdot e^{K \cos 2\psi} \tag{10.7}$$

within the range $\{-\pi/2 \leq \psi \leq \pi/2\}$. Here

$$I_0(K) = \sum_{r=0}^{\infty} (K/2)^{2r} (r!)^{-2} \tag{10.8}$$

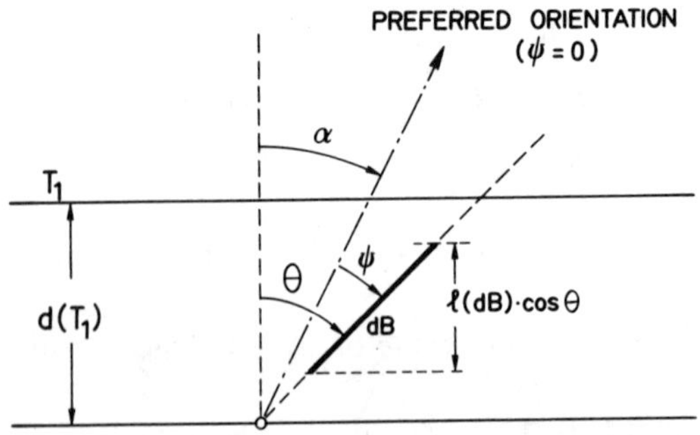

FIG. 10.5: Measures of needle orientation with respect to test lines.

is a modified Bessel function. K is the concentration parameter which determines the spread of the distribution: if $K=0$, $f(\psi)=1/\pi$ and the needle field is isotropic, but if $K\rightarrow\infty$ all needles become parallel to the preferred orientation; hence $K\geq 0$. In comparing eq. (10.7) with the formula describing the normal distribution we note that K is related to the variance σ^2.

The von Mises distribution is rather difficult to handle. Marriott (1971) has therefore proposed an approximation in the form of a cardioid distribution which is suitable for $\{-1\leqslant K\leqslant 1\}$; its probability density function is

$$f(\psi)=(1+K\cdot\cos 2\psi)/\pi \tag{10.9}$$

where $\{-\pi/2\leqslant\psi\leqslant\pi/2\}$. Some examples of this distribution are shown in Fig. 10.6a. We choose $\alpha=0$ so that $\psi=\theta$, for convenience. It is seen that the orientation pattern depends on the concentration parameter K: for $K=0$ there is no preferred orientation and $f(\psi)=1/\pi$; as K increases towards 1 the orientations around $\psi=0$ are more and more favoured, whereas orientations around $\psi=\pi/2$ are favoured if K is negative.

In Fig. 10.6c these distributions are plotted against ψ in the range $\{-\pi/2\leqslant\psi\leqslant\pi/2\}$; notice that in eq. (10.9) ψ can assume either positive or negative values. Figure 10.6b shows plots of $f(\psi)$ for the same cases on polar coordinates; the resulting plot, called "rose of compass", is quite popular and convenient in practice because it immediately shows the degree and, particularly, the direction of orientation (Underwood, 1968, p. 52; see Section 10.5).

With this orientation distribution at hand we can now develop the Buffon needle problem for non-random directions. Let a field of needles, as shown in Fig. 10.6, be cut by a set, T_1, of parallel test lines $d(T_1)$ apart and orientated at an angle α to the preferred orientation of the needles (Fig. 10.4). We obtain the probability of intersection between needles and test lines from eq. (10.1) deducing $g(\theta)$ from eq. (10.9). Because of eq. (10.6) (Fig. 10.5)

$$g(\theta)=[1+K\cdot\cos 2(\theta-\alpha)]/\pi \tag{10.10}$$

Since $g(\theta)$ is a normalized probability density function we find the probability of intersection between needles and lines from

$$\Pr\{dB\uparrow T_1\}=\frac{l(dB)}{d(T_1)}\cdot\frac{1}{\pi}\int_0^{\pi}[1+K\cdot\cos 2(\theta-\alpha)]\,|\cos\theta|\,d\theta \tag{10.11}$$

Performing the integration we obtain

$$\Pr\{dB\uparrow T_1;K,\alpha\}=\frac{2\cdot l(dB)}{\pi\cdot d(T_1)}\cdot[1+\tfrac{1}{3}K\cdot\cos 2\alpha] \tag{10.12}$$

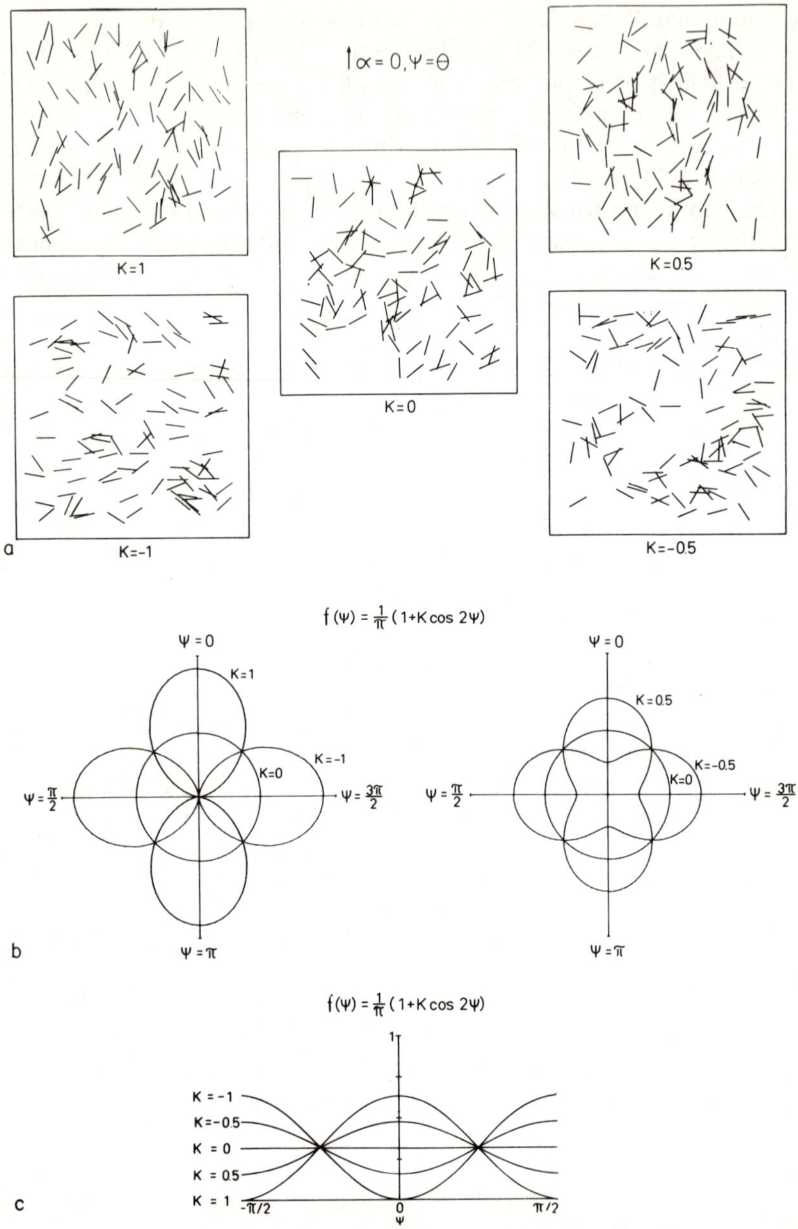

FIG. 10.6: Marriott orientation distributions for needles on plane: (a) plots of field with different concentration parameter K, (b) rose of compass for these examples, (c) plot of orientation distributions.

It is now seen that the probability of intersection depends solely on the orientation α of the preferred orientation (also called critical direction) to the grid lines, and on the concentration parameter K. If $K = 0$ eq. (10.12) reduces to eq. (10.5) for an isotropic field of needles.

This result evidently depends on the particular choice of orientation distribution as described by eq. (10.9) (Marriott, 1971). Indeed, the probability for having an intersection between needles and grid lines can only be solved if a particular orientation distribution is assumed: Buffon had assumed a uniform distribution with $g(\theta) = 1/\pi$. Streit (1978) has analysed other distributions, in particular two variants of the circular von Mises distribution, as it is described in Mardia (1972), and also discusses the statistical tests needed to decide on the appropriateness of such a stochastic model for orientation. For a broad discussion of the topic of non-random directions and their statistical implications the reader is referred to the book by Mardia (1972), "Statistics of Directional Data".

10.3.2. THE CASE OF ORIENTED LINES IN SPACE: BUFFON'S PROBLEM IN THREE DIMENSIONS

In Section 3.3 we have essentially dealt with Buffon's problem in three dimensions when we considered the probability that a section plane, T_2, intersects a short needle dJ of length $l(dJ)$ randomly oriented in space. Let us state this problem in a slightly different way, analogous to the Buffon problem.

Assume a set of needles of length $l(dJ)$ to be suspended in space (Fig. 10.7). This space is sectioned by a set of parallel planes, T_2, separated by

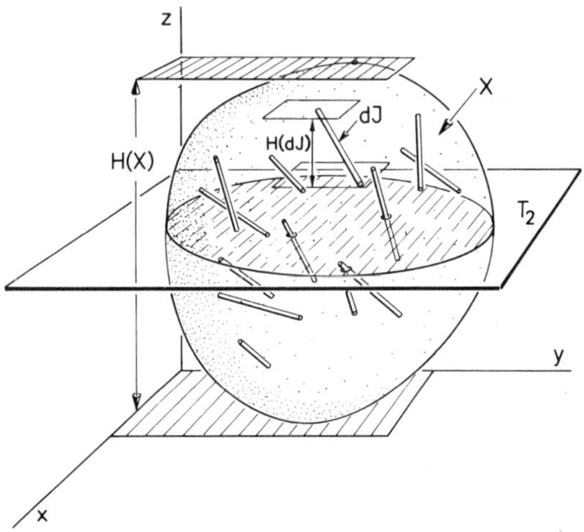

FIG. 10.7: Cutting spatial field of needles with plane.

a distance $d(T_2)$, and perpendicular to the z-axis. We seek the probability $\Pr\{T_2 \uparrow dJ\}$ which will evidently depend on the orientation of the needles measured by the angles θ and ϕ. Allowing for any orientation distribution with probability density function $g(\theta, \phi)$ the general statement is

$$\Pr\{T_2 \uparrow dJ\} = \int_0^{2\pi} \int_0^{\pi/2} \Pr\{T_2 \uparrow dJ \mid \theta, \phi\} \cdot g(\theta, \phi)\, d\theta\, d\phi \tag{10.13}$$

which is analogous to eq. (10.1) for the two-dimensional case. Integration is over spherical polar coordinates, as discussed in Section 2.3.

From the general principles of geometric probability theory (Section 2.2) we know that

$$\Pr\{T_2 \uparrow dJ \mid \theta, \phi\} = H(dJ \mid \theta, \phi)/d(T_2) \tag{10.14}$$

where $H(dJ \mid \theta, \phi)$ is the height of dJ perpendicular to T_2 (Fig. 10.7). We have found this to be invariant with respect to ϕ, (Fig. 3.13), so that

$$\Pr\{T_2 \uparrow dJ \mid \theta, \phi\} = l(dJ)\cos\theta/d(T_2) \tag{10.15}$$

Hence, eq. (10.13) becomes

$$\Pr\{T_2 \uparrow dJ\} = \frac{l(dJ)}{d(T_2)} \cdot \int_0^{2\pi} \int_0^{\pi/2} \cos\theta \cdot g(\theta, \phi)\, d\theta\, d\phi \tag{10.16}$$

Note that this probability can be stated in an alternative form if the needles dJ are contained within a specimen X (Fig. 10.7). Cutting this specimen with a plane T_2 we find

$$\Pr\{T_2 \uparrow dJ \mid T_2 \uparrow X\} = \frac{l(dJ)}{H(X)} \cdot \int_0^{2\pi} \int_0^{\pi/2} \cos\theta \cdot g(\theta, \phi)\, d\theta\, d\phi \tag{10.17}$$

where $H(X)$ is the caliper diameter of the specimen in the direction of T_2; this is analogous to eq. (10.16). Note that $d(T_2)$ is simply the caliper diameter of the slice of the specimen contained between two of the equidistant section planes. For convenience and more generality we shall use eq. (10.17) for the further developments.

We now need to make appropriate assumptions about the orientation distribution $g(\theta, \phi)$. Assuming first uniform distribution, i.e. isotropy, we have found in Section 2.3 that

$$g(\theta, \phi) = \sin\theta/2\pi \tag{10.18}$$

This leads to

$$\Pr\{T_2 \uparrow dJ \mid T_2 \uparrow X\} = \frac{1}{2} \cdot \frac{l(dJ)}{H(X)} \tag{10.19}$$

in analogy to eq. (3.139).

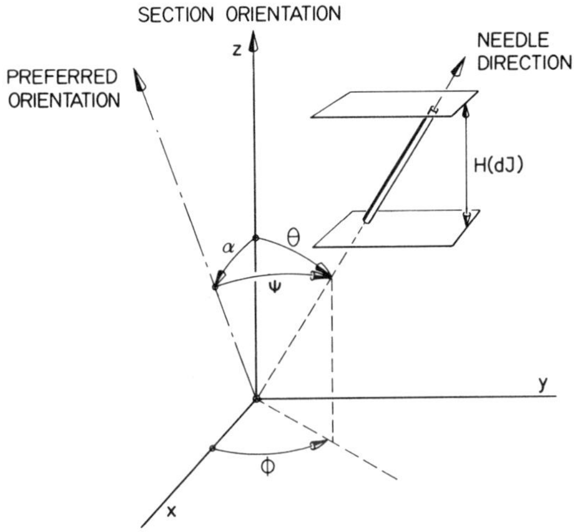

FIG. 10.8: Measures of orientation and height of needle.

In the general case we allow for the needles to prefer certain orientations, thus introducing anisotropy (see Section 10.2). We now need three angles to describe the orientation of the needles: in addition to θ and ϕ, the angle ψ between the needle and the preferred direction (Fig. 10.8). A further angle α describes the orientation of the preferred direction to the normal to T_2, i.e. the z-axis; for convenience we choose the coordinate system such that the preferred orientation α lies in the x–z plane (Fig. 10.8), which does not detract from generality. The relation between these angles is more complicated than in the two-dimensional case and reads

$$\cos\psi = \cos\theta\cos\alpha + \sin\theta\sin\alpha\cos\phi \tag{10.20}$$

Allowing for preferred orientation of the needles assumes that the angle ψ occurs with a certain probability density function $f(\psi)$ which is related to $g(\theta, \phi)$ by

$$g(\theta, \phi) = f(\psi)\sin\theta \tag{10.21}$$

as proposed by Baldwin *et al.* (1971). Note that $\sin\theta$, a well-known term, namely the probability density of θ (compare Section 2.3) occurs explicitly in all distributions that can be assumed. Solving eq. (10.16) evidently depends on an appropriate assumption for $g(\theta, \phi)$ or $f(\psi)$. We consider three cases.

(1) *Uniform distribution*: $f(\psi) = 1/(2\pi)$ (10.22)

Here evidently eq. (10.17) describes the isotropic case whose solution is given in eq. (10.19).

(2) *Fisher distribution*: $f(\psi) = c(K) \cdot e^{K \cos 2\psi}$ (10.23)

where $c(K)$ is a normalizing constant. Note that this distribution is analogous to the von Mises distribution for the planar case (eq. 10.7). The Fisher distribution allows for all degrees of anisotropy: as $K \to \infty$ the lines are all parallel to the preferred direction α, $\psi = 0$, as $K \to -\infty$ they lie parallel to the plane normal to α, i.e. they are all normal to α, $\psi = \pi/2$, and if $K = 0$ the cloud of needles is isotropic.

Computing the right hand side of eq. (10.17) with

$$g(\theta, \phi; K, \alpha) = c(K) \cdot \sin\theta \cdot e^{K \cos 2\psi} \tag{10.24}$$

as the orientation probability density function derived from the Fisher distribution is very awkward for the general case. It becomes manageable, however, for some particular cases, when the section plane is either normal or parallel to the preferred orientation, i.e. if $\alpha = 0$ or $\alpha = \pi/2$. Since the cases of cross-sections ($\alpha = 0$) or longitudinal sections ($\alpha = \pi/2$) through orientated structures are very important in practice (see Volume 1, Sections 3.6, 6.4, 8.4 and 8.5) we shall develop these two cases.

Case of transverse sections: $\alpha = 0$

In this case matters are simplified because $\psi = \theta$. With $c(K)$ as a normalization constant we have

$$g(\theta, \phi; K, 0) = c(K) \cdot \sin\theta \cdot e^{2K \cos^2\theta} \tag{10.25}$$

because $\cos 2\theta = 2\cos^2\theta - 1$. Accordingly eq. (10.17) becomes

$$\Pr\{T_2 \uparrow dJ; K, 0\} = \frac{l(dJ)}{H(X)} \cdot c(K) \int_0^{2\pi} \int_0^{\pi/2} \cos\theta \sin\theta \cdot e^{2K \cos^2\theta} \, d\theta \, d\phi \tag{10.26}$$

We first find the normalization constant by integrating $g(\theta, \phi; K, 0)$ over the ranges $\{0 \leq \phi \leq 2\pi\}$, $\{0 \leq \phi \leq \pi/2\}$ to be

$$c(K) = \left[2\pi \sum_{r=0}^{\infty} (2K)^r / \{r!\,(2r+1)\} \right]^{-1} \tag{10.27}$$

The integral becomes

$$\int_0^{2\pi} \int_0^{\pi/2} \cos\theta \sin\theta \cdot e^{2K \cos^2\theta} \, d\theta \, d\phi = 2\pi \left[-\frac{1}{4K} e^{2K \cos^2\theta} \right]_0^{\pi/2}$$

$$= \frac{2\pi}{4K} (e^{2K} - 1) \tag{10.28}$$

From this we obtain by appropriate substitution of eqs (10.27 and 10.28) into eq. (10.26)

$$\Pr\{T_2 \uparrow dJ;\, K, 0\} = \frac{l(dJ)}{H(X)} \cdot \frac{e^{2K} - 1}{4K \sum_{r=0}^{\infty} [(2K)^r/\{r!\,(2r+1)\}]}$$

$$= \gamma_1(K, 0) \cdot \frac{l(dJ)}{H(X)} \tag{10.29}$$

The probability of having an intersection depends only on the concentration parameter K. Note that if $K \to 0$

$$\Pr\{T_2 \uparrow dJ;\, 0, 0\} = \frac{l(dJ)}{H(X)} \cdot \frac{1}{2} \tag{10.30}$$

which describes the isotropic case (eq. 10.19), whereas if $K \to \infty$ (all lines normal to the section planes)

$$\Pr\{T_2 \uparrow dJ;\, \infty, 0\} = \frac{l(dJ)}{H(X)} \tag{10.31}$$

On the other hand, if all lines are parallel to the section plane ($K \to -\infty$), the probability of intersection vanishes to 0.

Case of longitudinal sections: $\alpha = \pi/2$
By eq. (10.20) we find, for $\alpha = \pi/2$,

$$\cos \psi = \sin \theta \cos \phi \tag{10.32a}$$

Noting furthermore that

$$\cos 2\psi = 2 \cos^2 \psi - 1 \tag{10.32b}$$

we obtain for the probability of orientation between line segments and test plane

$$g(\theta, \phi;\, K, \pi/2) = c(K) \sin \theta \cdot e^{2K \sin^2 \theta \cos^2 \phi} \tag{10.33}$$

The probability of intersection now is

$$\Pr\{T_2 \uparrow dJ;\, K, \pi/2\} = \frac{l(dJ)}{H(X)} \cdot c(K) \int_0^{2\pi} \int_0^{\pi/2} \cos \theta \sin \theta \cdot e^{2K \sin^2 \theta \cos^2 \phi}\, d\theta\, d\phi \tag{10.34}$$

The normalization constant $c(K)$ is the same as above (eq. 10.27), because it is valid for all values of α. The solution of the integral of eq. (10.34) is

$$\int_0^{2\pi} \int_0^{\pi/2} \cos \theta \sin \theta e^{2K \sin^2 \theta \cos^2 \phi}\, d\theta\, d\phi = \pi \sum_{r=0}^{\infty} [K^r \cdot (2r-1)!!/\{r!\,(r+1)!\}] \tag{10.35}$$

where $n!! = n(n-2)\cdot(n-4)\ldots 2$ if n is even, and $n!! = n(-2)\cdot(n-4)\ldots 3\cdot 1$ if n is odd.

From this we find the probability of having an intersection between the line segments dJ and a longitudinal section to be

$$\Pr\{T_2 \uparrow dJ;\, K,\, \pi/2\} = \frac{l(dJ)}{H(X)} \cdot \frac{\sum_{r=0}^{\infty} [K^r(2r-1)!!/\{r!\,(r+1)!\}]}{2\sum_{r=0}^{\infty} [(2K)^r/\{r!\,(2r+1)\}]}$$
$$= \gamma_1(K, \pi/2) \cdot \frac{l(dJ)}{H(X)} \tag{10.36}$$

Here again the probability of having an intersection depends only on the concentration parameter K, and again if $K \to 0$, i.e. if the segments are isotropic, eq. (10.36) becomes eq. (10.19). Note that in this case the probability of intersections vanishes to 0 when $K \to \infty$. If $K \to -\infty$

$$\Pr\{T_2 \uparrow dJ;\, -\infty,\, \pi/2\} = \frac{l(dJ)}{H(X)} \cdot \frac{2}{\pi} \tag{10.37}$$

which is equivalent to eq. (10.5) for an isotropic field of needles on a plane; note that when $K \to -\infty$ the needles are all parallel to the plane normal to α, but their projections onto a reference plane normal to α are isotropic.

(3) *Marriott distribution*: $f(\psi) = c(K)(1 + K\cdot\cos 2\psi)$ (10.38)

This distribution is an approximation to the Fisher distribution (eq. 10.23) which is valid for $\{-1 \leqslant K \leqslant 1\}$. It is analogous to eq. (10.9) which was presented as Marriott's approximation to the von Mises distribution in the plane. For positive values of K $\{0 < K \leqslant 1\}$ the needles have a tendency to group around the preferred orientation α, $\psi = 0$, whereas for negative values $\{0 > K \geqslant -1\}$ they are grouped around a plane normal to the preferred orientation. Using eq. (10.37) we have

$$g(\theta, \phi;\, K, \alpha) = c(K)\cdot\sin\theta\cdot(1 + K\cdot\cos 2\psi) \tag{10.39}$$

where the normalization constant is found by integrating $g(\theta, \phi;\, K, \alpha)$ over the ranges $\{0 \leqslant \phi \leqslant 2\pi\}$, $\{0 \leqslant \theta \leqslant \pi/2\}$ to be

$$c(K) = [2\pi(1 - K/3)]^{-1} \tag{10.40}$$

To compute eq. (10.17) we need to integrate the product of $g(\theta, \phi;\, K, \alpha)$ with $\cos\theta$, which is, in this case

$$\int_0^{2\pi}\int_0^{\pi/2} \cos\theta\,\sin\theta(1 + K\cos 2\psi)\,d\theta\,d\phi = \pi\{1 - (K/2)\sin^2\alpha\} \tag{10.41}$$

Accordingly we find

$$\Pr\{T_2 \uparrow dJ\} = \frac{l(dJ)}{H(X)} \cdot c(K) \cdot \int_0^{2\pi} \int_0^{\pi/2} \cos\theta \cdot g(\theta, \phi)\, d\theta\, d\phi$$

$$= \frac{l(dJ)}{H(X)} \cdot \frac{1}{2} \cdot \frac{1-(K/2)\sin^2\alpha}{1-(K/3)}$$

$$= \frac{l(dJ)}{H(X)} \cdot \gamma_1(K, \alpha) \tag{10.42}$$

This result was demonstrated by Baldwin *et al.* (1971). It is evident that the probability of intersection depends only on the concentration parameter K and on the orientation of the test planes to the preferred direction.

The two particular cases of transverse ($\alpha = 0$) and longitudinal ($\alpha = \pi/2$) sections yield interesting results. For $\alpha = 0$

$$\Pr\{T_2 \uparrow dJ;\ \alpha = 0\} = \frac{l(dJ)}{H(X)} \cdot \frac{1}{2}[1/\{1-(K/3)\}] \tag{10.43}$$

and for $\alpha = \pi/2$

$$\Pr\{T_2 \uparrow dJ;\ \alpha = \pi/2\} = \frac{l(dJ)}{H(X)} \cdot \frac{1}{2}[\{1-(K/2)\}/\{1-(K/3)\}] \tag{10.44}$$

We shall exploit these results when we attempt to estimate the concentration parameter K in Section 10.4.

At present we can derive another interesting result from these relations. It is evident that for $K = 1$ a section plane at $\alpha = 0$ yields the largest probability (P_1) of intersecting the needles, one at $\alpha = \pi/2$ the lowest probability (P_2); from eqs (10.43) and (10.44) the ratio of these probabilities is found to be

$$P_1/P_2 = 1/\{1-(K/2)\} = 2 \tag{10.45}$$

Similarly, for $K = -1$ we obtain the largest probability at $\alpha = \pi/2$ (P_2) and the lowest at $\alpha = 0$ (P_1), so that

$$P_2/P_1 = \{1-(K/2)\} = \tfrac{3}{2} \tag{10.46}$$

The ratio of probabilities of any two mutually perpendicular planes other than these particular cases is always smaller so that we can derive the following limits for the ratio of probabilities on perpendicular planes:

$$\begin{aligned} &\text{for} \quad K = 1 \qquad P_1/P_2 \leqslant 2 \\ &\text{for} \quad K = -1 \qquad P_2/P_1 \leqslant \tfrac{3}{2} \end{aligned} \tag{10.47}$$

10.3.3. THE CASE OF ORIENTED SURFACE ELEMENTS IN SPACE INTERSECTED BY PLANES AND LINES

The next case, where anisotropy is a frequent property, is that of surfaces: they can tend to form lamellar arrangements—i.e. prefer to lie more or less parallel to each other—or fascicules of tubules showing a longitudinal orientation. To deal with this case assume a model made of infinitesimally small surface elements, dS, of area $s(dS)$ and diameter $h(dS)$, essentially flat and, say, circular discs (Figs 10.9 and 10.10). They are contained in a specimen X and their orientation is determined by the angles θ and ϕ that the *normals* form with the x, y, z coordinate systems, as usual. We allow for non-random directions by considering an orientation distribution $g(\theta, \phi)$ and the possibility of a preferred direction α, as in the preceding section (Fig. 10.10), whereby α lies again in the x–z plane.

Let this specimen be cut (a) by a section plane T_2 and (b) by a test line T_1. We seek the probabilities that T_2, or T_1 respectively, intersect or hit dS.

(a) *Sectioning with a test plane T_2*

Let the specimen X be cut by a plane T_2 which is taken, for the sake of convenience, to be normal to the z-axis (Fig. 10.9). The probability that T_2 cuts dS is

$$\Pr\{T_2 \uparrow dS\} = \int_0^{2\pi}\int_0^{\pi/2} \Pr\{T_2 \uparrow dS \mid \theta, \phi\} g(\theta, \phi)\, d\theta\, d\phi \qquad (10.48)$$

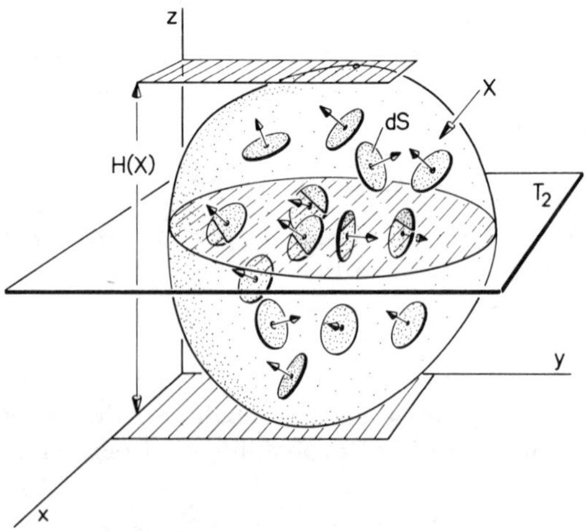

FIG. 10.9: Cutting spatial field of planar surface elements with plane.

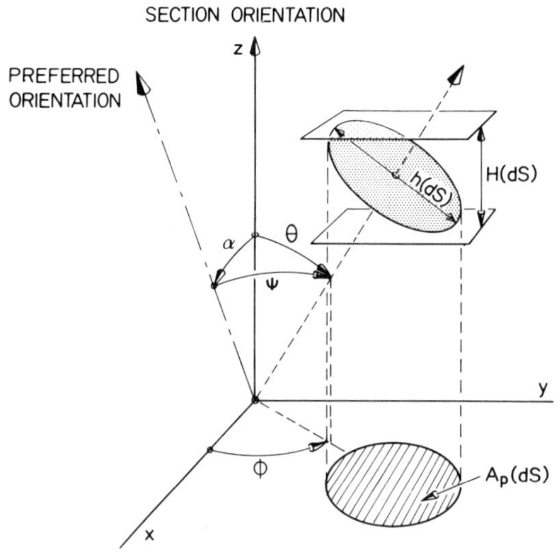

FIG. 10.10: Measures of orientation of planar surface element, together with caliper diameter and projected area.

Evidently (Figs 10.9 and 10.10)

$$\Pr\{T_2 \uparrow dS \mid \theta, \phi\} = H(dS \mid \theta, \phi)/H(X; 0, 0) \qquad (10.49)$$

where $H(X; 0, 0) = H(X)$ is the caliper diameter of the specimen in the direction $\theta = 0$, i.e. normal to the section plane, and

$$H(dS \mid \theta, \phi) = h(dS) \cdot \sin\theta \qquad (10.50)$$

so that

$$\Pr\{T_2 \uparrow dS \mid T_2 \uparrow X\} = \frac{h(dS)}{H(X)} \cdot \int_0^{2\pi} \int_0^{\pi/2} \sin\theta \cdot g(\theta, \phi)\, d\theta\, d\phi \quad (10.51)$$

It is noted that eq. (10.51) is analogous to eq. (10.17) except that $\cos\theta$ is replaced by $\sin\theta$ in the integral because the orientation is now determined by the direction of the normal of the surface element: the element is perpendicular to the section plane when $\theta = \pi/2$.

(b) *Sectioning with a test line T_1*
Let the specimen X be cut by a test line T_1 which is parallel to the z-axis. The probability that T_1 hits dS is

$$\Pr\{T_1 \uparrow dS\} = \int_0^{2\pi} \int_0^{\pi/2} \Pr\{T_1 \uparrow dS \mid \theta, \phi\} g(\theta, \phi)\, d\theta\, d\phi \qquad (10.52)$$

In analogy to eq. (3.99)

$$\Pr\{T_1 \uparrow dS \mid \theta, \phi\} = A_p(dS)/A_p(X) \tag{10.53}$$

where $A_p(dS)$ and $A_p(X)$ are the projected areas of dS and X onto the x–y plane, i.e. a plane normal to T_1 (Fig. 10.10). As in eq. (3.84) we have

$$A_p(dS) = s(dS) \cdot \cos\theta \tag{10.54}$$

so that

$$\Pr\{T_1 \uparrow dS \mid T_1 \uparrow X\} = \frac{s(dS)}{A_p(X)} \int_0^{2\pi} \int_0^{\pi/2} \cos\theta \cdot g(\theta, \phi)\, d\theta\, d\phi \tag{10.55}$$

Note that eq. (10.55) is analogous to eq. (10.17) except that the probability depends on a ratio of areal measures of dS and X.

Making appropriate assumptions for the orientation distribution of the surface elements, the integrals in eqs (10.51) and (10.55) can be computed. We proceed in the same way as for the needles (Section 10.2.2), allowing for a preferred direction α and an orientation distribution in terms of the angle ψ between the normals to dS and the preferred orientation, and using the same model distributions. The relation between the angles is given by eq. (10.20) and $g(\theta, \phi)$ by eq. (10.21).

(1) *Uniform distribution*: $f(\psi) = 1/(2\pi)$ (10.56)

This evidently describes the isotropic case. Solving eqs (10.51) and (10.55) yields

$$\begin{aligned}\Pr\{T_2 \uparrow dS\} &= \frac{h(dS)}{H(X)} \cdot \frac{1}{2\pi} \int_0^{2\pi} \int_0^{\pi/2} \sin^2\theta\, d\theta\, d\phi \\ &= \frac{\pi}{4} \cdot \frac{h(dS)}{H(X)}\end{aligned} \tag{10.57}$$

$$\begin{aligned}\Pr\{T_1 \uparrow dS\} &= \frac{s(dS)}{A_p(X)} \cdot \frac{1}{2\pi} \int_0^{2\pi} \int_0^{\pi/2} \cos\theta \sin\theta\, d\theta\, d\phi \\ &= \frac{1}{2} \cdot \frac{s(dS)}{A_p(X)}\end{aligned} \tag{10.58}$$

(2) *Fisher distribution*: $f(\psi) = c(K) \cdot e^{K \cos 2\psi}$ (10.59)

where the normalizing constant $c(K)$ is the same as that discussed in relation to eq. (10.23). The concentration parameter K determines the arrangement of the surface elements: with $K = 0$ they are isotropic, with $K \to \infty$ they tend to form a lamellar structure, whereas with $K \to -\infty$ they form a tubular structure, i.e. they are uniformly distributed with respect to an orientation cylinder around α (Figs 10.3, 10.11 and 10.12). The orientation probability density function is

$$g(\theta, \phi; K, \alpha) = c(K) \cdot \sin\theta \cdot e^{K\cos 2\psi} \tag{10.60}$$

As before, computing eqs (10.51) and (10.55) with this function is very awkward, but it becomes manageable for the two particular cases of $\alpha = 0$ and $\alpha = \pi/2$ when the preferred direction of the surface elements, i.e. that of their *normals*, is either parallel or perpendicular to the direction of the test probes; remember that the direction of a test plane is also given by that of its normal.

Case of $\alpha = 0$ (Fig. 10.11):
Preferred surface orientation parallel to section orientation
We have $\psi = \theta$ and, as in eq. (10.25)

$$g(\theta, \phi; K, 0) = c(K)\sin\theta e^{2K\cos^2\theta} \tag{10.61}$$

The normalization constant is as in eq. (10.27). The two probabilities become

$$\Pr\{T_2 \uparrow dS;\ K, 0\} = \frac{h(dS)}{H(X)} \cdot c(K) \int_0^{2\pi} \int_0^{\pi/2} \sin^2\theta e^{2K\cos^2\theta}\, d\theta\, d\phi \tag{10.62}$$

$$\Pr\{T_1 \uparrow dS;\ K, 0\} = \frac{s(dS)}{A_p(X)} \cdot c(K) \int_0^{2\pi} \int_0^{\pi/2} \cos\theta \sin\theta e^{2K\cos^2\theta}\, d\theta\, d\phi \tag{10.63}$$

The integral of eq. (10.63) is the same as in eq. (10.26) whose solution is given by eq. (10.28). We further find that, in relation to eq. (10.62),

$$\int_0^{2\pi} \int_0^{\pi/2} \sin^2\theta e^{2K\cos^2\theta}\, d\theta\, d\phi = 2\pi \cdot (\pi/4) \sum_{r=0}^{\infty} (2r-1)!![r!\,(r+1)!]^{-1} K^r \tag{10.64}$$

where $n!!$ is as in eq. (10.35). Using the results of eqs (10.27), (10.28) and (10.64) we obtain the probabilities of cutting the surface elements with planes or lines whose *directions* are *parallel* to the preferred orientations; they are:

$$\Pr\{T_2 \uparrow dS;\ K, 0\} = \frac{h(dS)}{H(X)} \cdot \frac{\pi \sum_{r=0}^{\infty} K^r[(2r-1)!!/\{r!\,(r+1)!\}]}{4\sum_{r=0}^{\infty} [(2K)^r/\{r!\,(2r+1)\}]} = \gamma_2(K, 0) \cdot \frac{h(dS)}{H(X)} \tag{10.65}$$

and for test lines,

$$\Pr\{T_1 \uparrow dS;\ K, 0\} = \frac{s(dS)}{A_p(X)} \cdot \frac{e^{2K} - 1}{4K \sum_{r=0}^{\infty} [(2K)^r/\{r!\,(2r+1)\}]}$$

$$= \gamma_1(K, 0) \cdot \frac{s(dS)}{A_p(X)} \tag{10.66}$$

Note that $\gamma_1(K, 0)$ is identical to that found in eq. (10.29) which is not surprising because in both cases lines and planes are interacting. The values of the coefficients $\gamma(K, 0)$ depend only on the concentration parameter K (Fig. 10.11). In the case of an *isotropic* structure ($K = 0$) we have

$$\gamma_2(0, 0) = \pi/4$$
$$\gamma_1(0, 0) = \tfrac{1}{2}$$

which reduces eqs (10.65) and (10.66) to eqs (10.57) and (10.58) respectively. For *lamellar* structures (Fig. 10.11a), as $K \to \infty$, i.e. as all surface elements lie parallel to each other,

$$\gamma_2(\infty, 0) \to 0$$
$$\gamma_1(\infty, 0) \to 1$$

And finally for *tubular* structures where $K \to -\infty$ (Fig. 10.11b)

$$\gamma_2(-\infty, 0) \to 1$$
$$\gamma_1(-\infty, 0) \to 0$$

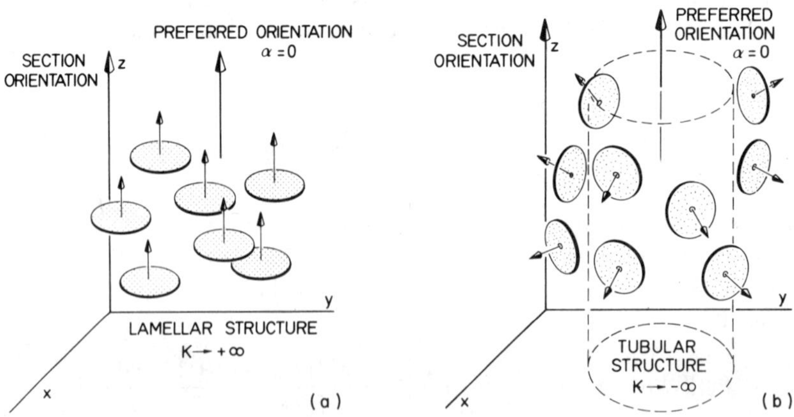

FIG. 10.11: Construction of lamellar and tubular structures with preferred orientation $\alpha = 0$.

In other words, the caliper diameter of discs in tubular arrangement measured in the orientation direction of horizontal planes is $h(dS)$ for all elements (Fig. 10.11b) making the probability to cut them with a plane directly dependent on $h(dS)/H(X)$, whereas such discs can evidently not be hit by a vertical test line. For discs in lamellar arrangement, on the other hand, it is found that the probability of hitting them by a vertical test line depends solely on $s(dS)/A_p(X)$, whereas they cannot be cut by a horizontal test plane.

Case of $\alpha = \pi/2$ (Fig. 10.12):
Preferred surface orientation perpendicular to section orientation
By eq. (10.20) we find, for $\alpha = \pi/2$

$$\cos \psi = \sin \theta \cos \phi \tag{10.67}$$

so that, as in eq. (10.33)

$$g(\theta, \phi; K, \pi/2) c(K) \sin \theta e^{2K \sin^2 \theta \cos^2 \phi} \tag{10.68}$$

where $c(K)$ is again the same in eq. (10.27). The two probabilities become

$$\Pr\{T_2 \uparrow dS;\ K, \pi/2\} = \frac{h(dS)}{H(X)} \cdot c(K) \int_0^{2\pi} \int_0^{\pi/2} \sin^2 \theta e^{2K \sin^2 \theta \cos^2 \phi}\, d\theta\, d\phi \tag{10.69}$$

$$\Pr\{T_1 \uparrow dS;\ K, \pi/2\} = \frac{s(dS)}{A_p(X)} \cdot c(K) \int_0^{2\pi} \int_0^{\pi/2} \cos \theta \sin \theta e^{2K \sin^2 \theta \cos^2 \phi}\, d\theta\, d\phi \tag{10.70}$$

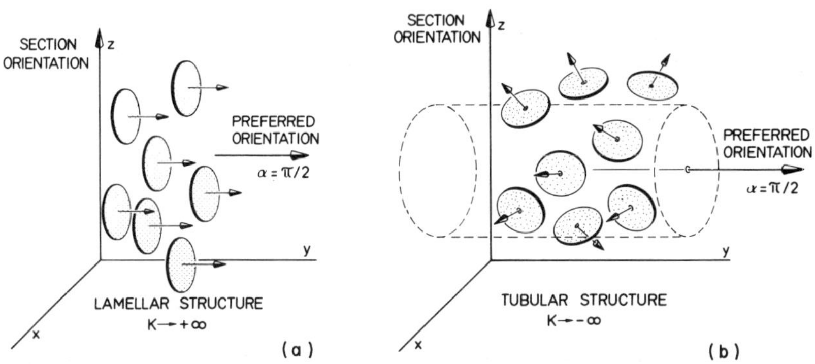

FIG. 10.12: Construction of lamellar and tubular structures with preferred orientation $\alpha = \pi/2$.

The integral of eq. (10.70) has the same value as in eq. (10.35). We further find that, for eq. (10.69),

$$\int_0^{2\pi}\int_0^{\pi/2} \sin^2\theta e^{2K\sin^2\theta\cos^2\phi}\, d\theta\, d\phi = 2\pi(\pi/4)\sum_{r=0}^{\infty}\frac{2r+1}{r+1}\cdot\frac{[(2r-1)!!]^2}{(r!)^3}\cdot\left(\frac{K}{2}\right)^r \quad (10.71)$$

Using the results of eqs (10.27), (10.35) and (10.71) we find the probabilities of cutting the surface elements with planes or lines whose *directions* are *perpendicular* to the preferred direction of the surface elements as for section planes:

$$\Pr\{T_2 \uparrow dS;\, K, \pi/2\} = \gamma_2(K, \pi/2)\cdot\frac{h(dS)}{H(X)} \quad (10.72)$$

where

$$\gamma_2(K, \pi/2) = \frac{\pi}{4}\cdot\frac{\sum_{r=0}^{\infty}\frac{2r+1}{r+1}\cdot\frac{[(2r-1)!!]^2}{(r!)^3}\cdot\left(\frac{K}{2}\right)^r}{\sum_{r=0}^{\infty}[(2K)^r/\{r!\,(2r+1)\}]} \quad (10.73)$$

and for test lines:

$$\Pr\{T_1 \uparrow dS;\, K, \pi/2\} = \gamma_1(K, \pi/2)\cdot\frac{s(dS)}{A_p(X)} \quad (10.74)$$

where

$$\gamma_1(K, \pi/2) = \frac{1}{2}\cdot\frac{\sum_{r=0}^{\infty}[K^r(2r-1)!!/\{r!\,(r+1)!\}]}{\sum_{r=0}^{\infty}[(2K)^r/\{r!\,(2r+1)\}]} \quad (10.75)$$

Note that, again, $\gamma_1(K, \pi/2)$ is the same as in eq. (10.36) for line segments intersected by test planes. The values of $\gamma(K, \pi/2)$ also depend exclusively on the concentration parameter K which describes the degree of alignment between the surface elements. For isotropic structures ($K=0$) they are identical to those given above for $\alpha=0$. If the structure is lamellar, $K\to\infty$ (Fig. 10.12a), and

$$\gamma_2(\infty, \pi/2)\to 1$$
$$\gamma_1(\infty, \pi/2)\to 0$$

because the probability of hitting surface elements by test lines (γ_1) which are perpendicular to the direction of the element (measured by its normal) is 0. For tubular structures (Fig. 10.12b), where $K\to-\infty$, we obtain

$$\gamma_2(-\infty, \pi/2) \rightarrow 2/\pi$$
$$\gamma_1(-\infty, \pi/2) \rightarrow 2/\pi$$

These results are interpreted as follows. The probability to cut lamellar structures by test planes whose *normal* is perpendicular to the preferred orientation axis of the surface elements ($\alpha = \pi/2$) depends directly on the ratio of the disc diameter to the diameter of the containing space, $h(dS)/H(X)$, so that $\gamma_2(\infty, \pi/2) = 1$, but such discs cannot be cut by vertical test lines whose direction is perpendicular to the disc orientation. However, to obtain the probability of cutting such discs in tubular arrangement we need to average their projected diameter (for cutting with a plane) or their projected area (for cutting with a test line) by rotating them around a circle perpendicular to $\alpha = \pi/2$, the master circle of the "orientation cylinder" of the tubular structure (Fig. 10.12b). The factor $2/\pi$ is the result of this averaging procedure.

(3) *Marriott distribution*: $f(\psi) = c(K) \cdot [1 + K \cdot \cos 2\psi]$

The nature of this distribution, which is an approximation of the Fisher distribution for small values of K, valid in the range $\{-1 \leq K \leq 1\}$, has been discussed in relation to eq. (10.38). Applied to discs this means that for $K > 0$ the discs tend to lamellar, for $K < 0$ to tubular arrangement, without ever reaching the totally ordered patterns shown in Figs 10.11 and 10.12. The orientation distribution $g(\theta, \phi;\ K, \alpha)$ is given in eq. (10.39):

$$g(\theta, \phi;\ K, \alpha) = c(K) \cdot \sin\theta(1 + K \cdot \cos 2\psi) \tag{10.76}$$

where $c(K)$ is the same as before (eq. 10.40).

The solutions to eqs (10.51) and (10.55) are now obtained from

$$\Pr\{T_2 \uparrow dS\} = \frac{h(dS)}{H(X)} \cdot c(K) \int_0^{2\pi} \int_0^{\pi/2} \sin^2\theta(1 + K\cos 2\psi)\, d\theta\, d\phi \tag{10.77}$$

$$\Pr\{T_1 \uparrow dS\} = \frac{s(dS)}{A_p(X)} \cdot c(K) \int_0^{2\pi} \int_0^{\pi/2} \cos\theta \sin\theta(1 + K\cos 2\psi)\, d\theta\, d\phi \tag{10.78}$$

The solution to the integral of eq. (10.78) is given in eq. (10.41). The integral in eq. (10.77) yields

$$\int_0^{2\pi} \int_0^{\pi/2} \sin^2\theta(1 + K\cos 2\psi)\, d\theta\, d\phi = \frac{\pi^2}{2}\{1 - (K/4)(1 + \cos^2\alpha)\} \tag{10.79}$$

Using the results of eqs (10.40), (10.41) and (10.79) we find the probabilities of cutting the surface elements with planes T_2 and test lines T_1 orientated at an angle α to the preferred orientation of the surface elements to be for section planes:

$$\Pr\{T_2 \uparrow dS;\ K, \alpha\} = \gamma_2(K, \alpha) \cdot \frac{h(dS)}{H(X)} \tag{10.80}$$

where

$$\gamma_2(K, \alpha) = \frac{\pi}{4} \cdot \frac{1-(K/4)(1+\cos^2 \alpha)}{1-K/3} \tag{10.81}$$

and for test lines:

$$\Pr\{T_1 \uparrow dS;\ K, \alpha\} = \gamma_1(K, \alpha) \cdot \frac{s(dS)}{A_p(X)} \tag{10.82}$$

where

$$\gamma_1(K, \alpha) = \frac{1}{2} \cdot \frac{1-(K/2)\sin^2 \alpha}{1-K/3} \tag{10.83}$$

It is again to be noticed that $\gamma_1(K, \alpha)$ is identical in eqs (10.42) and (10.83) which is not surprising if one considers that in both cases lines and planes are interacting; to exchange the roles between planes and lines as elements of the structure or of the section does not apparently matter. We find that both for cutting the structure with a section plane or with a test line, the probabilities depend, on the one hand, on the concentration parameter K, determining the orientation distribution pattern of the elements, and on the other hand, on the orientation α of the probe (T_2 or T_1) with respect to the preferred orientation of the structural elements.

Again the two particular cases of $\alpha = 0$ and $\alpha = \pi/2$ yield interestingly simple results which depend solely on K. For plane sections we find

$$\gamma_2(K, 0) = \frac{\pi}{4} \cdot \frac{1-K/2}{1-K/3} \tag{10.84}$$

$$\gamma_2(K, \pi/2) = \frac{\pi}{4} \cdot \frac{1-K/4}{1-K/3} \tag{10.85}$$

and for test lines

$$\gamma_1(K, 0) = \frac{1}{2} \cdot \frac{1}{1-K/3} \tag{10.86}$$

$$\gamma_1(K, \pi/2) = \frac{1}{2} \cdot \frac{1-K/2}{1-K/3} \tag{10.87}$$

Note that the coefficients given in eqs (10.86) and (10.87) for surface elements cut by lines are identical with those found in eqs (10.43) and (10.44) for needles cut by planes, as is actually to be expected.

Let us now look at some special cases of K. For $K = 1$ (tending to lamellar arrangement of the surface elements) a vertical plane section ($\alpha = \pi/2$) will yield the largest number of cuts (P_2), and a horizontal section ($\alpha = 0$) the smallest number (P_1). Taking sections at right angles to each other we find that the ratio of sectioning probabilities is

$$P_2/P_1 \leqslant \tfrac{3}{2} \tag{10.88}$$

the equality applying to exact transverse and longitudinal sections. For $K = -1$ where the elements tend to tubular arrangement the largest number of cuts occurs at $\alpha = 0$, and the smallest at $\alpha = \pi/2$. The ratio is

$$P_1/P_2 \leqslant \tfrac{6}{5} \tag{10.89}$$

Similar relations can be easily derived for lines intersecting the surface elements, and one will not be surprised to find them corresponding to those for lines cut by a plane given in eq. (10.47). For $K = 1$ a vertical line ($\alpha = 0$) gives the largest number of cuts (P_2), and a horizontal line the smallest; any two perpendicular lines will have a ratio of cuts

$$P_2/P_1 \leqslant 2 \tag{10.90}$$

whereas for $K = -1$ we find

$$P_1/P_2 \leqslant \tfrac{3}{2} \tag{10.91}$$

because the largest number of cuts occurs at $\alpha = \pi/2$, i.e. with lines which "traverse" the tubular structure crosswise.

10.4. Estimating average stereological parameters for anisotropic structures

In the first part of this chapter we have considered the geometric probability of hitting anisotropic structures with stereological probes such as test lines and planes. We should now ask how this will affect the estimation of average stereological parameters, for in the principles derived in previous chapters we had always assumed isotropic conditions to prevail. We shall sequentially look at the cases of anisotropic curves in a plane, anisotropic curves in space and anisotropic surfaces in space. The general result that will ensue is that the same basic measuring principles are valid, but that a coefficient needs to be introduced which depends exclusively on the degree and pattern of order or preferred orientation in the structure.

10.4.1. LENGTH DENSITY OF AN ANISOTROPIC CURVE IN A PLANE

Our specimen is a planar domain X of area $A(X)$ which contains a set of curves Y showing some preferred orientation; the total length of these curves is $B(Y)$. We would like to estimate the length density of Y in X

$$B_A(Y \subset X) = B(Y)/A(X) \tag{10.92}$$

from a count of intersections $I(Y \cap T_1)$ between the anisotropic curves and a test line T_1 of fixed orientation with respect to the specimen X; for convenience we set the orientation of the test line equal to $\pi/2$ (Fig. 10.13a). In analogy to Section 3.2.3 we should like to find B_A from the ratio of average intersection count to average length of test line enclosed in X (eq. 3.122).

To find the average or expected number of intersections we consider Y to be made of infinitesimally small essentially straight line elements dY of length $b(dY)$ (Fig. 10.1b) so that

$$E\{I(Y \cap T_1)\} = \int_Y E\{I(dY \cap T_1)\} \tag{10.93}$$

By definition $I(dY \cap T_1) = 1$ if $T_1 \uparrow dY$ and 0 otherwise, so that

$$E\{I(dY \cap T_1)\} = \Pr\{T_1 \uparrow dY\} \tag{10.94}$$

Now the infinitesimal line element dY is equivalent to the "needle" dB considered in Section 10.3.1. Therefore, according to eqs (10.1) to (10.3) we can write

$$\Pr\{T_1 \uparrow dY\} = \left\{ \int_0^{\pi} |\cos\theta|\, g(\theta)\, d\theta \right\} b(dY)/H(X;0) \tag{10.95}$$

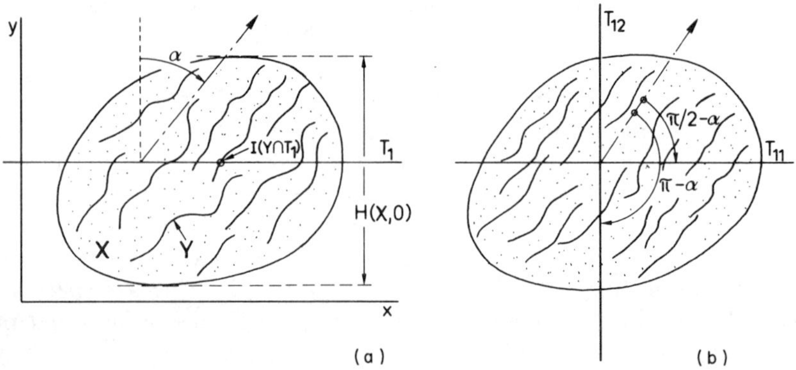

FIG. 10.13: Estimating length of anisotropic curves on planar specimen from intersection density with test line.

where $g(\theta)$ is a probability density function describing the orientation pattern of the curve elements and by that the orientation properties of Y; $H(X; 0)$ is the caliper diameter of X in the direction normal to T_1, representing the space within which T_1 can be randomly positioned (Fig. 10.13a).

According to eqs (10.93–10.94) the expected number of intersections is obtained by integrating eq. (10.95) over the curve Y which yields

$$E\{I(Y \cap T_1)\} = \left\{ \int_0^{\pi} |\cos\theta|\, g(\theta)\, d\theta \right\} \cdot B(Y)/H(X; 0) \tag{10.96}$$

On the other hand it is easily seen that the average length of test line contained in X is

$$E\{L(X \cap T_1)\} = A(X)/H(X; 0) \tag{10.97}$$

so that, by dividing eq. (10.96) by eq. (10.97) and rearranging we obtain

$$B(Y)/A(X) = \left\{ \int_0^{\pi} |\cos\theta|\, g(\theta)\, d\theta \right\}^{-1} \cdot E\{I(Y \cap T_1)\}/E\{L(X \cap T_1)\} \tag{10.98}$$

or, in the short-hand symbolism of stereology

$$B_A = \gamma_0^{-1} \cdot I_L \tag{10.99}$$

It is easily seen that this is equivalent to eqs (3.104) and (3.122) for the case of isotropic lines, except that the coefficient $\pi/2$ is replaced by a coefficient γ_0^{-1} which depends on the orientation distribution of the line elements. These coefficients have in fact been derived in Section 10.3.1. Thus, if a uniform orientation distribution is imposed we found $\gamma_0 = 2/\pi$ (eq. 10.5) which makes eq. (10.99) equal to eq. (3.104). If the curve's orientation pattern can be described by a Marriott distribution as given in eq. (10.9) and shown in Fig. 10.6 then γ is a function of the preferred orientation α and what has been described as the concentration parameter K of orientations about α (eq. 10.12):

$$\gamma_0(K, \alpha) = \frac{2}{\pi}\left[1 + \left(\tfrac{1}{3}\right) \cdot K \cos 2\alpha\right] \tag{10.100}$$

Note that the preferred orientation is measured as the angle to the normal to T_1, so that the angle between the test line and the preferred orientation is $\pi/2 - \alpha$.

In this particular case we have

$$B_A = \left\{ \frac{2}{\pi}\left[1 + \left(\tfrac{1}{3}\right) \cdot K \cdot \cos 2\alpha\right] \right\}^{-1} \cdot I_L \tag{10.101}$$

We should note a few particular cases: (1) if $K=0$ the curves are isotropic (Fig. 10.6) and eq. (10.101) reduces to eq. (3.104); (2) if the test line is normal to the preferred orientation ($\alpha=0$) then

$$B_{\mathrm{A}}(\text{normal}) = \left\{\frac{2}{\pi}[1+K/3]\right\}^{-1} \cdot I_L \tag{10.102}$$

We have discussed the case of estimating the length density of curves in a plane in detail because in the case of anisotropic surfaces in space it may often appear desirable to estimate simply the line density of surface profiles on a section of known orientation with respect to the specimen. In order to obtain an estimate of B_{A} in such instances we need to know the direction of preferred orientation α with respect to the test line and to have an estimate of K. How to obtain these estimates will be discussed in Section 10.5.

A case, interesting in practice, has been worked out by Cruz-Orive (personal communication). If we use two mutually perpendicular test lines, T_{11} and T_{12}, to estimate intersection densities I_{L1} and I_{L2}, respectively (Fig. 10.13b), we will have

$$\begin{aligned} B_{\mathrm{A}} &= \left\{\frac{2}{\pi}[1+(\tfrac{1}{3})K\cos 2\alpha]\right\}^{-1} I_{L1} \\ &= \left\{\frac{2}{\pi}[1-(\tfrac{1}{3})K\cos 2\alpha]\right\}^{-1} I_{L2} \end{aligned} \tag{10.103}$$

because the angles between test line and preferred orientation are now $(\pi/2-\alpha)$ for T_{11} and $(\pi-\alpha)$ for T_{12}. In this case we find

$$(\tfrac{1}{3})K\cos 2\alpha = (I_{L1}-I_{L2})/(I_{L1}+I_{L2}) \tag{10.104}$$

so that

$$B_{\mathrm{A}} = (\pi/4)(I_{L1}+I_{L2}) \tag{10.105}$$

irrespective of K and of α. Thus with this model one can find B_{A} from counts on test lines of a square lattice, introducing the average of the intersection counts $(I_{L1}+I_{L2})/2$ into the basic eq. (3.104) without considering the orientation distribution at all. Note, however, that I_{L1} and I_{L2} have to be obtained as the ratio $E\{I\}/E\{L\}$ as in eqs (10.98) and (3.122).

In a general sense it should be noted that for the basic stereological principles, as developed in Chapter 3, to hold, all that counts is a uniform distribution of the angles between test line and curve Y. This can be approximated with an arbitrary degree of precision by using n sets of parallel test lines a distance d apart and choosing the angles between them to be π/n (Moran, 1966). We then have

$$B_A = (\pi/2) \cdot \left(\sum_{i=1}^{n} I_{Li} \right) \Big/ n \qquad (10.106)$$

Note that eq. (10.105) is eq. (10.106) for $n = 2$.

Finally, an alternative in the same sense is to use a totally isotropic test line system, such as that made of semicircles (Merz, 1968) where the intersection counts are directly entered into eq. (3.104); for the practical method of using this test system see Volume 1, Section 4.2.3.3.

Thus having gone through this development we note that it is not really necessary to consider anisotropy of the structure if we are only interested in estimating the length of the curve and if proper precautions are taken in setting up the test line system used for intersection counting. The simplest way is to do the count on square lattices at arbitrary orientation and to introduce the average of the counts in both directions, but note that it is not necessary to orientate the lines at the angles of 19° and 71° to the preferred orientation, as it has been proposed (Sitte, 1967); it is not even correct unless the elements dY are strictly parallel.

10.4.2. LENGTH DENSITY OF AN ANISOTROPIC CURVE IN SPACE

We consider a specimen X of volume $V(X)$ which contains an anisotropic curved line Y of total length $J(Y)$. We need to estimate the length density of the curve within the specimen

$$J_V = J(Y)/V(X) \qquad (10.107)$$

from a count of transections $Q(Y \cap T_2)$ on the specimen profile of area $A(X \cap T_2)$ obtained by cutting the specimen with a plane T_2 held at fixed orientation; in fact, for convenience we choose the coordinate system such that the polar coordinates of the normal to T_2, θ and ϕ, are both 0, and that the preferred orientation α lies in the x–z plane, as before (Figs 10.8 and 10.14).

To find the average or expected number of transections we consider Y to be made of infinitesimally short essentially straight line elements dY of length $l(dY)$ so that

$$E\{Q(Y \cap T_2)\} = \int_Y E\{Q(dY \cap T_2)\} \qquad (10.108)$$

As with respect to eq. (10.93), $Q(dY \cap T_2) = 1$ if $T_2 \uparrow dY$ and 0 otherwise so that

$$E\{Q(dY \cap T_2)\} = \Pr\{T_2 \uparrow dY\} \qquad (10.109)$$

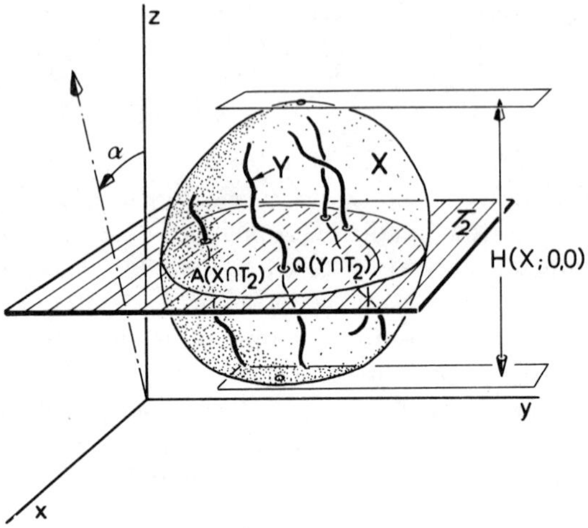

FIG. 10.14: Estimating length of anisotropic curves in spatial specimen from transection density on test plane.

Taking $\Pr\{T_2 \uparrow dY\}$ from eq. (10.17), noting that dY is equivalent to the "needle" dJ, and integrating over Y, according to eq. (10.108) we obtain immediately

$$E\{Q(Y\cap T_2)\} = \left\{\int_0^{2\pi}\int_0^{\pi/2} \cos\theta \cdot g(\theta, \phi)\, d\theta\, d\phi\right\} \cdot J(Y)/H(X; 0, 0) \tag{10.110}$$

where $H(X; 0, 0)$ is the caliper diameter of the specimen in the direction of the normal to T_2 (Fig. 10.14).

The average profile area of X at this orientation of T_2 is evidently

$$E\{A(X\cap T_2)\} = V(X)/H(X; 0, 0) \tag{10.111}$$

so that by dividing eq. (10.110) by eq. (10.111) and rearranging we find

$$J(Y)/V(X) = \left\{\int_0^{2\pi}\int_0^{\pi/2} \cos\theta \cdot g(\theta, \phi)\, d\theta\, d\phi\right\}^{-1} \cdot E\{Q(Y\cap T_2)\}/E\{A(X\cap T_2)\} \tag{10.112}$$

or, in the short-hand symbolism of stereology

$$J_V = \gamma_1^{-1} \cdot Q_A \tag{10.113}$$

Note, on the one hand, that this development is identical to the one in Section 10.4.1, and on the other hand, that eqs (10.112) and (10.113) are equivalent to eqs (3.142) and (3.128) respectively, except that the coefficient 2 is replaced by γ_1^{-1} which depends on the orientation distribution of the curve elements.

In Section 10.3.2 we have, in fact, derived γ_1 for a particular orientation distribution, the Mariott distribution which is an approximation to the more general Fisher distribution. From eq. (10.42) we find

$$\gamma_1(K, \alpha) = \{\tfrac{1}{2}[1-(K/2)\sin^2\alpha]/[1-(K/3)]\} \tag{10.114}$$

where α is the preferred orientation and K the concentration parameter of the distribution (see Section 10.3.2 for further detail). Note that α is the angle between the preferred orientation and the normal to T_2. In this particular case we hence have

$$J_V(K, \alpha) = \{2[1-(K/3)]/[1-(K/2)\sin^2\alpha]\} \cdot Q_A \tag{10.115}$$

Let us note a few interesting cases deriving from this general relation. First, if $K=0$ the curve is isotropic, and $\gamma_1^{-1}=2$, as in eq. (3.128); the same result was obtained in eq. (10.19) by introducing uniform orientation distribution (eq. 10.22). Secondly, by looking at the particular case of transverse ($\alpha = 0$) and longitudinal sections ($\alpha = \pi/2$) we find (eqs 10.43 and 10.44):

$$\gamma_1(K, 0) = \tfrac{1}{2}[1/\{1-(K/3)\}] \tag{10.116}$$

$$\gamma_1(K, \pi/2) = \tfrac{1}{2}[\{1-(K/2)\}/\{1-(K/3)\}] \tag{10.117}$$

We can hence write for counts obtained on transverse and longitudinal sections respectively:

$$J_V = 2[1-(K/3)] \cdot Q_{A1} \quad \text{for} \quad \alpha = 0 \tag{10.118}$$

$$J_V = 2[\{1-(K/3)\}/\{1-(K/2)\}] \cdot Q_{A2} \quad \text{for} \quad \alpha = \pi/2 \tag{10.119}$$

Since both estimates of J_V should be the same we can derive K by dividing eq. (10.119) by eq. (10.118) to yield

$$K = 2[1-(Q_{A2}/Q_{A1})] \tag{10.120}$$

Introducing this into eq. (10.118) we derive

$$J_V = \tfrac{2}{3}(Q_{A1}+2Q_{A2}) \tag{10.121}$$

The conclusion is that one can estimate J_V without knowledge of the orientation distribution pattern, measured by K, if one obtains transection densities on transverse and longitudinal sections of the specimen.

Note that this simplification is essentially due to some nice properties of the Mariott distribution. For the more general Fisher distribution, which allows more marked degrees of anisotropy, things are more difficult. In Section 10.3.2 we have evaluated γ_1 for the cases of transverse and

Table 10.1
Multiplicative coefficients for stereological formulae applicable to anisotropic structures, based on Fisher's orientation distribution, as a function of the concentration parameter K

$K>0$	$\gamma_1(K,0)^{-1}$	$\gamma_1(K,\pi/2)^{-1}$	$\gamma_2(K,0)^{-1}$	$\gamma_2(K,\pi/2)^{-1}$
0.0	2.0000	2.0000	1.2732	1.2732
0.3	1.8420	2.0920	1.3318	1.2463
0.5	1.7025	2.2025	1.4021	1.2193
0.8	1.5820	2.3325	1.4849	1.1930
1.0	1.4803	2.4820	1.5801	1.1682
1.3	1.3961	2.6498	1.6869	1.1456
1.5	1.3274	2.8331	1.8036	1.1256
1.8	1.2720	3.0286	1.9280	1.1083
2.0	1.2278	3.2319	2.0575	1.0937
3.0	1.1251	4.0509	2.5789	1.0560
4.0	1.0819	4.7856	3.0466	1.0382
5.0	1.0608	5.4207	3.4509	1.0289
6.0	1.0485	5.9823	3.8085	1.0234
7.0	1.0405	6.4918	4.1328	1.0196
8.0	1.0348	6.9622	4.4323	1.0169
9.0	1.0305	7.4019	4.7122	1.0149
10.0	1.0272	7.8163	4.9760	1.0133
12.0	1.0223	8.5845	5.4651	1.0110
16.0	1.0164	9.9430	6.3299	1.0081
20.0	1.0130	11.1363	7.0896	1.0064
∞	1	∞	∞	1
$K<0$				
0.0	2.0000	2.0000	1.2732	1.2732
−0.3	2.1746	1.9246	1.2253	1.2992
−0.5	2.3629	1.8637	1.1864	1.3234
−0.8	2.5616	1.8148	1.1553	1.3456
−1.0	2.7671	1.7758	1.1305	1.3656
−1.3	2.9757	1.7447	1.1107	1.3833
−1.5	3.1847	1.7199	1.0949	1.3988
−1.8	3.3914	1.7001	1.0823	1.4123
−2.0	3.5941	1.6840	1.0721	1.4240
−3.0	4.3502	1.6444	1.0468	1.4581
−4.0	5.0147	1.6244	1.0341	1.4789
−5.0	5.6051	1.6128	1.0268	1.4928
−6.0	6.1401	1.6055	1.0221	1.5028
−7.0	6.6320	1.6002	1.0187	1.5103
−8.0	7.0897	1.5964	1.0163	1.5162
−9.0	7.5201	1.5935	1.0144	1.5209
−10.0	7.9265	1.5910	1.0129	1.5248
$-\infty$	∞	$\pi/2$	1	$\pi/2$

longitudinal sections; they are given in eqs (10.29) and (10.36) respectively. It is difficult to conceive at present a practical case where the use of such difficult coefficients would be justified. For sake of completeness Table 10.1 lists some of these coefficients, calculated numerically to four decimal places; they may be of interest particularly in dealing with highly oriented structures where $K \gg 1$ or $K \ll -1$. Note the particular cases of "perfectly" parallel lines ($K \rightarrow \infty$) where for a transverse section $\gamma_1 = 1$, and for curves "perfectly" parallel to a plane ($K \rightarrow -\infty$) where for a "longitudinal" section (i.e. a section perpendicular to the plane of orientation) $\gamma_1 = \pi/2$ (see eqs 10.31 and 10.37).

10.4.3. SURFACE DENSITY OF AN ANISOTROPIC SURFACE IN SPACE

The case of an anisotropic surface in space is a very frequent one in nature. One can cite a large number of examples from both biology and materials sciences, e.g. the membrane system of muscle cells, the inner surface of bone, the xylem walls in wood, or the lamellar structures in wrought alloys. In practical terms this is therefore an important problem to solve.

The model specimen (Fig. 10.15) is now again a structure X of volume $V(X)$ which contains surfaces (or interfaces between phases) Y of total

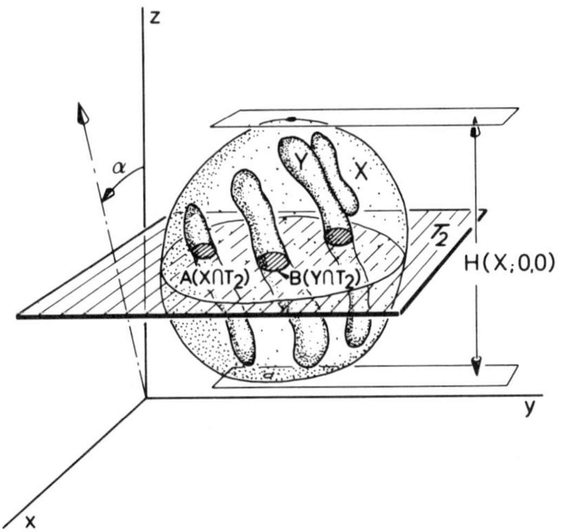

FIG. 10.15: Estimating surface of anisotropic objects from boundary length density on test plane.

area $S(Y)$. We need to estimate the surface density

$$S_V = S(Y)/V(X) \tag{10.122}$$

from sections. There are two approaches, as in Section 3.2. (1) The specimen can be sectioned by a plane section T_2 (Fig. 10.15) and we seek to find S_V from the boundary length density

$$B_A = E\{B(Y \cap T_2)\}/E\{A(X \cap T_2)\} \tag{10.123}$$

where $B(Y \cap T_2)$ is the length of the surface trace and $A(X \cap T_2)$ the area of the profile of the specimen. (2) The specimen can be probed with a test line T_1 (Fig. 10.16) and we attempt to estimate S_V from the intersection density

$$I_L = E\{I(Y \cap T_1)\}/E\{L(X \cap T_1)\} \tag{10.124}$$

where $I(Y \cap T_1)$ is the number of intersections of the surface with the test line and $L(X \cap T_1)$ is the length of test line enclosed in the specimen. For the case of an isotropic surface the relationships between S_V, B_A and I_L were worked out in Section 3.2; we are now interested in finding similar relations when the surface is anisotropic.

First, consider the *case of a plane section* T_2 (Fig. 10.15). To find the average length of the surface trace, $E\{B(Y \cap T_2)\}$, consider the surface to be subdivided into infinitesimally small essentially planar surface elements, dY, of area $s(dY)$; if these are cut by T_2 a trace of length $b(dY \cap T_2)$ is obtained. As before we find the average length of surface

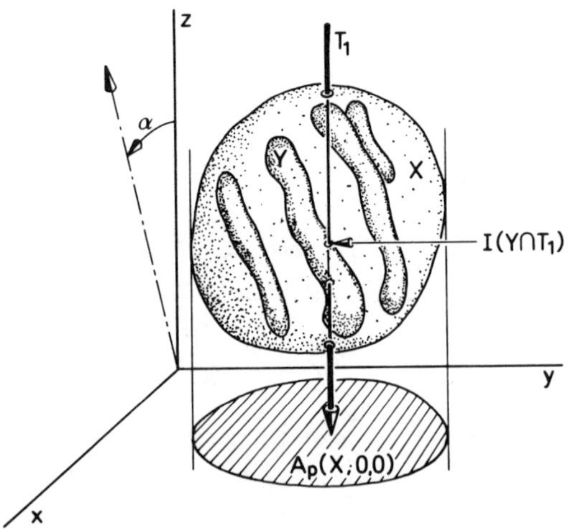

FIG. 10.16: Estimating surface of anisotropic objects from intersection density on test line.

trace by integrating over the surface the average trace length of the elements:

$$E\{B(Y \cap T_2)\} = \int_Y E\{b(dY \cap T_2)\} \tag{10.125}$$

In estimating the average trace length of the element we must first note that a trace is only obtained if $T_2 \uparrow dY$ and that this may depend on the orientation of the element; let us call this length $dB(\theta, \phi)$. Thus $b(dY \cap T_2) = dB(\theta, \phi)$ if $T_2 \uparrow dY$ and 0 otherwise. We therefore find, allowing for an orientation distribution,

$$E\{b(dY \cap T_2)\} = \int_0^{2\pi} \int_0^{\pi/2} dB(\theta, \phi) \cdot \Pr(T_2 \uparrow dY \mid \theta, \phi) \cdot g(\theta, \phi)\, d\theta\, d\phi \tag{10.126}$$

In Section 3.2.1 we have shown (eq. 3.71) that for any orientation

$$dB = s(dY)/h(dY) \tag{10.127}$$

where $h(dY)$ is the diameter of the element dY. On the other hand the probability to hit the element is

$$\Pr\{T_2 \uparrow dY \mid \theta, \phi\} = \sin\theta \cdot h(dY)/H(X; 0, 0) \tag{10.128}$$

where $H(X; 0, 0)$ is, as before, the caliper diameter of the specimen normal to the section plane. Hence

$$E\{b(dY \cap T_2)\} = \left\{ \int_0^{2\pi} \int_0^{\pi/2} \sin\theta \cdot g(\theta, \phi)\, d\theta\, d\phi \right\} \cdot s(dY)/H(X; 0, 0) \tag{10.129}$$

Integrating over Y according to eq. (10.125) we find

$$E\{B(Y \cap T_2)\} = \left\{ \int_0^{2\pi} \int_0^{\pi/2} \sin\theta \cdot g(\theta, \phi)\, d\theta\, d\phi \right\} \cdot S(Y)/H(X; 0, 0) \tag{10.130}$$

The average profile area of the specimen is, as in eq. (10.111),

$$E\{A(X \cap T_2)\} = V(X)/H(X; 0, 0) \tag{10.131}$$

so that we obtain by dividing eq. (10.130) by eq. (10.131) and rearranging

$$S(Y)/V(X) = \left\{ \int_0^{2\pi} \int_0^{\pi/2} \sin\theta \cdot g(\theta, \phi)\, d\theta\, d\phi \right\}^{-1} \cdot E\{B(Y \cap T_2)\}/E\{A(X \cap T_2)\} \tag{10.132}$$

or in short-hand symbolism

$$S_V = \gamma_2^{-1} \cdot B_A \tag{10.133}$$

It is seen that an orientation dependent coefficient γ_2^{-1} replaces the coefficient $4/\pi$ obtained for the isotropic case in eqs (3.66) and (3.76).

If we now look at the *case of a line section* T_1 (Fig. 10.16) we proceed through similar steps. The average number of intersections is obtained by

$$E\{I(Y \cap T_1)\} = \int_Y E\{I(dY \cap T_1)\} \tag{10.134}$$

and we note that $I(dY \cap T_1) = 1$ if $T_1 \uparrow dY$ and 0 otherwise, so that, simply,

$$E\{I(dY \cap T_1)\} = \int_0^{2\pi} \int_0^{\pi/2} \Pr\{T_1 \uparrow dY \mid \theta, \phi\} \cdot g(\theta, \phi)\, d\theta\, d\phi \tag{10.135}$$

The probability of hitting dY has been worked out in Section 3.2.2 to be the ratio of projected area of dY, $A_p(dY \mid \theta, \phi)$, to the projected area of the specimen $A_p(X; 0, 0)$, both normal to the test line; according to eqs (3.84) and (10.54) we have

$$A_p(dY \mid \theta, \phi)/A_p(X; 0, 0) = s(dY) \cdot \cos\theta / A_p(X; 0, 0) \tag{10.136}$$

Consequently,

$$E\{I(dY \cap T_1)\} = \left\{ \int_0^{2\pi} \int_0^{\pi/2} \cos\theta \cdot g(\theta, \phi)\, d\theta\, d\phi \right\} \cdot s(dY)/A_p(X; 0, 0) \tag{10.137}$$

and, by integrating over the entire surface

$$E\{I(Y \cap T_1)\} = \left\{ \int_0^{2\pi} \int_0^{\pi/2} \cos\theta \cdot g(\theta, \phi)\, d\theta\, d\phi \right\} \cdot S(Y)/A_p(X; 0, 0) \tag{10.138}$$

As developed in Section 3.2.2 the average length of test line in the specimen is

$$E\{L(X \cap T_1)\} = V(X)/A_p(X; 0, 0) \tag{10.139}$$

so that, finally, by dividing and rearranging we find

$$S(Y)/V(X) = \left\{ \int_0^{2\pi} \int_0^{\pi/2} \cos\theta \cdot g(\theta, \phi)\, d\theta\, d\phi \right\}^{-1} \cdot E\{I(Y \cap T_1)\}/E\{L(X \cap T_1)\} \tag{10.140}$$

or

$$S_V = \gamma_1^{-1} \cdot I_L \tag{10.141}$$

It should be noted that the orientation-dependent coefficient γ_1^{-1}, replacing the coefficient 2 for the isotropic case derived in eqs (3.80) and (3.95), is the same as that found above in eq. (10.112) with respect to a space curve cut by a plane. We have commented on this in Section 10.3.3.

The coefficients γ_1 and γ_2 depend on the orientation distributions describing the surface, measured by the distribution of the normals to the surface elements. We have dealt with this in some detail in Section 10.3.3 assuming certain orientation probability density functions $g(\theta, \phi)$. We had found that the most general Fisher distribution (eq. 10.60) was difficult to handle, except for special cases, whereas the Marriott distribution (eq. 10.76) gave some useful estimates of γ_1 and γ_2, as given in eqs (10.83) and (10.81) respectively. The Marriott distribution is an approximation of the Fisher distribution for small values of the concentration parameter K $(-1 \leq K \leq 1)$. If we know the direction of preferred orientation α, which is the angle of the preferred orientation to the test line or to the normal to the test plane, we can estimate S_V of anisotropic surfaces by

$$S_V = \frac{4}{\pi} \{[1-(K/3)]/[1-(K/4)(1+\cos^2\alpha)]\} \cdot B_A \tag{10.142}$$

$$S_V = 2\{[1-(K/3)]/[1-(K/2)\sin^2\alpha]\} \cdot I_L \tag{10.143}$$

Note that both these formulas convert to the isotropic case if $K=0$.

Cruz-Orive (personal communication) observed that one can estimate S_V without knowledge of the orientation distribution if B_A is measured on transverse $(\alpha = 0)$ and longitudinal $(\alpha = \pi/2)$ sections of the structures. In analogy to the development followed through above in eqs (10.116) to (10.121) one finds, with B_{A1} and B_{A2} measured on longitudinal and transverse sections, respectively,

$$K = 4(B_{A2} - B_{A1})/(2B_{A2} - B_{A1}) \tag{10.144}$$

so that

$$S_V = \frac{4}{3\pi}(B_{A1} + 2B_{A2}) \tag{10.145}$$

This should prove to be a very useful procedure in practice, but it depends essentially on whether the model, i.e. the assumption of a Marriott distribution, holds. In that respect we must note that this distribution allows only a moderate degree of anisotropy, for which the boundary conditions can be formulated as follows. For the case of *elongated* structures ($0<K\leq 1$) the ratio of measures should be

$$B_{A2}/B_{A1}\leq\tfrac{3}{2} \tag{10.146}$$

whereas for *flattened* structures ($-1\leq K<0$)

$$B_{A1}/B_{A2}\leq\tfrac{6}{5} \tag{10.147}$$

This can evidently be tested; all we need to have is a general notion on the nature of anisotropy.

Cruz-Orive has also demonstrated that one can, in fact, estimate S_V without any knowledge of the pattern of anisotropy—provided it is moderate and can be described by a Marriott distribution—by obtaining an estimate of B_A on three mutually perpendicular but otherwise arbitrary sections. It makes use of the fact that

$$\cos^2\alpha_1+\cos^2\alpha_2+\cos^2\alpha_3=1 \tag{10.148}$$

where α_1, α_2 and α_3 are the angles between the anisotropy axis and the normals to the three mutually perpendicular planes. It is then found that

$$S_V=\frac{4}{3\pi}(B_{A1}+B_{A2}+B_{A3}) \tag{10.149}$$

Similar results could be obtained for the estimation of S_V through I_L; this is not further pursued because it is probably not of great value. The reason is that with anisotropic surfaces one cannot assume test lines deposited on a section to be equivalent to the test line considered in this model. In estimating S_V from an intersection count—still a most practical procedure—one would have to go through a two-step procedure, combining what we have derived here for the relation between B_A and S_V, with that between I_L and B_A on a flat section, as described in Section 10.4.1. Clearly, cutting an anisotropic surface with a plane may yield an anisotropic image of the profiles and this requires appropriate precautions in estimating B_A.

Because of the limitations of these methods to "mild" degrees of anisotropy it would be useful to derive similar methods for the more general case as described by the Fisher distribution. These are very difficult to handle, except for the special cases of longitudinal and transverse sections where $\alpha=\pi/2$ and $\alpha=0$ respectively. The coefficients γ_2 and γ_1 are given in eqs (10.65), (10.66), (10.73) and (10.75) for these cases. Table 10.1 lists some of these coefficients; they may be of some interest when dealing with highly oriented surfaces where $K\gg 1$ (very elongated) or $K\ll -1$ (very flattened). Note particularly that for a structure equivalent to parallel straight tubules (Fig. 10.12b) where $K\to -\infty$, $\gamma_2=1$ for a cross-section ($\alpha=0$) and $\gamma_2=2/\pi$ for a longitudinal section ($\alpha = \pi/2$), for lamellar structures ($K \rightarrow \infty$), $\gamma_2 = 1$ for a section parallel to the anisotropy axis, i.e. perpendicular to the lamellae and $\gamma_2 = 0$ for a section parallel to the lamellae, evidently.

10.5. Estimating direction and degree of anisotropy

In the preceding section we have shown how the problems posed by anisotropy can be overcome when attempting to estimate average stereological parameters. It now remains to show how direction and degree of anisotropy can be estimated by stereological methods.

10.5.1. THE PLANAR CASE

Assume we are given a picture with some linear trace for which we need to determine direction and degree of anisotropy (Fig. 10.17a). The

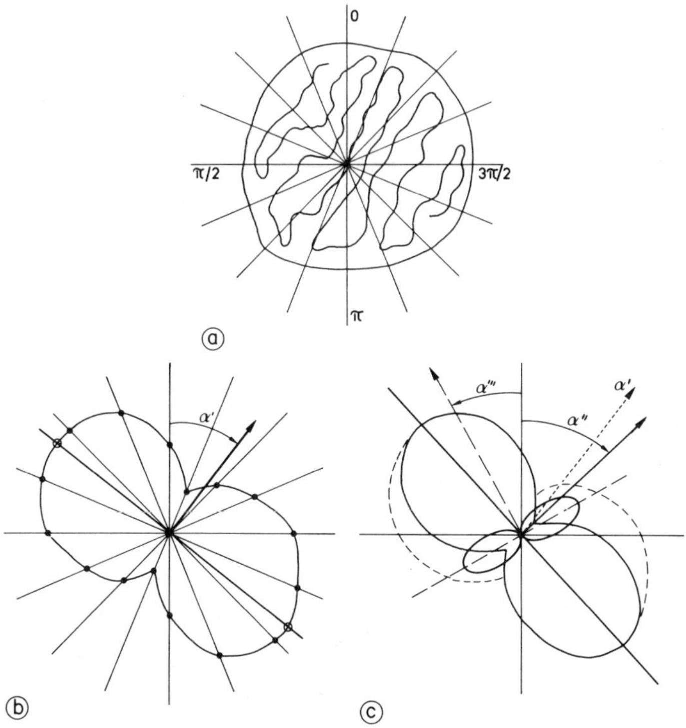

FIG. 10.17: Estimating degree and direction of anisotropy for plane curve by means of rose of compass.

procedure we may choose derives from the preceding discussions: we overlay a grid of parallel lines and count intersections; as we turn the grid stepwise through 180°, $I_L(\alpha)$ will vary depending on α, the angle between the normal to the test lines and the preferred orientation. The easiest way to proceed is to plot $I_L(\alpha)$ on polar coordinates, as shown in Fig. 10.17b. The result is a "rose of orientation" on which it is now fairly easy to determine the direction of anisotropy: by smoothing the curve we seek the maximal value of I_L; the anisotropy direction is now at right angles to the line connecting the two symmetrical maxima, indicated as α' in Fig. 10.17b.

Plotting the rose of orientation is actually a very informative procedure, as discussed in detail by Underwood (1968) and Mardia (1972). In the example of Fig. 10.17b it is revealed that the curve of Fig. 10.17a probably contains more than just one anisotropy axis. Referring back to Fig. 10.6 we note that the rose should show bilateral symmetry about the anisotropy axis if the line elements deviate randomly from the preferred orientation. This is not true of the rose in Fig. 10.17b. The interpretation of this finding could be this: the elements of the curve clearly favour the orientation α', but they favour also deviations from α' to one side, namely to negative values of ψ (Fig. 10.5); the probability density function of ψ is therefore not symmetrical but skewed. Alternatively, one could say that there are two anisotropy axes. In Fig. 10.17c we have graphically derived two symmetrical roses which, by addition, result in the empirical rose of Fig. 10.17b. The two anisotropy axes, α'' and α''' fall to either side of the overall main anisotropy axis α'. But they clearly have different weight since it is seen that α' is closer to α'' than to α'''; the relative weight of the two superimposed anisotropies is roughly represented by the relative areas of the two roses.

In order to estimate the *degree* of anisotropy we need to estimate the concentration parameter K as it was introduced in the various model distributions (eqs 10.7 and 10.9; Fig. 10.6). K is a measure of the concentration about the preferred orientation α (Fig. 10.5). How can K be estimated? Marriott (1969) proposes a method for estimating K based on a maximum likelihood technique. An alternative is as follows.

In eq (10.104) we have shown that a relationship exists between K and the number of intersections counted on test lines that are perpendicular to each other. If we make the orientation of one of these test lines coincide with the direction of the preferred orientation we will have $\alpha = 0$, and for the one at right angles to it $\alpha = \pi/2$. In this case $\cos 2\alpha = 1$ for both test lines and we have, from eq. (10.104),

$$K = 3[(I_{L1} - I_{L2})/(I_{L1} + I_{L2})] \quad (10.150\,a)$$

$$= 3[(I_{L1}/I_{L2}) - 1]/[(I_{L1}/I_{L2}) + 1] \quad (10.150b)$$

where $I_{L1} > I_{L2}$ are intersection densities measured on test lines perpendicular and parallel to the direction of preferred orientation, respectively. It is also easily seen that I_{L1} and I_{L2} are the largest and smallest intersection counts obtained as the test lines are rotated through 180° (Fig. 10.17b). Thus, in principle, it should be possible to estimate K simply from the ratio of largest to smallest values of I_L. Unfortunately eq. (10.150) holds strictly only for the model of the Marriott distribution which allows only small values for K, namely values in the range $\{-1 \leqslant K \leqslant 1\}$. Whether this distribution adequately describes the system can be tested by looking at the ratio I_{L1}/I_{L2} for which the following limits are derived from eq. (10.150b):

$$1 \leqslant I_{L1}/I_{L2} \leqslant 2 \tag{10.151}$$

The lower limit corresponds to $K=0$, i.e. isotropy where we would expect intersection counts in all directions to be equal; the upper limit corresponds to $K=1$. For sake of completeness we should mention that for negative values of K the limits would be

$$1 \geqslant I_{L2}/I_{L1} \geqslant \tfrac{1}{2} \tag{10.152}$$

but this is not a practically important distinction for the planar case.

If we wanted to estimate K for the structure in Fig. 10.17a we would have $I_{L1}/I_{L2} = 3.7$. This orientation distribution hence does not conform to the Marriott distribution and, strictly speaking, we cannot estimate a meaningful value of K by eq. (10.150). A rough estimate indicates that $K \sim 2$. This demonstrates the limitations of the analysis of anisotropies through the Marriott distribution: it is valid only for mild degrees of anisotropy, and for cases in which the distribution of orientation is symmetrical about α. A pattern which conforms to these limits is shown in Fig. 10.18: the rose is almost symmetrical, the ratio $I_{L1}/I_{L2} = 2$, so that

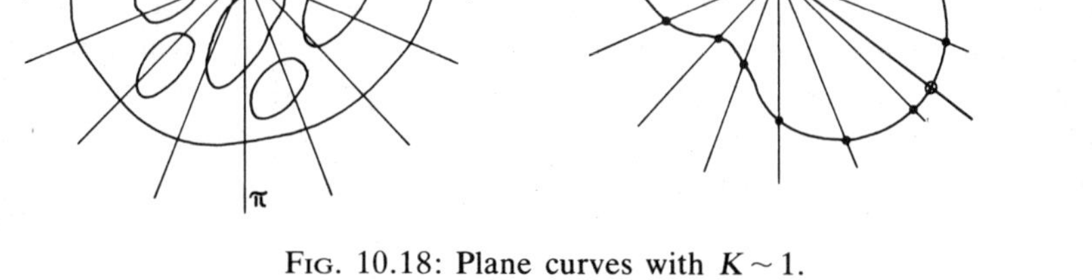

FIG. 10.18: Plane curves with $K \sim 1$.

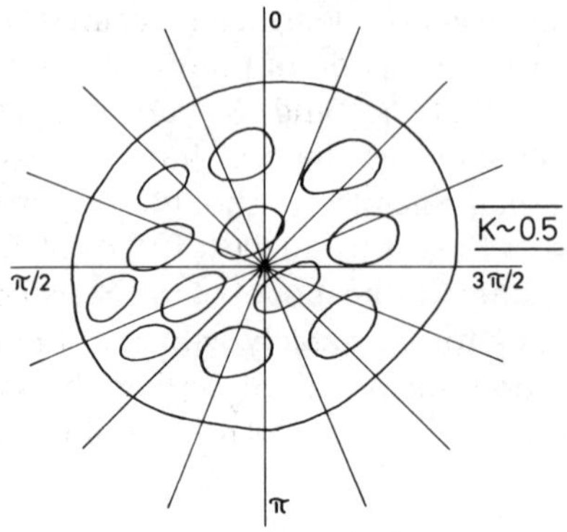

FIG. 10.19: Plane curves with $K \sim 0.5$.

it is described by a Marriott distribution with $K = 1$. A similar pattern for which $K = 0.5$ is shown in Fig. 10.19.

10.5.2. THE SPATIAL CASES

The procedures for estimating direction and degree of anisotropy are basically the same for spatial curves and surfaces. In analogy to the planar case we can seek to determine the direction of anisotropy by cutting sections in various directions and plotting some kind of a three-dimensional rose of orientation. For the case of spatial curves the number of transections per unit section area, $Q_A(\alpha)$, will depend on the orientation of the section with respect to the preferred orientation in the specimen; for spatial surfaces the density of boundary lengths, $B_A(\beta)$, will vary with orientation of the section (Fig. 10.20c).

In attempting to implement this procedure one will meet with a number of practical difficulties which are easy to imagine: (1) it is not easy to measure precisely the orientation of the section with respect to a three-dimensional polar coordinate system; (2) once a section is cut the specimen block is partly destroyed, rendering repeated sections at different angles difficult to obtain. It might thus be more practical to use a trial and error procedure which should be practicable at least with respect to surfaces. If one cuts an arbitrary plane section, T_{2a}, of the specimen (Fig. 10.20a) the anisotropy of the spatial surface will be reflected in the anisotropy of the profiles on the section (Fig. 10.20b) for which we can estimate the direction of anisotropy, α', by the procedure outlined above in Section 10.5.1. We now have a first indicator of the direction of spatial anisotropy, α, which must be contained in a plane normal to the arbitrary section plane, T_{2a}, and either parallel or normal to α': parallel to α' if the

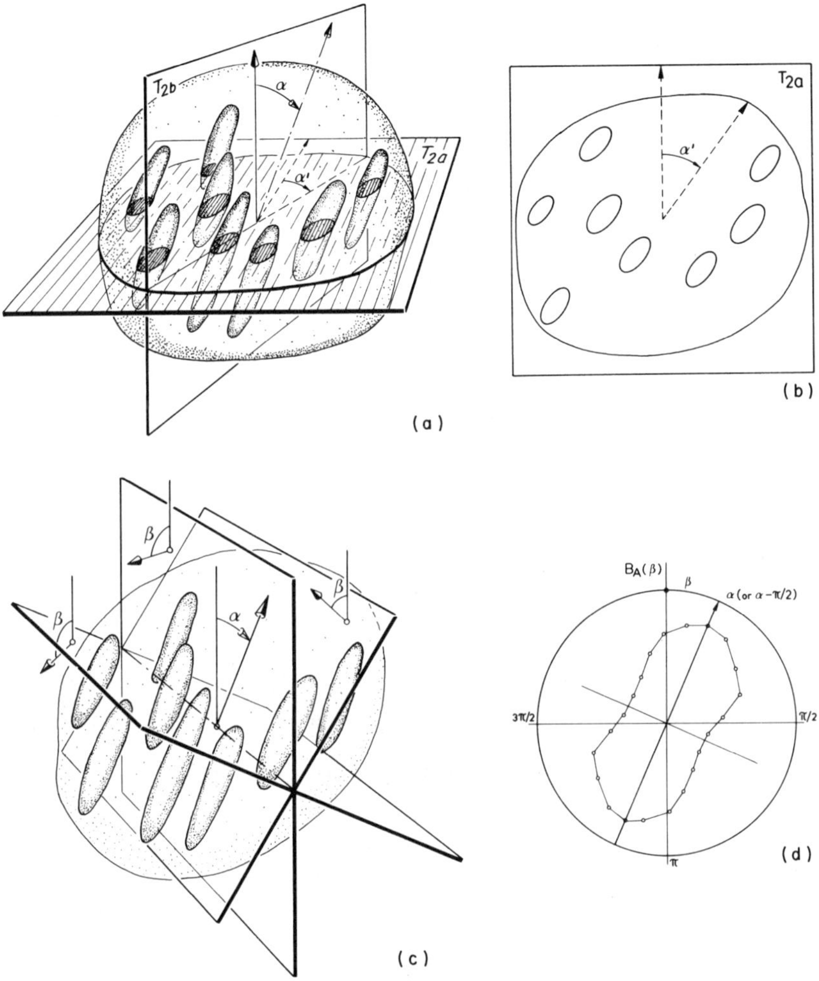

FIG. 10.20: Estimating direction and degree of anisotropy for spatial objects: (a) and (b) arbitrary plane section, T_{2a}, reveals direction of plane T_{2b} containing preferred orientation; (c) cutting sections at varying angles β allows α to be estimated from rose of compass (d).

structure is elongated ($K>0$, prolate ellipsoids or tubules); normal to α' if the structure is flattened ($K<0$, oblate ellipsoids or lamellae). It may be difficult to decide which kind of anisotropy prevails: sometimes one has independent information on the shape of the objects enclosed.

Assuming that we have decided on the type of anisotropy (in fact, on the sign of the concentration parameter K; see Section 10.3.3) we can now find α by cutting a series of sections of varying inclination β with respect to the plane of T_{2a} but all passing through a reference line which is either parallel (if $K<0$) or normal (if $K>0$) to the direction α' on the section (Fig. 10.20c). Estimating on each of these sections the trace length per unit area, $B_A(\beta)$, we can now plot a two-dimensional rose of orientation (Fig. 10.20d). We find α, as in Fig. 10.7, with respect to the line connecting the largest values of $B_A(\beta)$: in the case of elongated structures ($K>0$) α is the direction parallel to this line, whereas in the case of flattened structures ($K<0$) α is in the direction normal to the line connecting the maximal values of $B_A(\beta)$.

If the arbitrary section had been cut with respect to some identifiable orientation marks on the specimen we can now identify the axis of anisotropy from α' and α; to make the structure conform with the models used in the previous sections we must simply turn the specimen around the z-axis to let α' and α lie in the x–z plane.

If we are undecided whether the structure is elongated or flattened the best procedure appears to be to cut a section T_{2b} *parallel* to α' and normal to T_{2a}, and a second one, T_{2c}, *normal* to α' and to T_{2a}. Estimating the ratio of largest to smallest intersection densities $I_{L1}/I_{L2}=R$ (see eq. 10.150) we should have

$$\text{for} \quad K>0: \quad R(T_{2b})>R(T_{2c}) \tag{10.153}$$

$$\text{for} \quad K<0: \quad R(T_{2b})<R(T_{2c}) \tag{10.154}$$

The problem with this procedure is that the difference between these ratios will depend on the direction of T_{2a} with respect to α; it will be greater the larger α. The estimation of α' will also be more reliable the larger α. The conclusion therefore is that one should try to cut the first arbitrary section, T_{2a}, as close as possible to a plane that one guesses to contain α, if one has some outside indication of anisotropy.

When dealing with spatial curves this procedure is not as easy to implement. However, if the curves or filaments are contained in a translucent matrix, one can cut a thick slice and estimate the direction of anisotropy from the orientation of the projections of the small segments of filaments on to the observation plane. On T_{2a} α' is estimated as the direction normal to that of test lines yielding the largest intersection counts. The spatial anisotropy axis is then found by cutting a series of sections inclined at an angle β, as in Fig. 10.20c, passing through a reference line normal to α'; the rose of orientation is constructed by plotting the density of transections, $Q_A(\beta)$, which yields α as the direction normal to the maximal values of $Q_A(\beta)$, as in Fig. 10.20d. Here again, we have to turn the specimen around the z-axis until α' and α lie in the x–z plane to make the situation conform with that of models used in Section 10.3.2.

The data obtained by this procedure also allow an estimation of the degree of anisotropy through the concentration parameter K. In eq. (10.144) we had found K for anisotropic surfaces to be related to the ratio of the boundary density on transverse and longitudinal sections, B_{A1} and B_{A2}, respectively, such that, in slightly modified version,

$$K = 4[(B_{A2}/B_{A1}) - 1]/[2(B_{A2}/B_{A1}) - 1] \tag{10.155}$$

Note that if $B_{A2} > B_{A1}$ the structure is elongated and $K > 0$, whereas if $B_{A2} < B_{A1}$ we are dealing with a flattened structure and $K < 0$. B_{A1} and B_{A2} are the values of $B_a(\beta)$ obtained on the series of sections with varying inclination β which are, respectively, parallel and perpendicular to α; they can, in fact, be read off the rose of orientations, as constructed in Fig. 10.20d.

Similarly, for the case of spatial curves of filaments we have, according to eq. (10.120)

$$K = 2[1 - (Q_{A2}/Q_{A1})] \tag{10.156}$$

where Q_{A1} and Q_{A2} are the transection densities estimated on transverse and longitudinal sections of the structure; they are likewise obtained from the rose of orientation plotted for the series of sections with varying inclination β.

The crux with eqs (10.155) and (10.156) is that they apply only for cases in which the orientation distribution can be described by the Marriott distribution for the spatial case (eqs 10.39 and 10.76) for which $\{-1 < K < 1\}$. As in the planar case this means that only mild degrees of anisotropy can be evaluated properly.

APPENDIX 1

Alphabetic list of symbols used in this text and their definition

Note: the same letter may occasionally have to be used for different meanings, but context, actual definition, and subscripts should avoid ambiguities. See also Table 1.1.

A:	area of profiles
A_A:	areal density
A_T:	test area
a:	area of individual profile
B:	boundary length of profiles
B_A:	boundary length density on area
b:	boundary length (perimeter) of individual profile
C:	curvature of plane figure
C_A:	curvature density of plane figures on area
D:	diameter (of spherical particles)
d:	diameter of profile
$d(T)$:	spacing of test planes or test lines
$E\{x\}$:	expected or average value of x
$\exp\{a\}$:	$= e^a$
$F(x)$, $f(x)$:	function of x, specifically frequency distribution
G:	total or Gaussian curvature
H:	caliper diameter (tangent diameter) in space
$\bar{H}$:	mean caliper diameter (averaged for orientation)
h:	caliper diameter of plane figure
h_0:	height of lost cap section
I:	intersections between test line and surface or boundary
I_L:	intersection density on test line
J:	length of linear feature in space
J_V:	length density of linear features in space
i:	size class for profiles
j:	size class for particles

$\underline{K}$: mean curvature of surface (integrated)
$\bar{K}$: average (or mean) mean curvature
K_V: density of integral mean curvature per unit volume
K_S: density of integral mean curvature per unit surface
K_t: correction coefficient for section thickness
k: constants or coefficients
L: length of line or intecept (straight line)
L_L: length density of intercepts on test line
L_T: length of test line
l: length of intercept or chord
M_k: kth moment
N: number of discrete objects or profiles
n: number of profiles
N_A: numerical density of profiles on test area
N_L: numerical density of intercepts on test line
N_N: relative number (numerical frequency)
N_V: numerical density of particles in volume
P: point number (test points)
P_T: test point set
P_P: test point density
$\Pr\{x\}$: probability of event x
p: element of probability measure
Q: transection of spatial lineal features with plane
Q_A: transection density on unit area
R: radius of sphere
r: radius of circle
S: surface area of spatial object
S_S: relative surface
S_V: surface density in unit volume
T: tangent count (sweeping line on area)
T: section or probe
T_3: slice or slab section
T_2: plane section
T_1: line probe or "section"
T_0: point set as probe or "section"
T_+ (T_-): positive (negative) tangents
T_A: tangents per unit area
t: section or slice thickness
V: volume of components or structure
V_T: test volume
V_V: volume density
v: volume of individual object
X: structure set
Y: object set or components of structure
Y_3: 3D object set (solid)
Y_2: 2D object set (surface or interface)
Y_1: 1D object set (filaments or needles)
Y_0: 0D object set (points, corners)
$|Y|_1, |X|_1$: measure of Y or X (one-dimensional)
Z: profile ($Y \cap T$)

Z_3:	3D profile $(Y_3 \cap T_3)$
Z_2:	2D profile $(Y_3 \cap T_2)$
Z_1:	1D profile $(Y_3 \cap T_1)$ or $(Y_2 \cap T_2)$
Z_0:	0D profile $(Y_3 \cap T_0)$ or $(Y_2 \cap T_1)$ or $(Y_1 \cap T_2)$.
α, β, γ:	constants or coefficients, angles
δ, λ, ρ:	shape factors
Δ:	diameter class interval
θ:	orientation angle between vector and z-axis (co-altitude)
κ:	local curvature
λ_a:	mean free distance of components a
ϕ:	orientation angle between x-axis and projection of vector into x–y plane (latitude)
τ:	sweeping line for tangent count
τ:	barrier (sheet or foil) thickness
τ_h:	harmonic mean sheet thickness

Symbols for set operations

$a \cup b$:	union of sets a and b
$a \cap b$:	intersection of sets a and b
$a \in b$:	a is element of set b, or a belongs to b
$a \bar{\cap} b$:	set a is tangent to set b
$a \uparrow b$:	set a hits set b $[(a \cap b) \neq \varnothing]$
$\varnothing$	empty set $[(a \cap b) = \varnothing$ means "no intersection"$]$

APPENDIX 2

Glossary of terms

anisotropic: exhibiting different properties in different directions of space.

arithmetic mean: sum of the sample values divided by the number of samples (= average).

average: sum of the sample values divided by the number of samples (= arithmetic mean).

biased estimate: the estimate deviates from the true value by a certain amount and in a certain direction.

boundary: outline of a profile (= perimeter).

chord: straight line segment between two surface points (see intercept).

complement: (set theory) a subset required to fill a set, to make it complete, e.g. if a structure X contains components Y embedded in a matrix this matrix is said to be the complement to Y.

component: distinguishable part of a structure.

consistent estimate: a method of estimation is called consistent, if the estimate becomes equal to the population value when the sample consists of the entire population.

containing space: structure space to which the parameters describing the internal components are related.

convex solid: a solid for which any straight line connecting any of two points of the solid lies totally within the solid.

curvature: deviation of curved surface (line) from plane (straight line) measured by reciprocal of radius of the largest tangent circle.

density: the quantity per unit volume, unit area or unit length.

element of probability measure: element of a set of N possible events for which the probability of occurrence is $1/N$, and equal for all elements of the set.

frequency distribution: distribution of frequencies of occurrence of an event over the range of all possible events, e.g. the frequency of occurrence of a certain size class of particles relative to all size classes.

harmonic mean: sum of the reciprocals of the sample values divided by sample number.

intercept: segment or chord of test line contained within object and extending between two points on the object's surface.

intersection: intersection between test line and object surface or profile boundary; set theory, $(a \cap b)$.

isotropic: having the same properties in all directions of space.

isotropic uniform random (*IUR*): all orientations and all positions in space are equally likely.

morphometry: quantitative morphology; the measurement of structures by any method, including stereology.

normal: line perpendicular to a surface (or line), indicates orientation of surface.

object: discrete element of structure.

orientation: direction in space (e.g. of normal to surface) measured by polar coordinates (angles θ, ϕ to coordinate system).

orthogonal complement: complement to test set of dimension s with respect to an n-dimensional space has dimension $s^c = n - s$, e.g. in three-dimensional space the orthogonal complement to a test line is a plane normal to the test line (Fig. 2.14).

particle: discrete element of structure.

perimeter: outline of a profile (= boundary).

phase: aggregate of all parts of structure which are identical in nature.

probability: relative frequency of the occurrence of an event, measured by the ratio between its occurrence and the total number of cases necessary to ensure its occurrence when such cases are viewed as indefinitely extended.

probability density function: a function of a discrete (or continuous) variate that gives the probability that a specified value will occur.

probe: any geometric entity (plane, line, point, slice etc.) used to penetrate into interior of structure.

profile: sharply outlined plane figure resulting from sectioning (or projecting) an object with a plane (on to a plane).

random sample: sample obtained by random process (by means of random numbers), i.e. without regard to the properties of structure.

random section: section of structure obtained by random sampling, whereby all regions of the structure and all orientations must have equal chance of being cut.

reference system: quantity to which feature or component characteristics are related (containing space, test system etc.).

section: plane intersecting a structure; sometimes thin slice of thickness t; also used for any "probe" such as line or point sets that intersect structure.

set: a collection of things belonging together which are called the elements of the set (a set of points, a set of lines etc.).

slice: "section" of finite thickness t.

stereology: set of mathematical methods relating three-dimensional parameters of structure to measurements obtainable on sections.

structure: something made of interdependent parts in a definite pattern of organization.

test system: set of geometric elements (points, lines, areas) used to obtain stereological estimates.

uniform random sample: all points in space are equally likely to be included in the sample.

union: set of all elements of set a together with those of set b are called the union of a and b ($a \cup b$).

Bibliography

The bibliography is given in two parts. The first part lists a number of references to general works on stereology and related sciences, such as books and proceedings of conferences and symposia. The second part compiles individual references as quoted in the text. A list of journals where papers on stereology appear regularly is given at the end.

I. General references to stereology and related sciences

Cochran, W. C. (1953). "Sampling Techniques". Wiley and Sons, New York.

De Hoff, R. T. and Rhines, F. N. (1968). "Quantitative Microscopy". McGraw-Hill, New York.

Elias, H. (1967). "Stereology". Proceedings of the Second International Congress for Stereology. Springer Verlag, New York.

Hadwiger, H. (1955). "Altes und Neues über konvexe Körper". Birkhäuser, Basel.

Hadwiger, H. (1957). "Vorlesungen über Inhalt, Oberfläche und Isoperimetrie". Springer Verlag, Berlin–Göttingen–Heidelberg.

Haug, H. (1963). "Proceedings First International Congress for Stereology Vienna". (1963). Congressprint, Wien.

Kendall, M. G. and Moran, P. A. P. (1963). "Geometrical Probability". Charles Griffin, London.

Mardia, K. V. (1972). "Statistics of Directional Data". Academic Press, London and New York.

Matheron, G. (1967). "Eléments pour une Théorie des Milieux Poreux". Masson, Paris.

Matheron, G. (1975). "Random Sets and Integral Geometry". Wiley and Sons New York.

Miles, R. E. and Serra, J. (1978). "Geometrical Probability and Biological Structures: Buffon's 200th Anniversary". Vol. 23: Lecture Notes in Biomathematics. Springer Verlag, Berlin–Heidelberg–New York.

Nicholson, W. L. (1972). "Proceedings of the Symposium on Statistical and Probabilistic Problems in Metallurgy". Special Supplement to Advances in Applied Probability. Applied Probability Trust, Seattle, Washington.

Saltykov, S. A. (1958). "Stereometric Metallography" [in Russian]. 2nd edition 3rd edition 1970). State Publishing House for Metals and Sciences, Moscow. (German translation of 3rd edition: VEB Deutscher Verlag für Grundstoffindustrie, Leipzig, 1974).

Santaló, L. A. (1976). "Integral Geometry and Geometric Probability". Vol, 1 of Encyclopedia of Mathematics and its Applications. Addison-Wesley, Reading, Mass.

Underwood, E. E. (1970). "Quantitative Stereology". Addison-Wesley, Reading, Mass.

Underwood, E. E., de Wit, R. and Moore, G. A. (1976). "Proceedings of the Fourth International Congress for Stereology". National Bureau of Standards, Special Publication Nr. 431, U. S. Government Printing Office, Washington.

Weibel, E. R. (1963). "Morphometry of the Human Lung". Springer Verlag, Berlin, and Academic Press, New York and London.

Weibel, E. R. and Elias, H. (1967). "Quantitative Methods in Morphology". Springer Verlag, Berlin–Heidelberg–New York.

Weibel, E. R., Meek, G., Ralph, B., Echlin, P. and Ross, R. (1972). "Stereology 3. Proceedings of the Third International Congress for Stereology". Blackwell Scientific Publications, Oxford–London–Edinburgh–Melbourne. (Reprinted from J. Microscopy, Vol. 95, Parts 1 and 2).

II. Individual references

Abercrombie, M. (1946). Estimation of nuclear population from microtomic sections. *Anat. Rec.* **94,** 239.

Aherne, W. (1967). Methods of counting discrete tissue components in microscopical sections. *J. Royal Micr. Soc.* **87,** 493.

Anderssen, R. S. (1976). Stable procedures for the inversion of Abel's equation. *J. Inst. Maths. Applics.* **17,** 329.

Anderssen, R. S. and Jakeman, A. J. (1975). Abel type integral equations in stereology. II. Computational methods of solution and the random spheres approximation. *J. Microsc.* **105,** 135.

Bach, G. (1959). Ueber die Grössenverteilung von Kugelschnitten in durchsichtigen Schnitten endlicher Dicke. *Z. wiss. Mikrosk.* **64,** 265.

Bach, G. (1967). Kugelgrössenverteilung und Verteilung der Schnittkreise; ihre wechselseitigen Beziehungen und Verfahren zur Bestimmung der einen aus der andern. *In* "Quantitative Methods in Morphology" (E. R. Weibel and H. Elias, eds), p. 23. Springer Verlag, Berlin–Heidelberg–New York.

Baldwin, J. P., Tinker, P. B. and Marriott, F. H. C. (1971). The measurement of length and distribution of onion roots in the field and in the laboratory. *J. Appl. Ecology* **8,** 543.

Bertrand, J. (1899). "Calcul des Probabilités". Gauthier-Villars, Paris.

Binet, J. L. and Roger, J. (1977). "Un autre Buffon". Collection Savoir. Hermann, Paris.

Bolender, R. P., Paumgartner, D., Losa, G., Muellener, D. and Weibel, E. R. (1978). Integrated stereological and biochemical studies on hepatocytic membranes. I. Membrane recoveries in subcellular fractions. *J. Cell Biol.* **77,** 565.

Buffon, G. L. L. (1777). Essai d'arithmétique morale. *Suppl. à l'Histoire Naturelle (Paris)* **4**.

Cahn, J. W. (1967). The significance of average mean curvature and its determination by quantitative metallography. *Trans AIME* **239,** 610.

Cahn, J. W. and Fullman, R. L. (1956). On the use of lineal analysis for obtaining particle size distribution functions in opaque samples. *Trans AIME* **206,** 610.

Cahn, J. W. and Nutting, J. (1959). Transmission quantitative metallography. *Trans. Am. Inst. Min. metall. Engrs.* **215,** 526.

Casley-Smith, J. R. and Davy, P. (1978). The estimation of distances between parallel membranes on thick sections. *J. Microsc.* **114,** 249.

Cauchy, A. (1908). Mémoire sur la rectification des courbes et de la quadrature des surfaces courbes. *Mém. Acad. Sci. Paris* **22,** 3, 1850; Oeuvres complètes, Vol. 1 (1908).

Chalkley, H. W. (1943). Methods for quantitative morphological analysis of tissue. *J. Nat. Cancer Inst.* **4,** 47.

Chalkley, H. W., Cornfield, J. and Park, H. (1949). A method for estimating volume–surface ratios. *Science* **110,** 295.

Chayes, F. (1956). "Petrographic Modal Analysis". John Wiley, New York.

Coleman, R. (1969). Random paths through convex bodies. *J. Appl. Prob.* **6,** 430.

Coleman, R. (1973). Random paths through rectangles and cubes. *Metallography* **6,** 103.

Coleman, R. (1978). The stereological analysis of two-phase particles. *In* "Geometrical Probability and Biological Structures: Buffon's 200th Anniversary". Vol. 23 Lecture Notes in Biomathematics (R. E. Miles and J. Serra, eds.), p. 37. Springer Verlag, Berlin–Heidelberg–New York.

Corrsin, S. (1954). A measure of the area of a homogeneous random surface in space. *Quart. Appl. Math.* **12,** 404.

Courant, R. and Hilbert, D. (1953). "Methods of Mathematical Physics", Vol. 1. Wiley (Interscience), New York.

Courant, R. and Robbins, H. (1941). "What is Mathematics?". Oxford University Press, New York.

Courant, R. and Robbins, H. (1962). "Was ist Mathematik?" Springer Verlag, Berlin–Göttingen–Heidelberg.

Crofton, M. W. (1885). Probability. *In* "Encyclopedia Brittanica". 9th edition, Vol. 19, p. 768.

Cruz-Orive, L. M. (1976). Particle size-shape distributions: the general spheroid problem. I: Mathematical model. *J. Microsc.* **107,** 235.

Cruz-Orive, L. M. (1976b). Quantifying "pattern": a stereological approach. *J. Microscopy* **107,** 1.

Cruz-Orive, L. M. (1977). Size distributions. Lecture Notes, Stereology Course. Bern, March 1977. Department of Anatomy, Univ. of Bern.

Cruz-Orive, L. M. (1978a). Particle size-shape distributions: the general spheroid problem. II. Stochastic model and practical guide. *J. Microsc.* **112,** 153.

Cruz-Orive, L. M. (1978b). Stereological analysis of spheroidal particles of variable size and shape. *Practical Metallography*, special issue **8,** 350.

Cruz-Orive, L. M. (1979). Estimation of sheet thickness distributions from linear and plane sections. *Biom. J.* **21,** 717.

De Hoff, R. T. (1967). The quantitative estimation of mean surface curvature. *Trans AIME* **239,** 617.

De Hoff, R. T. (1968). Curvature and topological properties of interconnected phases. *In* "Quantitative Microscopy". (R. T. De Hoff and F. N. Rhines, eds) p. 291. McGraw-Hill, New York.

De Hoff, R. T. and Bousquet, P. (1970). Estimation of the size distribution of triaxial ellipsoidal particles from the distribution of linear intercepts. *J. Microsc.* **92,** 119.

De Hoff, R. T. and Rhines, F. N. (1961). Determination of the number of particles per unit volume from measurements made on random plane sections: the general cylinder and the ellipsoid. *Trans AIME* **221,** 975.

De Hoff, R. T. and Rhines, F. N. (1968). "Quantitative Microscopy". McGraw-Hill, New York.

De Hoff, R. T., Aigeltinger, E. H. and Craig, K. R. (1972). Experimental determination of the topological properties of three-dimensional microstructures. *J. Microsc.* **95,** 69.

Delesse, M. A. (1847). Procédé mécanique pour déterminer la composition des roches. *C. R. Acad. Sci. (Paris)* **25,** 544.

Duffin, R. J., Meussner, R. A. and Rhines, F. N. (1953). Statistics of particle measurement and of particle growth. Carnegie Inst. Technol., Pittsburgh, Technical Report No. 32, C. I. T. AF8A-TR32.

Ebbeson, S. O. E. and Tang, D. (1965). A method for estimating the number of cells in histological sections. *J. Royal Micr. Soc.* **84,** 449.

Elias, H. and Pauly, J. E. (1966). "Human Microanatomy". 3rd edition, Davis, Philadelphia, Pennsylvania.

Fischmeister, H. F. (1968). Scanning methods in quantitative metallography. *In* "Quantitative Microscopy" (R. T. De Hoff and F. N. Rhines, eds), p. 336. McGraw-Hill, New York.

Floderus, S. (1944). Untersuchungen über den Bau der menschlichen Hypophyse mit besonderer Berücksichtigung der quantitativen mikromorphologischen Verhältnisse. *Acta path. microbiol. scand. Suppl.* **53,** 1.

Freedman, L. S. (1974). A note on Aherne's method of counting tissue components in relatively thick sections. *J. Microsc.* **100,** 219.

Fullman, R. L. (1953). Measurement of particle sizes in opaque bodies. *Trans. Am. Inst. Min. Metall. Engrs.* **197,** 447.

Gibbs, J. W. (1873). "A method of geometrical representation of the thermodynamic properties of substances by means of surfaces". Trans. Connecticut Academy II, 382 (reproduced in J. W. Gibbs, Scientific Papers, Vol. 1, p. 33. Dover–New York, 1961).

Giger, H. (1967). A system of basic stereologic formulae. *In* "Stereology". Proceedings of the Second International Congress for Stereology (H. Elias, ed.), p. 257. Springer Verlag, New York.

Giger, H. (1970). Grundgleichungen der Stereologie. *Metrika* **16,** 43.

Glagoleff, A. A. (1933). On the geometrical methods of quantitative mineralogic analysis of rocks. *Trans. Inst. Econ. Min.*, Moscow, **59,** 1.

Goldsmith, P. L. (1967). The calculation of true particle size distributions from the sizes observed in a thin slice. *Brit. J. Appl. Phys.* **18,** 813.

Green, J. A. (1965). "Sets and Groups". Routledge and Kegan Paul, London–Boston–Henley.

Gundersen, H. J. and Østerby, R. (1973). Statistical analysis of transformations leading to normal distributions of the peripheral glomerular basement membrane. *J. Microsc.* **97,** 293.

Gundersen, H. J., Jensen, T. B. and Østerby, R. (1978). Distribution of membrane thickness determined by lineal analysis. *J. Microsc.* **113,** 27.

Hadwiger, H. (1955). "Altes und Neues über konvexe Körper". Birkhäuser, Basel.

Hadwiger, H. (1957). "Vorlesungen über Inhalt, Oberfläche und Isoperimetrie". Springer Verlag, Berlin.

Haug, H. (1967). Probleme und Methoden der Strukturzählung im Schnittpräparat. *In* "Quantitative Methods in Morphology" (E. R. Weibel and H. Elias, eds), p. 58. Springer Verlag, Berlin–Heidelberg–New York.

Hennig, A. (1956). Bestimmung der Oberfläche beliebig geformter Körper mit besonderer Anwendung auf Körperhaufen im mikroskopischen Bereich. *Mikroskopie* **11,** 1.

Hennig, A. (1957). Zur Geometrie von Schnitten. *Z. wiss. Mikr.* **63,** 362.

Hennig, A. (1963a). Länge eines räumlichen Linienzuges. *Z. wiss. Mikr.* **65,** 193.

Hennig, A. (1963b). Length of a three-dimensional linear tract. *In* "Proceedings First International Congress for Stereology Vienna" (H. Haug, ed.), p. 44. Congressprint, Wien.

Hennig, A. (1969). Fehler in der Volumenermittlung aus der Flächenrelation in dicken Schnitten (Holmes-Effekt). *Mikroskopie* **25,** 154.

Hilliard, J. E. (1967). The calculation of the mean caliper diameter of a body for use in the analysis of the number of particles per unit volume. In "Stereology". Proceedings of the Second International Congress for Stereology (H. Elias, ed.), p. 211. Springer Verlag, New York.

Hilliard, J. E. (1968). Direct determination of the moments of the size distribution of particles in an opaque sample. *Trans AIME* **242,** 1373.

Holmes, A. (1927). "Petrographic Methods and Calculations". Murby, London.

Horikawa, E. (1954). On a new method of representation of the mixture of several austhenite grain sizes. *Tetsu to Hagane* **40,** 991.

Hull, F. C. and Houk, W. J. (1953). Statistical grain structure studies: plane distribution curves of regular polyhedrons. *Trans AIME* **197,** 565.

Jakeman, A. J. and Anderssen, R. S. (1975). Abel type integral equations in stereology. I. General discussion. *J. Microsc.* **105,** 121.

Keiding, N., Jensen, S. T. and Ranek, L. (1972). Maximum likelihood estimation of the size distribution of liver cell nuclei from the observed distribution in a plane section. *Biometrics* **28,** 813.

Kendall, M. G. and Moran, P. A. P. (1963). "Geometrical probability". Griffin, London.

Krumbein, W. C. (1935). Thin-section mechanical analysis of indurated sediments. *J. Geol.* **43,** 482.

Lagerkrantz, C. (1948). Photo-electric counting of individual microscopic plant and animal cells. *Nature* **161,** 25.

Lord, G. W. and Willis, T. F. (1951). Calculation of air bubble size distribution from results of a Rosiwal traverse of aerated concrete. *A.S.T.M. Bull.* **56,** 177.

Mardia, K. V. (1972). "Statistics of Directional Data". Academic Press, London.

Marriott, F. H. C. (1969). Associated directions. *Biometrics* **25,** 775.

Marriott, F. H. C. (1971). Buffon's problem for non-random directions. *Biometrics* **27,** 233.

Matheron, G. (1967). "Eléments pour une Théorie des Milieux Poreux". Masson, Paris.

Matheron, G. (1972). Random sets theory and its applications to stereology. *J. Microsc.* **95,** 15.

Matheron, G. (1975). "Random Sets and Integral Geometry". John Wiley, New York.

Mayhew, T. M. (1972). A comparison of several methods for stereological determination of the numbers of organelles per unit volume of cytoplasm. *J. Microsc.* **96,** 37.

Merz, W. A. (1968). Streckenmessung an gerichteten Strukturen im Mikroskop und ihre Anwendung zur Bestimmung von Oberflächen-Volumen-Relationen im Knochengewebe. *Mikroskopie* **22,** 132.

Miles, R. E. (1969). Poisson flats in Euclidean spaces. *Adv. Appl. Probab.* **1**, 211.

Miles, R. E. (1972). Multidimensional perspectives on stereology. *J. Microsc.* **95,** 181.

Miles, R. E. (1976). Estimating aggregate and overall characteristics from thick sections by transmission microscopy. *J. Microsc.* **107,** 227.

Miles, R. E. (1978). The importance of proper model specification in stereology. *In* "Geometrical Probability and Biological Structures: Buffon's 200th Anniversary" (R. E. Miles and J. Serra, eds). Vol. 23 Lecture Notes in Biomathematics, p. 115. Springer Verlag, Berlin–Heidelberg–New York.

Miles, R. E. and Davy, P. (1976a). Probabilistic foundations of stereology. *In* "Proceedings of the Fourth International Congress for Stereology" (E. E. Underwood, R. de Wit and G. A. Moore, eds), p. 443. National Bureau of Standards, Special Publication Nr. 431, U.S. Government Printing Office, Washington.

Miles, R. E. and Davy, P. (1976b). Precise and general conditions for the validity of a comprehensive set of stereological formulae. *J. Microscopy* **107,** 211.

Miles, R. E. and Davy, P. (1978). Particle number or density can be stereologically estimated by wedge sections. *J. Microsc.* **113,** 45.

Miles, R. E. and Serra, J. (1978). "Geometrical Probability and Biological Structures: Buffon's 200th Anniversary". Vol. 23, Lecture Notes in Biomathematics. Springer Verlag, Berlin–Heidelberg–New York.

Minkowski, H. (1903). Volumen und Oberfläche. *Math. Ann.* **57,** 447.

Moore, G. A. (1972). Recent progress in automatic image analysis. *J. Microsc.* **95,** 105.

Moran, P. A. P. (1966). Measuring the length of a curve. *Biometrika* **53,** 359.

Moran, P. A. P. (1972). The probabilistic basis of stereology. *Adv. in Applied Probability*, Spec. Suppl. (Proc. Symp. Probabilistic Problems in Metallurgy, Seattle, 1971), p. 69.

Myers, E. J. (1967). Size distribution of cubic particles. *In* "Stereology", Proceedings of the Second International Congress for Stereology (H. Elias, ed.), p. 187. Springer Verlag, New York.

Nicholson, W. L. (1970). Estimation of linear properties of particle size distributions. *Biometrika* **57,** 273.

Nicholson, W. L. (1976). Estimation of linear functionals by maximum likelihood. *J. Microsc.* **107,** 323.

Plummer, H. C. (1940). "Probability and Frequency", p. 54. MacMillan, London.

Preston, K. D. (1973). Automated microscopy for cytological analysis. *In* "Advances in Optical and Electron Microscopy", Vol. 5, p. 43. Academic Press, London and New York.

Reid, W. P. (1955). Distribution of size of spheres from a study of slices of the solid. *J. Math. Phys.* **34,** 95.

Rosiwal, A. (1898). Ueber geometrische Gesteinsanalysen. *Verh. K. K. Geol. Reichsanst. Wien*, 143.

Salmon, G. (1954). "Treatise on Conics". Chelsea, New York.

Saltykov, S. A. (1945). "Stereometric Metallography". 1st edition. State Publishing House for Metals Sciences, Moscow.

Saltykov, S. A. (1958). "Stereometric Metallography". 2nd edition (3rd edition

1970). State Publishing House for Metals and Sciences. Moscow. (German translation of 3rd edition: VEB Deutscher Verlag für Grundstoffindustrie, Leipzig, 1974.)

Saltykov, S. A. (1967). The determination of the size distribution of particles in an opaque material from a measurement of the size distribution of their sections. *In* "Stereology". Proceedings of the Second International Congress for Stereology (H. Elias, ed.), p. 163. Springer Verlag, New York.

Santaló, L. A. (1936). Integralgeometrie 5: Ueber das kinematische Mass im Raum. *Actualités Sci. Industr.* **357,** 1.

Santaló, L. A. (1943). Integral geometry on surfaces of constant negative curvature. *Duke Math. J.* **10,** 687.

Santaló, L. A. (1953). "Introduction to Integral Geometry". Hermann, Paris.

Santaló, L. A. (1955). Sobre la distribucion de los tamanos de los corpusculos contenidos en un cuerpo a partir de la distribucion en sus secciones. *Trabajos Estadist.* **6,** 181.

Santaló, L. A. (1976). "Integral Geometry and Geometric Probability". Vol. 1 of Encyclopedia of Mathematics and its Applications. Addison Wesley, Reading, Mass.

Scheil, E. (1931). Die Berechnung der Anzahl der Grössenverteilung kugelförmiger Kristalle in undurchsichtigen Körpern mit Hilfe der durch einen ebenen Schnitt erhaltenen Schnittkreise. *Z. Anorg. allg. Chem.* **27,** 201.

Schwartz, H. A. (1934). The metallographic determination of the size distribution of temper carbon nodules. *Metals and Alloys* **5,** 139.

Serra, J. (1969). Introduction à la morphologie mathématique. Cahiers du Centre de Morphol. mathémat. Fasc. 3, Ecole des Mines de Paris.

Serra, J. (1972). Stereology and structuring elements. *J. Microsc.* **95,** 93.

Serra, J. (1978). One, two, three, ... infinity. *In* "Geometrical Probability and Biological Structures: Buffon's 200th Anniversary" (R. E. Miles and J. Serra, eds.), Vol. 23 Lecture Notes in Biomathematics. p. 137. Springer Verlag, Berlin–Heidelberg–New York.

Shand, S. J. (1916). A recording micrometer for geometrical rock analysis. *J. Geol.*, **24,** 394.

Sitte, H. (1967). Morphometrische Untersuchungen an Zellen. *In* "Quantitative Methods in Morphology" (E. R. Weibel and H. Elias, eds.), p. 167. Springer Verlag, Berlin–Heidelberg–New York.

Smith, C. S. and Guttman, L. (1953). Measurement of internal boundaries in three-dimensional structures by random sectioning. *Trans AIME* **197,** 81.

Sneddon, I. N. (1966). "Mixed Boundary Value Problems in Potential Theory". North-Holland, Amsterdam.

Sommerville, D. M. Y. (1958). "An Introduction to the Geometry of N Dimensions". Dover, New York.

Spektor, A. G. (1950). Analysis of distribution of spherical particles in non-transparent structures. *Zav. Lab.* **16,** 173.

Streit, F. (1978). Anisotropy and distance from the centre. Probabilistic models and statistical analysis. *In* "Geometrical Probability and Biological Structures: Buffon's 200th Anniversary" (R. E. Miles and J. Serra, eds), Vol. 23 Lecture Notes in Biomathematics. p. 203. Springer Verlag, Berlin–Heidelberg–New York.

Struik, D. J. (1961). "Lectures on Classical Differential Geometry". 2nd ed. Addison-Wesley, Reading, Mass.

Tallis, G. M. (1970). Estimating the distribution of spherical and elliptical bodies in conglomerates from plane sections. *Biometrics* **26,** 87.

Thomson, E. (1930). Quantitative microscopic analysis. *J. Geol.* **38,** 193.

Tomkeieff, S. L. (1945). Linear intercepts, areas and volumes. *Nature* **155,** 24.

Underwood, E. E. (1968). Particle size distribution. *In* "Quantitative Microscopy" (R. T. De Hoff and F. N. Rhines, eds), p. 149. McGraw-Hill, New York.

Underwood, E. E. (1970). "Quantitative Stereology". Addison-Wesley, Reading, Mass.

Underwood, E. E. (1972). The stereology of projected images. *J. Microsc.* **95,** 25.

Warren, R. (1976). The geometry and stereology of "cubo-spherical" particles. *In* "Proceedings of the Fourth International Congress for Stereology" (E. E. Underwood, R. de Wit and G. A. Moore, eds), p. 25. National Bureau of Standards, Special Publication Nr. 431, U.S. Government Printing Office, Washington.

Warren, R. and Naumovich, N. (1977). Relative frequencies of random intercepts through convex bodies. *J. Microsc.* **110,** 113.

Watson, G. S. (1971). Estimating functionals of particle size distributions. *Biometrika* **58,** 483.

Weibel, E. R. (1963). "Morphometry of the Human Lung". Springer Verlag, Berlin–Göttingen–Heidelberg, and Academic Press, New York and London.

Weibel, E. R. (1967). Structure in space and its appearance on sections. *In* "Stereology". Proceedings of the Second International Congress for Stereology (H. Elias, ed.), p. 15. Springer Verlag, New York.

Weibel, E. R. (1973). Morphological basis of alveolar-capillary gas exchange. *Physiol. Rev.* **53,** 419.

Weibel, E. R. and Gomez, D. M. (1962). A principle for counting tissue structures on random sections. *J. Appl. Physiol.* **17,** 343.

Weibel, E. R. and Knight, B. W. (1964). A morphometric study on the thickness of the pulmonary air-blood barrier. *J. Cell Biol.* **21,** 367.

Weibel, E. R. and Paumgartner, D. (1978). Integrated stereological and biochemical studies on hepatocytic membranes. II. Correction of section thickness effect on volume and surface density estimates. *J. Cell Biol.* **77,** 584.

Wicksell, D. D. (1925). The corpuscle problem I. *Biometrika* **17,** 84.

Wicksell, S. D. (1926). The corpuscle problem II. *Biometrika* **18,** 152.

Young, J. Z. and Roberts, F. (1951). A flying-spot microscope. *Nature* **167,** 231.

III. List of journals publishing articles on stereology

Advances in Applied Probability {Adv. Appl. Prob.} (Sheffield, England).

Biometrical Journal {Biom. J.} (Berlin, DDR), formerly Biometrische Zeitschrift {Biometr. Zschr.}.

Biometrics {Biometrics} (Washington).

Biometrika {Biometrika} (Cambridge, England).

Journal of Applied Probability {J. Appl. Prob.} (Sheffield, England).

Journal of Microscopy {J. Microsc.} (Blackwell's Oxford). Official journal of the International Society for Stereology, published with Royal Microscopical Society, London/Oxford; formerly (until 1968) Journal of the Royal Microscopical Society {J. Roy. Microsc. Soc.}.

Micron: Quarterly journal of electron microscopy {Micron} (West Watford, Herts.).

Microscopia Acta {Microsc. Acta} (Stuttgart), formerly (until 1971) Zeitschrift für wissenschaftliche Mikroskopie und mikroskopische Technik {Z. wiss. Mikrosk.}.

Mikroskopie {Mikroskopie} (Wien).

Transactions of the American Institute of Mining, Metallurgical (and Petroleum) Engineers {Trans AIME} (New York), particularly branch of Metals Society.

Author Index

T

U

W

Y

Subject Index

I

L

M

T

U